Ai6X6 RESEARCH LAB

Transformation Blueprint Series

Published by EDGE Publish, www.EdgePublish.com

Leadership Transformation for the AI Era, First Edition, 2026
ISBN-13: 978-1-971148-00-7, Printed in the United States of America.

Ai6X6
RESEARCH LAB

Transformation Blueprint Series

LEADERSHIP TRANSFORMATION FOR THE AI ERA

Designing the Vision, Not Just Directing It.
The 6x6 Blueprint for Strategic Trust and Decision Intelligence.

SUBRATA MUKHERJEE

EDGE
PUBLISH
www.EdgePublish.com

"For the grace of God, who lights the path.
And, in loving memory of my mother Rupali and my father Dilip, who taught me the importance of kindness and perseverance. This work is born from the love you gave, the sacrifices you made, and the quiet wisdom that still echoes in my life. This one is for you."

Personal message from our Founder & Executive Director

SUBRATA MUKHERJEE

For over twenty-five years, I have sat at the heart of global digital transformations—from shifting a 180-year-old institution like *The Economist* into a digital powerhouse, to architecting hyper-automation for Fortune 500 leaders. I have seen the rise of the internet, the mobile revolution, and the cloud. But none of these shifts compare to the era we have just entered.

The Ai6x6 Research Lab was born out of a singular, urgent observation: Most organizations are trying to solve tomorrow's problems with yesterday's mental models. Traditional transformation is often treated as a "project"—a linear path with a start and an end. But in an era of Agentic AI and self-directing systems, transformation is a living organism. It is a system of behaviors, decisions, and capabilities that must evolve in unison. I founded this Lab to move beyond the "hype" and the "tools." My motivation is to provide leaders with the Cohesion, Schema, and Custody required to navigate a world where the boundary between human intuition and machine intelligence is disappearing. People often ask why "6×6." Throughout history, the number six has been the sacred geometry of our existence—from the sexagesimal systems of ancient Sumerian astronomy to the hexagonal patterns found in the farthest reaches of our cosmos. Mathematically, six is the first "perfect number" because its divisors (1, 2, and 3) add up to itself, making it uniquely balanced. In my journey across the U.S., Europe, and Asia, I realized that this ancient symmetry holds the key to modern enterprise. By distilling organizational complexity into six fundamental pillars and viewing them through six transformative lenses, we activate a collective "sixth sense" for the organization—the ability to perceive and act on what others miss. This architecture forms the foundation of our three flagship methodologies—the "*guided navigation system every transformation needs*".

My journey to this blueprint was forged in the engine rooms of the world's most complex organizations. I have led wholesale business transformations for *The Economist*, reinvented Bloomberg's digital presence following the *Businessweek* acquisition, revitalized the iconic *Dummies* brand within Wiley, and optimized the global ratings processes within S&P Global and McGraw-Hill. This strategic work was balanced by deep technical execution: implementing RPA, AI, IoT, and Digital Twins within Raytheon, and architecting CRM, BPM, and digital workplace automations for the global clients of Cognizant and Unisys.

This perspective was further refined through rigorous academic pursuit, drawing on my MBA in Marketing from Jadavpur University, my Master's in Technology Management from Columbia University, and executive certifications from Harvard University and the Indian Institute of Management (IIM-C) to bridge the gap between high-level theory and industrial-scale execution.

Today, through the Ai6x6 Research Lab, I lead the development of the frameworks that are now being adopted by executive teams globally. At the center of this work is three methodologies: 6X6 Blueprint, IGNITE Methodology and REVEAL Methodology our multidimensional navigation system that unifies architecture, execution, and evaluation across the enterprise.

However, my background as a Tech Journalist for *The Telegraph* for over a decade and a columnist for *CIO Magazine* has taught me that true insight requires total independence. I have built this Lab as a space where my team and I can express our opinions freely, without the interference of corporate sponsors. Our goal is not just to report the news, but to "fold" it into the Ai6x6 Blueprints—giving you the critical "so-what" and the "what-next" to make global shifts actionable for your organization.

We are democratizing this knowledge because I believe that the ability to thrive in the AI era should not be reserved only for those who can afford expensive global consulting firms. Our mission is to empower businesses of all sizes with the targeted, practical, and battle-tested information that I often wish I had at my disposal during my own leadership assignments. Through the Ai6x6 Transformation Blueprints and our Executive Mentor Network, we are scaling AI literacy to ensure that as systems become more autonomous, leaders become more intentional. We are not just adopting new tools; we are reimagining the nature of the enterprise itself.

Ai6X6 RESEARCH LAB

NEW YORK | LONDON | KOLKATA

To our esteemed Industry Thought Leaders: Thank you!

Your generous time and profound insights have been the bedrock of this book. Each interview has been converted into sections and one-page cards that have been sprinkled across six books. You have provided a vital perspective, transforming our work from theory into a deeply practical guide rooted in real-world leadership in the AI-Era. Your willingness to share your hard-won wisdom, navigating complex challenges and revealing the future of your industries, enriches every page. You all have successfully navigated the AI transition in your careers, your companies and your industries. This book is a testament to your expertise. We are immensely grateful for your unwavering support and belief in this project. Thanks a lot, from me and my team at Ai6x6 and Executive Mentor Network!

TABLE OF CONTENTS

guided navigation for transformation

BLUEPRINT SERIES

SECTION A

Introduction

Ai6x6™ Transformation Blueprint

Ai6x6 Research Lab is a global intellectual institution dedicated to advancing the science and practice of organizational transformation. We develop original frameworks, diagnostic systems, capability playbooks, and research models that help leaders navigate complexity with clarity, precision, and confidence. Our work spans **Leadership, Workforce, Data, Customer, Product, and Operations**—the six pillars that define modern enterprise maturity. We are based in New York, London & Kolkata.

OUR MISSION FOR ORGANIZATIONAL TRANSFORMATION

To build a globally recognized body of knowledge that helps organizations think clearly, act decisively, and transform responsibly in an age of AI acceleration. We essentially play **six roles** for the industry:

THE CATALYST
BRING ALIGNMENT TO LEADERSHIP

THE STRATEGIST
BRING STRUCTURE TO AMBIGUITY

THE ARCHITECT
BRING CLARITY TO COMPLEXITY

THE ENGINEER
BRING DISCIPLINE TO TRANSFORMATION

THE ETHICIST
BRING INTEGRITY TO OUTCOMES

THE GUARDIAN
BRING OVERSIGHT TO AUTONOMY

WHAT WE DO?

1. Build Transformation Frameworks: We architect the Ai6x6 Transformation Blueprints— a structured, multi-lens transformation model that helps orgs decode patterns, risks, behaviors, and decision pathways.
2. Develop Diagnostic Systems: Our maturity diagnostics and radar models translate organizational signals into actionable insights, identifying Growth, Risk, Kill, and Shift zones.
3. Publish Research & Methodologies: We codify best practices in enterprise design through multi-volume series and research playbooks, available in bookstores or on our website ai6x6.com.
4. Create Executive Tools: From RACI matrices to transformation roadmaps, we build high-utility models for immediate executive application. We have many accelerated off-the-shelf offerings on our website.
5. Support Global Organizations: Our frameworks empower enterprises, consulting teams, and innovation labs. Our advisory team worldwide including our Executive Mentor Network is here to help.

Meet the Founder & Executive Director : Subrata Mukherjee

ACCOMPLISHMENTS

Subrata Mukherjee has delivered enterprise-scale transformation across the Americas, Europe, and Asia, shaping modernization programs for some of the world's most influential organizations. His work spans AI/ML, automation, digital product strategy, enterprise architecture, and large-scale organizational transformation. He has led digital reinvention in global publishing, driven automation and AI adoption in aerospace and defense, modernized ratings and compliance processes in financial services, and guided product and data transformation across multiple industries. In 2016, he was recognized as one of Folio's Top 100 People in Publishing for his impact on digital innovation.

EDUCATION

Subrata holds a Master's in Technology Management from Columbia University, an MBA in Marketing from Jadavpur University, and a Business Communications certification from Harvard University. He also holds a Management by Values certification from the Indian Institute of Management, Calcutta, deepening his grounding in leadership behavior, organizational culture, and values-driven transformation. He is a Certified Product Owner. His academic foundation spans technology, strategy, communication, and organizational leadership, informing his multidisciplinary approach to AI-era transformation.

INDUSTRIES

His career spans aerospace and defense, financial services, publishing, healthcare, consulting, education, manufacturing, and technology services. He has delivered transformation in highly regulated environments, mission-critical engineering contexts, global media organizations, and fast-moving digital businesses. This cross-industry exposure gives him a multidimensional perspective on how AI reshapes strategy, operations, and organizational identity.

COMPANIES

Across his career, Subrata has delivered transformation programs for organizations including The Economist, Raytheon Technologies, S&P Global, Bloomberg, Wiley, Cognizant, Unisys, and Gujarat Ambuja Cement. His work has ranged from leading The Economist's digital transformation and modernizing Bloomberg's subscription and data infrastructure, to launching Raytheon's enterprise-wide Citizen Development program for Automation and AI, transforming the Dummies brand at Wiley, modernizing S&P's Ratings Process after the sub-prime mortgage crisis, scaling CRM and Automation/AI consulting portfolios at Cognizant and Unisys, and establishing the technology backbone and smart-building systems for the Gujarat Ambuja conglomerate.

EXPERTISE

Subrata's expertise spans the full arc of enterprise transformation, beginning with early work in CAD/CAM for steel plants and smart-city infrastructure, which shaped his systems-thinking foundation. He then became a specialist in enterprise CRM and BPM transformation, leading large-scale modernization programs that re-architected customer operations, workflow orchestration, and enterprise process design. Over time, he transitioned into a strategist, product thinker, and transformation architect, guiding organizations on designing, adopting, and scaling next-generation digital capabilities. Today, he is recognized as a leading voice in AI and Emerging Technologies, with deep expertise across AI/ML, Generative AI, LLMs, RAG architectures, intelligent automation, hyper-automation, conversational AI, chatbots, agentic systems, and decision-automation frameworks. His work is rooted in strategy, product design, operating-model transformation, and enterprise-scale adoption—not coding or engineering execution—and focuses on translating emerging AI capabilities into practical, scalable, enterprise-grade systems that reshape how organizations operate and compete.

THOUGHT LEADERSHIP

Subrata is a prolific speaker who has presented at more than 100 global conferences and events. His writing has appeared in CIO Magazine, and he served as a weekly technology columnist for The Telegraph. Through the Ai6x6 Research Lab, he publishes frameworks, diagnostics, and transformation blueprints that help organizations navigate the AI economy. In 2026–27, he will launch the Executive Mentor Network to develop a global community of AI Strategy–certified coaches and expand AI literacy among senior leaders. Beyond his professional work, he is an avid traveler who has explored more than 40 countries and an accomplished portrait artist.

www.linkedin.com/in/sm360

1. GIVE **VALUE**, REPEATEDLY, BECOME A GOTO BOOK

The 18-book series brings in a lot of content from a practitioner perspective. The advice we are providing have worked for us in our past projects across industries; so we are preaching what we practiced before

2. BE **BRIEF**, PROVIDE CONTENT IN SMALL NUGGETS

I have never read a book cover to cover. I have short attention span, and more and more people are suffering from the same syndrome. Hence each micro-topic we have introduced is covered in two to three pages

3. BE **COMPREHENSIVE** FOR REAL WORLD IMPACT

Transformation is not easy. In the real world there are a lot of inter-connected challenges. We have proposed detailed frameworks that will connect the dots and provide comprehensive solutions to your team/client

4. BE **PRACTICAL**, PROVIDE EXAMPLES TO DEMYSTIFY

Most companies and leaders know they and their companies need to transform, but don't know where to start. I wanted to provide a step-by-step guide or a MOSCOW priority list on getting started

5. PUSH **BOUNDARIES** AND DARE LEADERS TO DREAM

Real transformation is not tinkering around the edges and playing safe; its often about self disrupting and pushing boundaries and reinventing oneself. Applies to individuals, teams and organizations.

6. PUSH LEADERS TO **INTROSPECT** & ASK QUESTIONS

Since the world around us is moving so fast, leaders often have a fear of been exposed about their lack of knowledge and ask "stupid" questions. These book educates you enough, so you can ask intelligent questions

7. BE **EXCITING**, DON'T BE BORING

I have tried to keep the book interesting, small text, lot of structured tables, quite a few images, bold formatting, case studies, Tips and Tricks, Interviews, RACI matrix, buying guides, assignment questions etc.

8. BRING **OTHERS**, HEAR FROM MY COMMUNITY

Tech sector worldwide is very collaborative. We learn from each other every day and we use the knowledge to impact lives positively. This book attempts to bring a few voices from my network and hear their perspective

9. PROMOTE **AI ECONOMY**, USE AI TOOLS YOURSELF

Generative AI has been used extensively for formatting content in a repeatable format & page size control. Without GenAI such an extensive book series would not have been possible in a short time.

10. BE **DIFFERENT**, CHARTER YOUR OWN COURSE

Most management self-help books are pages and pages of long text with guidelines provided by publishers that are a bin antiquated. Modern learning happens byte sized and is multi-modal and multi-channel..

GUIDED NAVIGATION FOR TRANSFORMATION – 18 BOOK SERIES

The **Guided Navigation for Transformation** framework is a comprehensive, end-to-end methodology designed to de-risk and accelerate AI adoption within an organization. It moves away from "random acts of digital" by providing a disciplined, three-stage approach that balances human capability with technical execution. The core of the system is the **6x6 Lens Sequence**, which ensures that every strategic move is evaluated through six critical dimensions: **Leadership, Workforce, Data, Customer, Product, and Operations**.

By applying these lenses consistently across the Blueprint, Reveal, and Ignite phases, the framework creates a "Golden Thread" that links high-level strategy to on-the-ground deliverables. This is not just a consulting model, but an educational ecosystem with an 18-book series that ensures that AI transformation is both measurable and sustainable.

"Guided Navigation for Transformation" in the AI Era

We provide guided navigation through three distinct phases: First, we establish your **Transformation Blueprint** to align skills. Second, we run the **Reveal Diagnostic** to assess your current maturity. Finally, we execute the **Consulting Playbook** to turn the 'Art of the Possible' into a 'Science of Deliverables'.

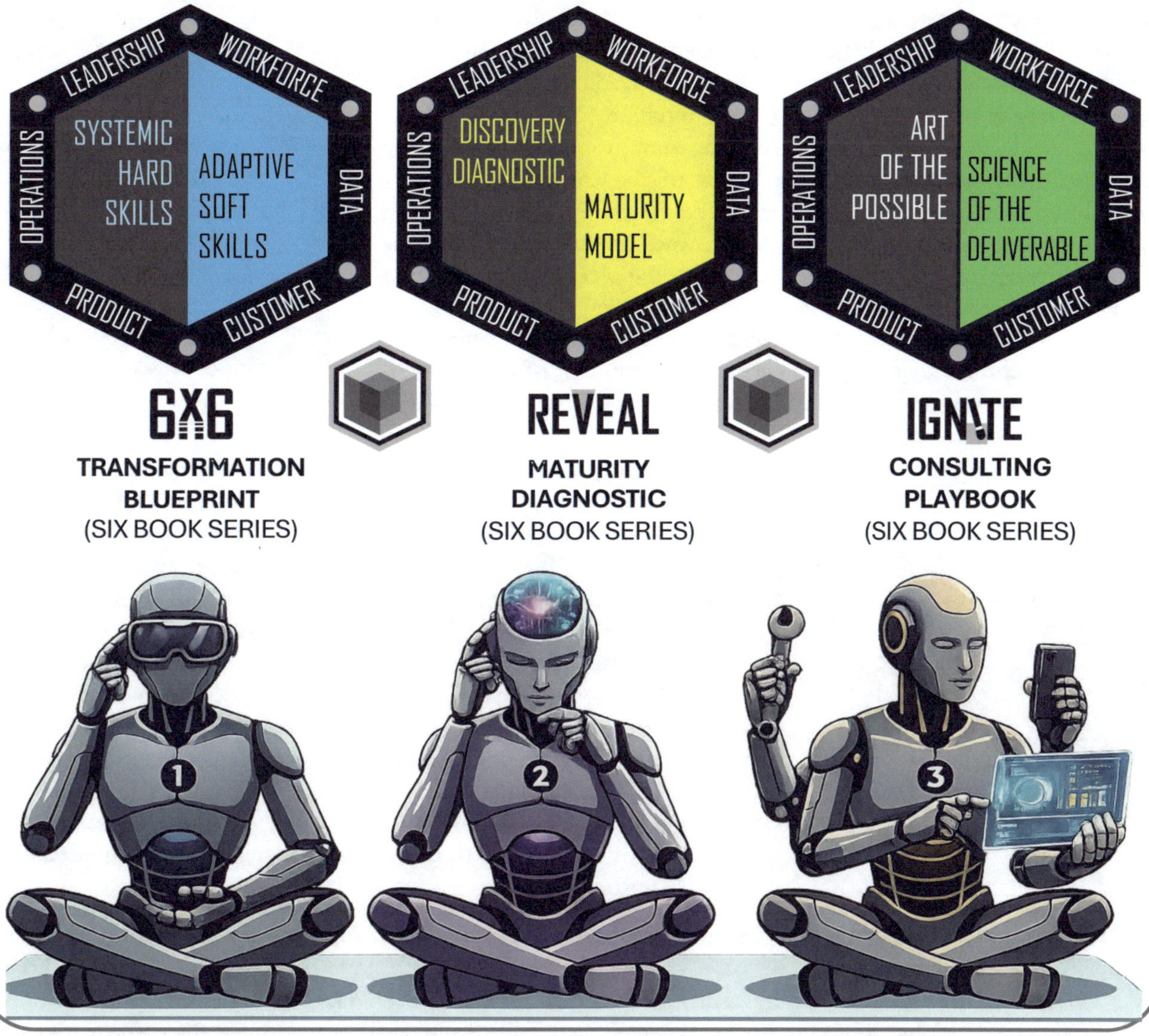

The **AI6x6 Transformation Engine™** is the flagship construct of the AI6x6 Research Lab—a precision-engineered ecosystem that serves as the definitive "Operating System" for the modern enterprise. Built on a proprietary 6×6 geometry, it unifies architecture, execution, and calibration into a single, high-performance unit. The Engine provides leaders with the high-torque machinery and **Guided Navigation** required to convert organizational inertia into scalable momentum in the era of autonomous systems.

The **Transformation Blueprint Series** defines the engine's architecture and establishes the structural logic of the entire system. By introducing the **Six Pillars** and **Six Lenses**, it forms the foundational geometry of enterprise modernization. Written for CEOs, Boards, and Strategy Leaders, this series acts as the **Navigation Deck**, providing the "Architectural North Star." It ensures that every move is strategically calibrated and ethically anchored before the gears are engaged, offering a shared mental model for aligning leadership teams. This intelligence layer removes the guesswork from transformation, providing a rigid yet adaptive framework that allows organizations to visualize their future state with absolute clarity. It is the vital starting point where complex, high-stakes ambition is codified into a disciplined, institutional reality that can withstand the pressures of rapid technological shift.

The **Maturity Diagnostic Series** provides the engine's accountability and calibration system, ensuring peak performance through proprietary **REVEAL ™** methodology. . Acting as the **Sensor-Driven Intake**, it utilizes 36 diagnostic cylinders to quantify organizational readiness and prioritize investments based on evidence rather than intuition. This series allows executive teams to identify hidden gaps and measure progress with scientific clarity, functioning as the **Measuring Layer** of the entire ecosystem. It ensures the engine remains tuned for resilience through a data-driven maturity model that tracks evolution over time. By providing rigorous assessment instruments and scoring logic, it offers the accountability required to sustain a long-term transformation journey. This layer acts as the engine's nervous system, constantly scanning the internal and external environment to calibrate the Blueprint's trajectory and ensure that every unit of energy spent drives the organization toward sustained excellence and institutional stability.

The **Consulting Playbook Series** translates architecture into action by operationalizing the engine through the proprietary **IGNITE™** methodology. It serves as the **Heavy-Duty Gearbox**, providing the repeatable consulting patterns, execution models, and directional force required to activate transformation across diverse domains. Designed for PMOs and delivery leads, these playbooks convert high-level strategy into tangible enterprise-scale outcomes, ensuring that the "torque" of the engine is effectively applied to the road. They function as the essential doing layer, reducing internal friction and shifting the organization from abstract concept to deep-seated conviction. By providing step-by-step pathways and intervention models, the series bridges the gap between executive intent and operational reality. This transmission layer ensures that momentum is not just created, but sustained and scaled, allowing the enterprise to accelerate through complex transitions with precision and measurable impact.

The **Transformation Engine** is built for scale across the global economy. Future **Industry Evolution** volumes will apply this 6×6 engine to specific sectors—such as Healthcare, Banking, and Manufacturing—revealing how the model drives specialized, high-velocity transformation. By adapting the Pillars and Lenses to sector-specific demands, these extensions ensure the Engine remains the definitive standard for navigating the complexities of the age of intelligent systems.

SERIES 1 6 BOOKS

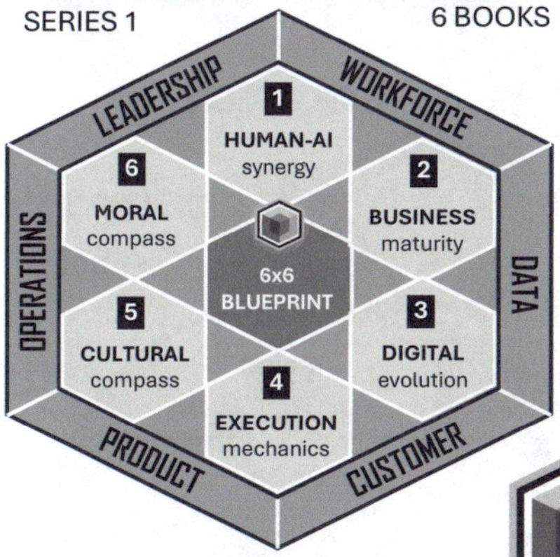

6X6 TRANSFORMATION BLUEPRINT

see

SERIES 2 6 BOOKS

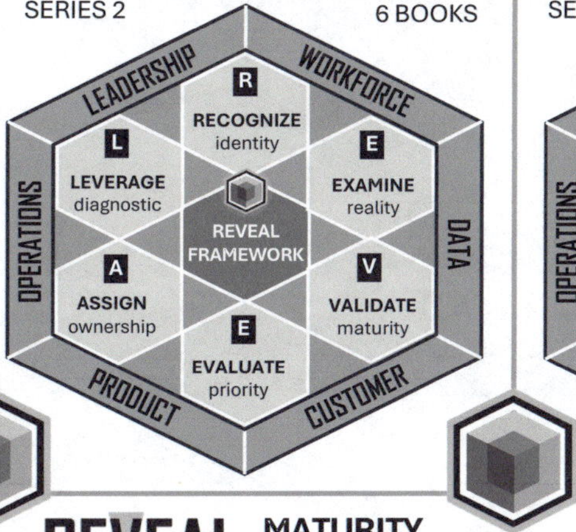

REVEAL MATURITY DIAGNOSTIC

think

SERIES 3 6 BOOKS

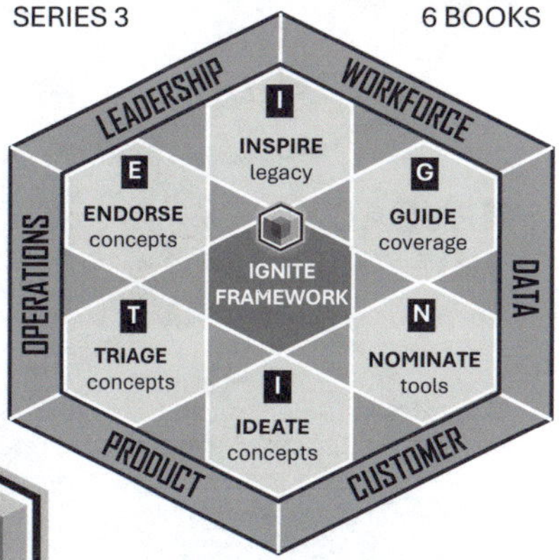

IGNITE CONSULTING PLAYBOOK

act

The Ai6x6 palette is a semantic system, not a decorative choice. Each color encodes the emotional and operational signature of its lens: **pink** captures the human spark that ignites potential; **green** reflects the strategic discipline required for maturity; **purple** signals the depth and computational power of digital evolution; **orange** embodies the kinetic urgency of execution; **blue** anchors the cultural soul in trust and cohesion; and **yellow** illuminates the ethical clarity of the moral audit. Together, these colors form a coherent visual language that guides the reader through the transformation cycle with intuitive, emotionally aligned cues.

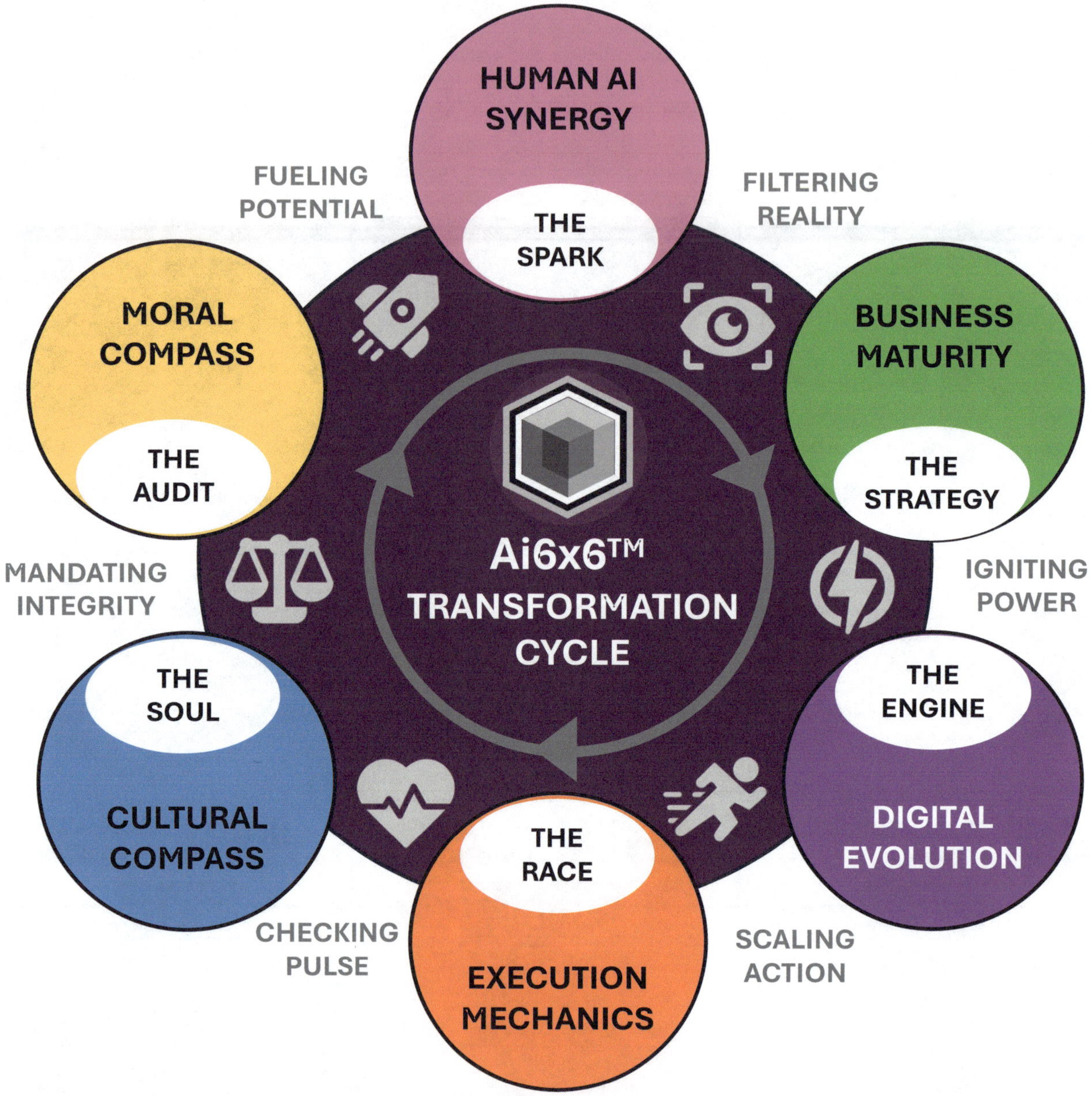

The **COLOR TABS ALONG THE PAGE EDGES IN THE MIDDLE** serve as the physical navigation system of the Ai6x6 series. Each tab mirrors the chromatic identity of its lens, allowing readers to locate themselves instantly within the transformation cycle. This creates a spatial reading experience where orientation is continuous and intuitive. The tabs turn our book into a modular interface—one where Human–AI Synergy, Business Maturity, Digital Evolution, Execution Mechanics, Cultural Compass, and Moral Compass become distinct territories with their own visual signatures. The result is a playbook that reads like a designed system, not a linear text.

6x6™ Transformation Blueprint

The Prologue to a Purpose

"The only way to go forward is to let go of the past."
The Myth of Orpheus and Eurydice

The tragic lesson of this myth is a cautionary tale for business. Orpheus's failure to trust his journey and looking back results in losing everything, underscoring the importance of fully committing to a forward-looking transformation.

The mission of this book series is simple: to transform leaders from passive observers of the AI Era into its active pioneers. We will ignite your thinking with exciting, highly practical content that pushes boundaries and provides the essential tools and frameworks you need to build real-world impact. We won't just tell you what to do; we'll show you how, using a diverse community's experiences and real-world examples that bring the abstract to life. This isn't a passive read; it's an active partnership and a dynamic dialogue designed to turn your natural curiosity into a definitive competitive advantage. This book is the first of a six-part series : Leadership Transformation; which governs your journey on how you should lead from the top and how you should inspire others to take on the AI Era and excel in it.

The **AI6x6 Transformation Engine** is a precision-engineered ecosystem designed to deliver **Guided Navigation** for enterprises facing high-stakes digital evolution. By integrating three proprietary systems—the **Transformation Blueprint**, the **Consulting Playbook**, and the **Maturity Diagnostic**—the Engine provides the robust "Operating System" required to dominate the AI era.

The architecture begins with the **Transformation Blueprint**, the Engine's intelligence layer. Here, the system utilizes **Six Lenses**—including the Cultural and Moral Compasses—to calibrate your trajectory and codify a master architectural plan. This strategic map then engages the **Consulting Playbook**, where the gears of the **IGNITE** methodology translate executive intent into immediate directional force through rapid-response execution cycles. Finally, the **Maturity Diagnostic** acts as a sensor-driven intake, utilizing six diagnostic cylinders to compress raw organizational data into high-pressure strategic insights that validate the transformation's impact. Bolted to a rigid **Chassis of Six Pillars**, the Engine ensures the enterprise remains resilient, balanced, and unshakeable even at peak acceleration. It is more than a strategy; it is a proprietary architecture for sustained institutional excellence.

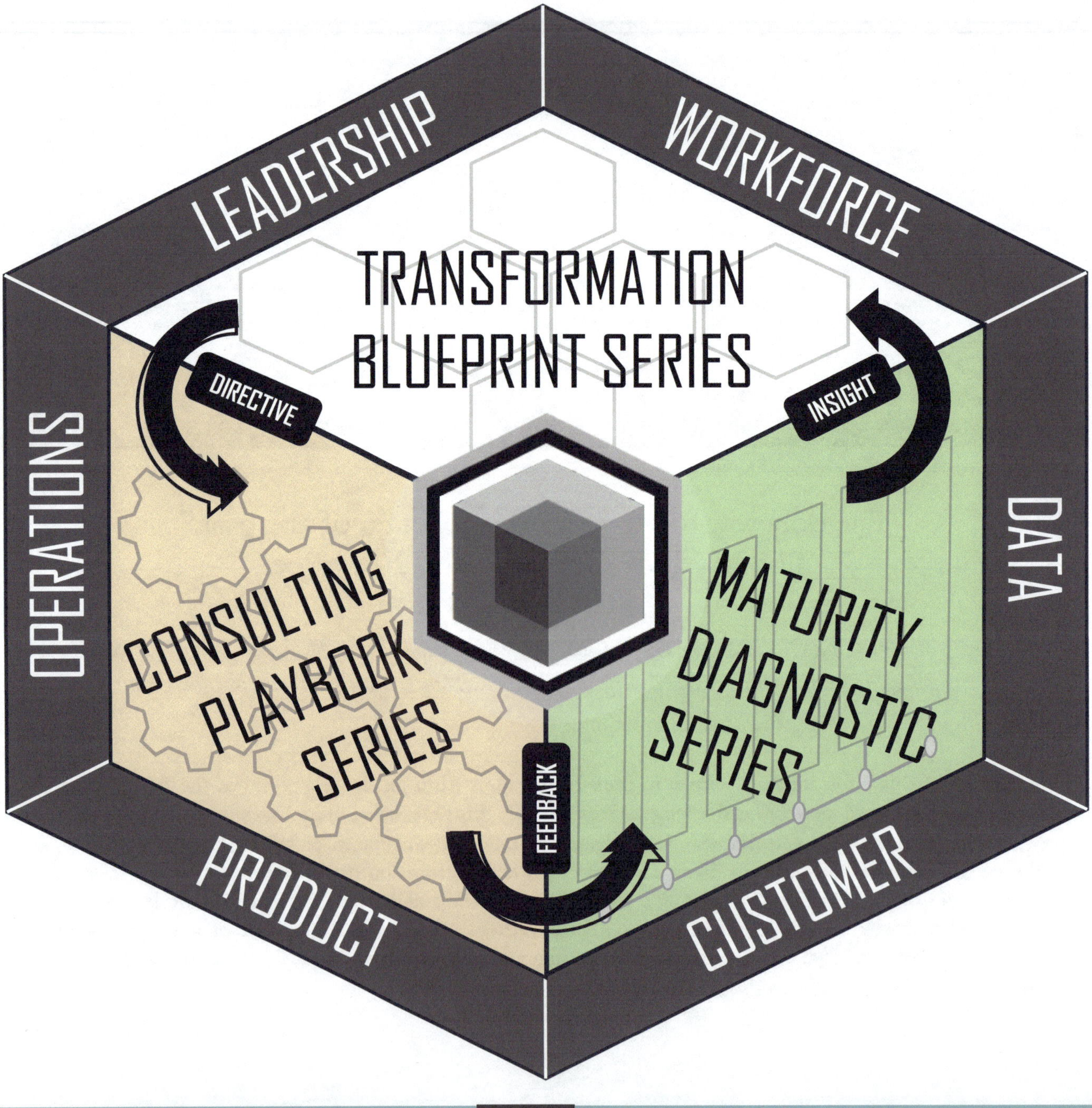

The AI6x6 Transformation Engine is a precision-engineered ecosystem that replaces guesswork with **Guided Navigation**. It is designed to convert organizational inertia into scalable momentum through a disciplined, mechanical approach to digital evolution.

CHASSIS (SIX PILLARS)
The entire system is anchored to a rigid structural frame. This chassis ensures that even as you reach peak acceleration, your enterprise remains resilient and unshakeable.

NAVIGATION DECK (SIX LENSES):
This is the intelligence layer. By applying our **Six Lenses**, we provide the **Guided Navigation** required to calibrate your trajectory, preventing costly drifts in strategy.

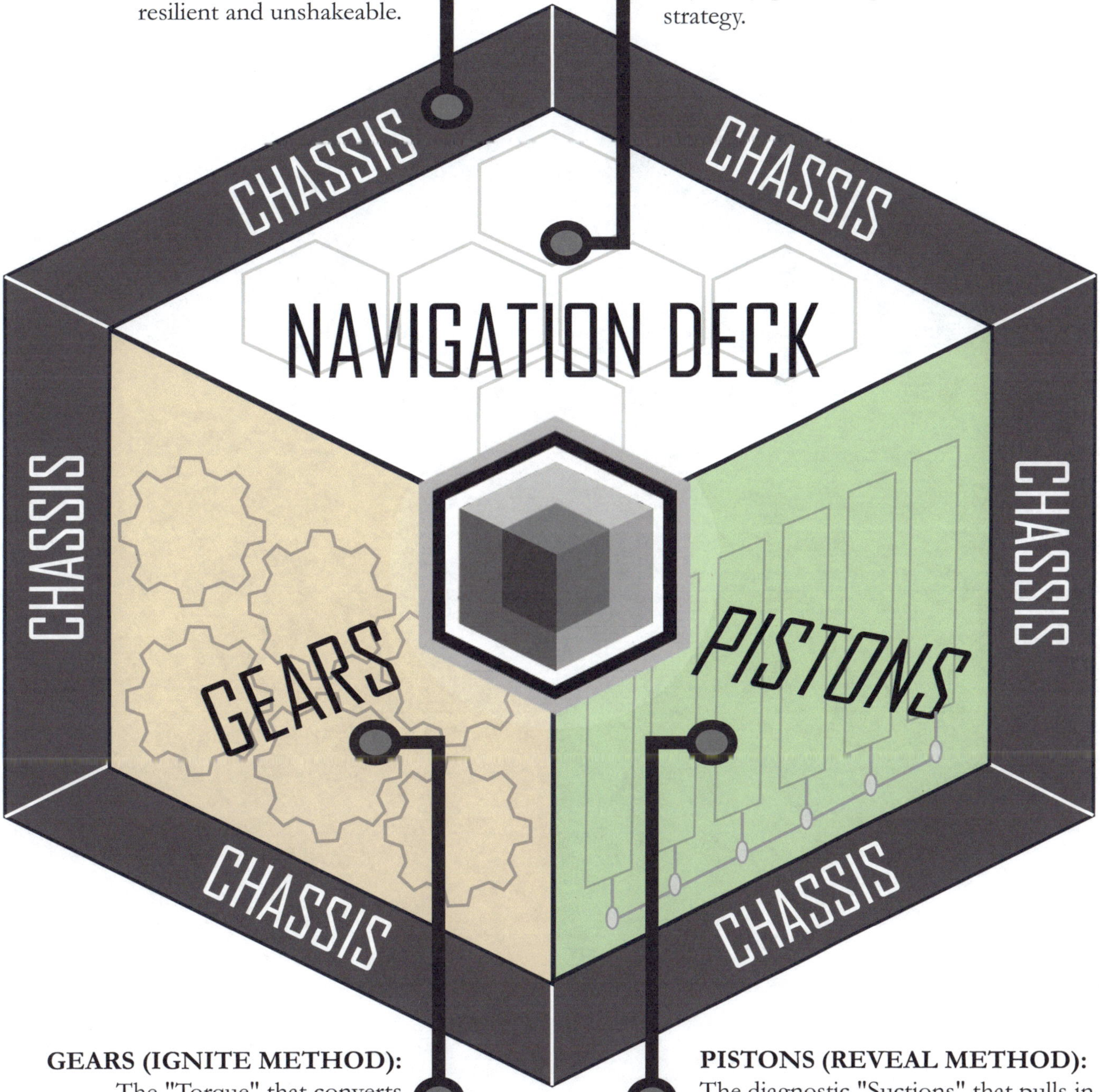

GEARS (IGNITE METHOD):
The "Torque" that converts diagnostic energy into immediate directional force through pilot programs and playbooks.

PISTONS (REVEAL METHOD):
The diagnostic "Suctions" that pulls in raw organizational data and compresses it into high-pressure actionable insights.

A car works best when its actions are coordinated through systems that generate power, transmit motion, and guide direction. Each component plays a precise mechanical role that keeps the vehicle stable, responsive, and moving with purpose.

CHASSIS
Supports vehicle structure, absorbs road stress, anchors mechanical systems, maintains stability, and ensures safe, balanced handling across varying speeds, terrains, and driving conditions.

NAVIGATION DECK
Displays real-time driving information, maps, routes, and alerts, helping the driver maintain direction, anticipate conditions, and make confident, informed decisions.

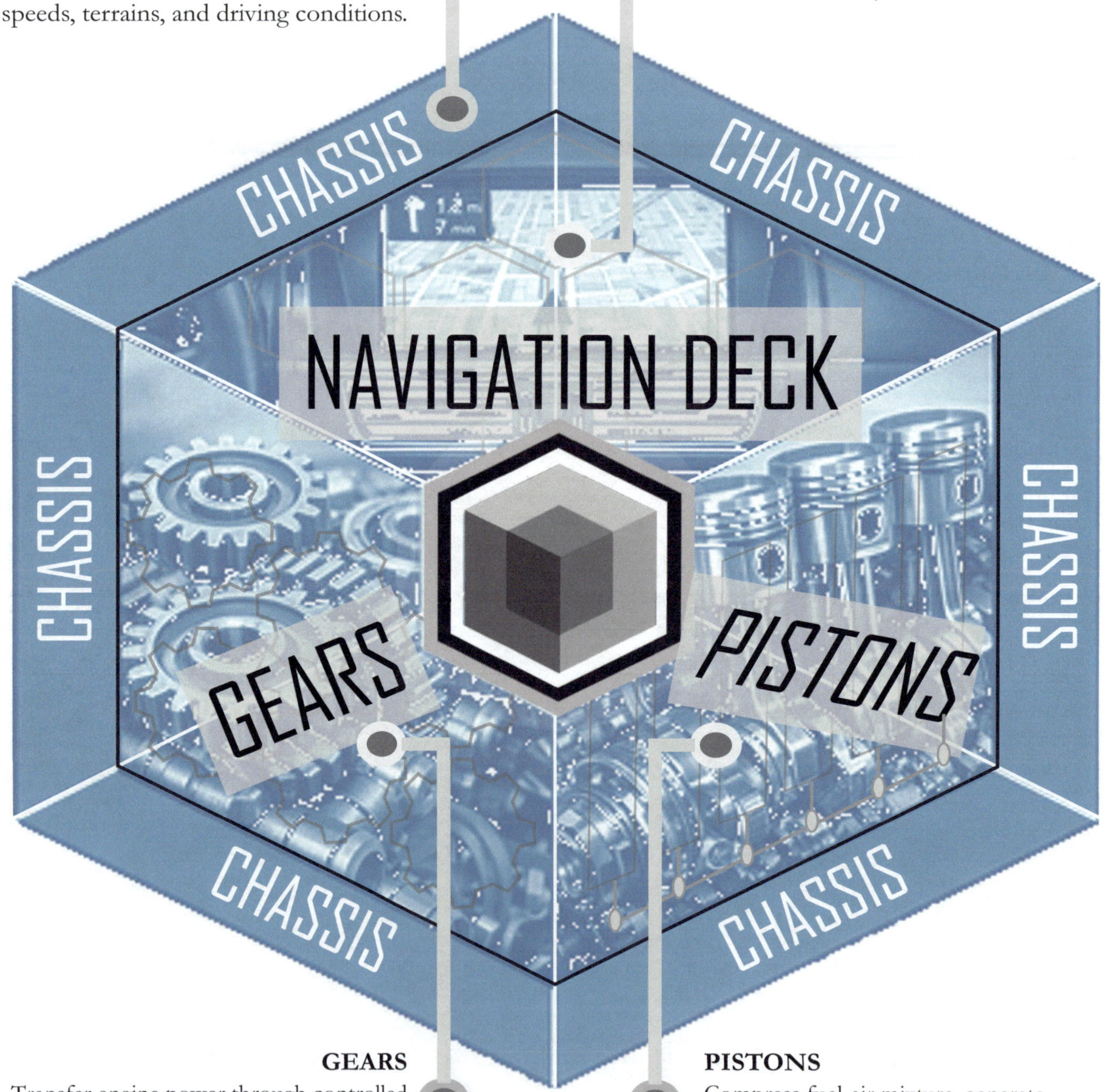

GEARS
Transfer engine power through controlled ratios, regulating speed and torque, ensuring smooth acceleration, efficient performance, and seamless coordination between mechanical components.

PISTONS
Compress fuel-air mixture, generate combustion force, drive the crankshaft, & convert explosive energy into continuous motion powering the vehicle's entire drivetrain system.

The **AI6x6 Transformation Engine** replaces the ambiguity of traditional consulting with mechanical certainty, converting organizational inertia into scalable momentum. By integrating three proprietary systems—the **REVEAL** diagnostic intake, the **IGNITE** consulting transmission, and the **6x6 Blueprint** navigation deck—the Engine provides the robust "Operating System" required to dominate the AI era. The **REVEAL** methodology acts as a sensor-driven intake, utilizing 6x6 diagnostic cylinders to compress raw organizational data into high-pressure strategic insights. This "data fuel" then powers the **IGNITE** phase, where consulting playbooks serve as a heavy-duty gearbox, translating executive intent into immediate directional force through rapid-response execution cycles. This entire power plant is governed by the **6x6 Blueprint**, the intelligence layer that uses **Six Lenses**—including the Cultural and Moral Compasses—to calibrate your trajectory and ensure ethical, operational alignment. Bolted to a rigid **Chassis** of **Six Pillars**, the Engine ensures the enterprise remains resilient, balanced, and unshakeable even at peak acceleration. It is more than a strategy; it is a proprietary architecture for sustained institutional excellence.

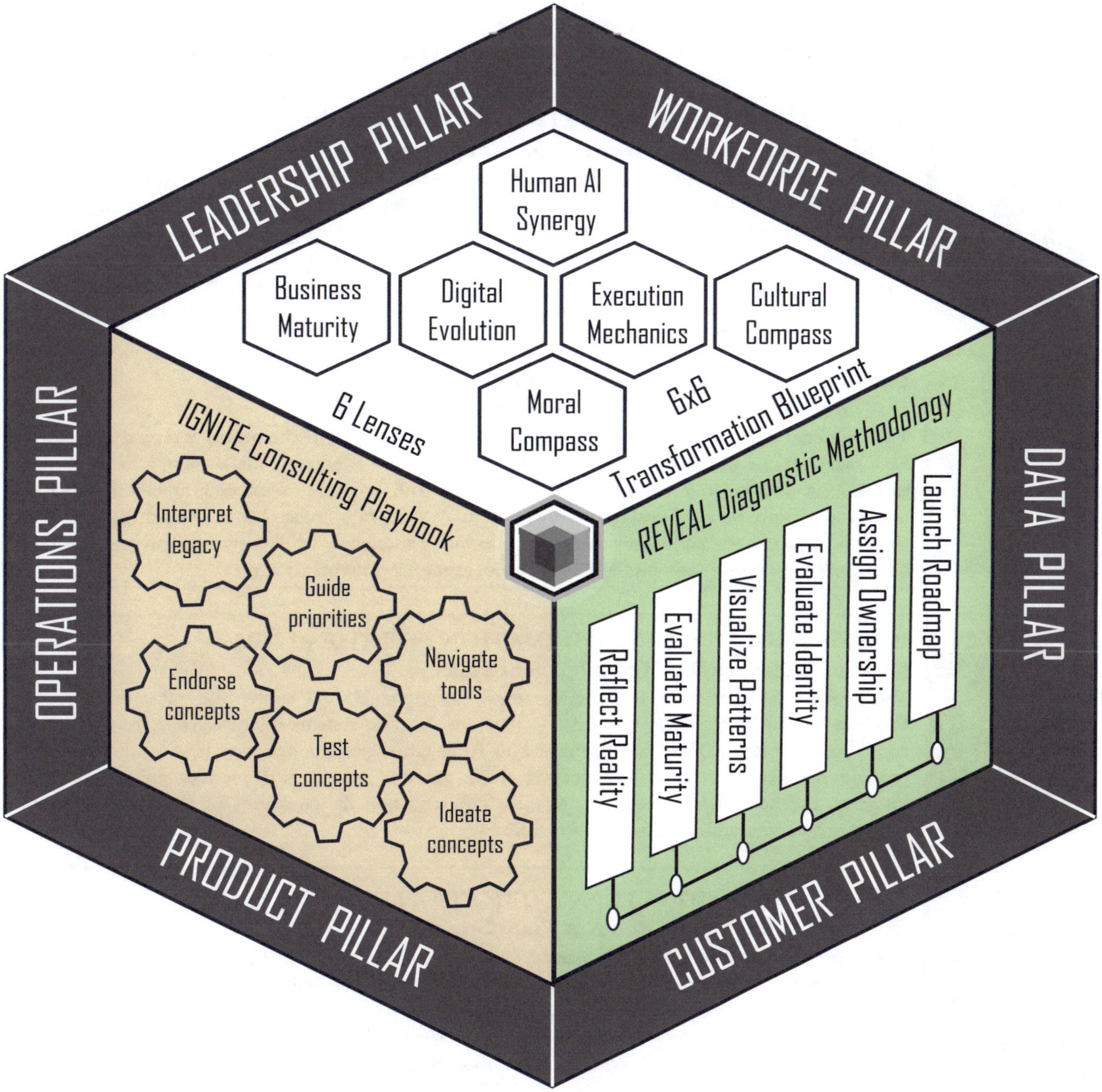

6 BOOKS OF THE Ai6x6 TRANSFORMATION BLUEPRINT SERIES

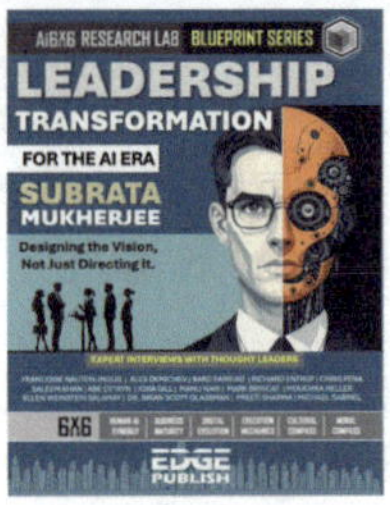

BOOK 1: LEADERSHIP TRANSFORMATION FOR THE AI ERA

Designing the Vision, Not Just Directing It

The Leadership book reframes what it means to lead in an era defined by intelligence, acceleration, and ambiguity. It explores how leaders think, decide, communicate, and shape meaning across the enterprise, emphasizing that leadership is not a role but an architecture. The narrative examines how leaders make decisions systemically, exercise judgment in an AI-augmented world, create alignment through narrative clarity, and uphold ethical stewardship as the foundation of trust. This book sets the tone for the entire series by positioning leadership as the integrator of all transformation forces.

BOOK 2: WORKFORCE TRANSFORMATION FOR THE AI ERA

Elevating Human Capability in a Technology-Driven Era

The Workforce book examines the human system of the enterprise — the skills, roles, mindsets, structures, and experiences that determine how people show up to the work of transformation. It explores how organizations support people as AI reshapes every function, and how leaders build environments where individuals feel empowered rather than displaced. The narrative dives into the evolution of adaptive talent ecosystems, the emergence of AI-augmented roles, the importance of team cohesion and psychological safety, and the recognition of human capability as a strategic asset.

BOOK 3: DATA TRANSFORMATION FOR THE AI ERA

Building the Intelligence Fabric of the Enterprise

The Data book establishes the informational backbone of the organization. It explores how data is collected, governed, interpreted, and transformed into intelligence, emphasizing that data is no longer a technical asset but the raw material of decision-making, personalization, automation, and innovation. The narrative examines how trustworthy data governance is built, how intelligence flows across systems, how data architectures are designed, and how data becomes a source of enterprise-level advantage. This book gives leaders the clarity to build an intelligent, insight-driven organization.

BOOK 4: CUSTOMER TRANSFORMATION FOR THE AI ERA

Designing Experiences That Earn Trust and Loyalty

The Customer book anchors the enterprise in the lived experience of the people it serves. It explores how organizations understand, anticipate, and respond to customer needs — not as a marketing exercise but as a cultural and operational discipline. The narrative examines how emotion shapes experience design, how AI enables personalization at scale, how transparency strengthens customer trust, and how end-to-end journey coherence becomes a competitive differentiator. This book ensures the organization evolves around the customer, not around its internal structures.

BOOK 5: PRODUCT TRANSFORMATION FOR THE AI ERA

Creating Value in a World of Intelligent Systems

The Product book explores how organizations design, build, and evolve the value they deliver. Products are no longer static offerings — they are dynamic systems of intelligence, behavior, and continuous adaptation. This book examines how AI enhances product behavior, how continuous value creation replaces episodic feature delivery, how product strategy and product operations must integrate, and how digital ecosystems redefine differentiation. It transforms product thinking from output to outcome, and from features to value.

BOOK 6: OPERATIONS TRANSFORMATION FOR THE AI ERA

Engineering the Engine Room of Transformation

The Operations book addresses the systems, workflows, and processes that determine whether strategy becomes reality. It explores how organizations orchestrate complexity, manage risk, and maintain resilience in environments defined by acceleration and uncertainty. The narrative examines how intelligent workflows are designed, how operational resilience is built, how AI optimizes processes, and how execution scales with precision. This is the book where transformation becomes tangible.

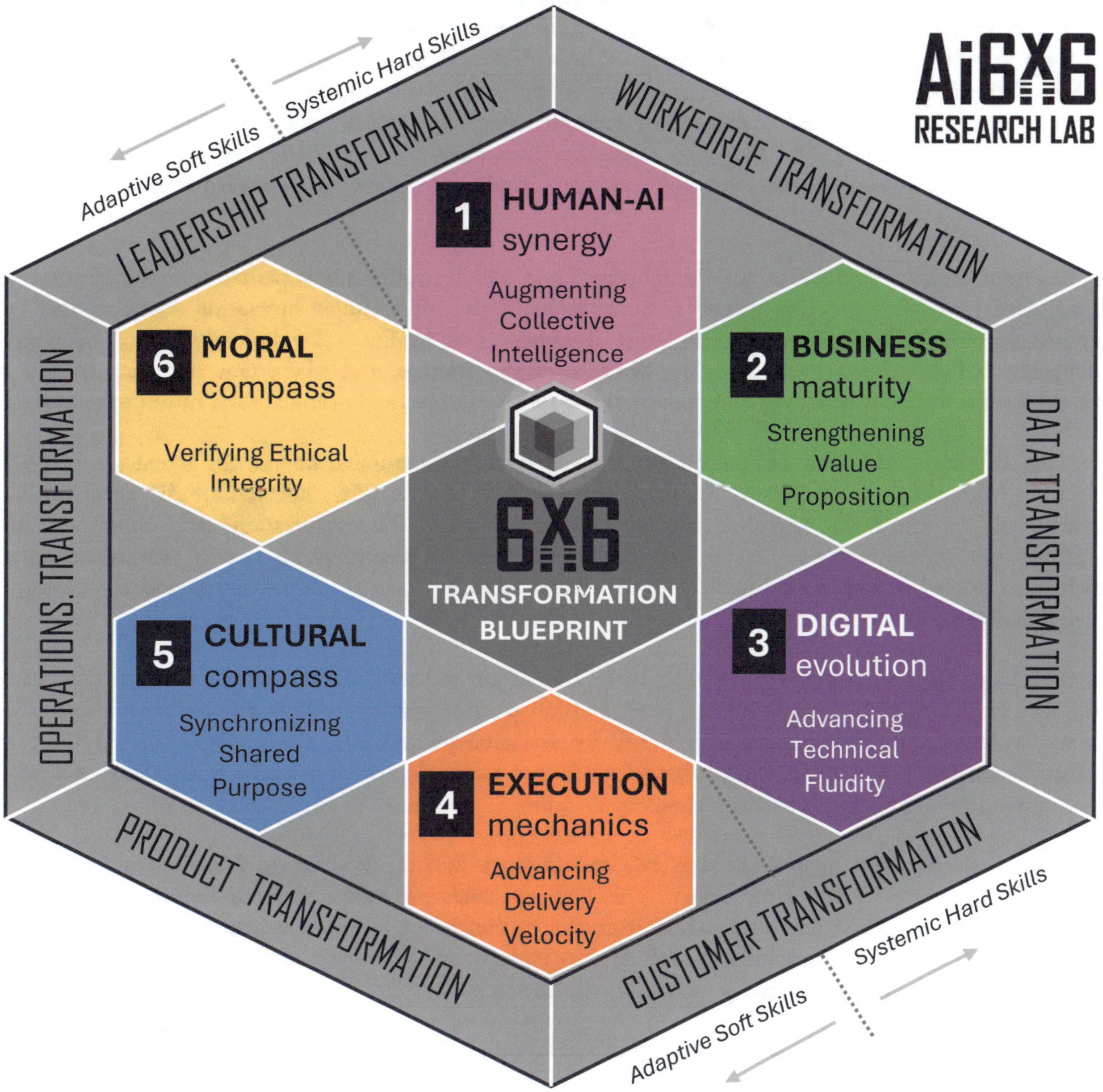

The **Ai6x6 TRANFORMATION BLUEPRINT™** provides the multidimensional structural integrity required for sustainable, high-stakes evolution. Serving as the **Navigation Deck** and **Chassis** of the Transformation Engine, the BLUEPRINT™ moves beyond episodic change to establish a permanent, scalable "Operating System" for the enterprise. The architecture is defined by the intersection of **Six Pillars** and **Six Lenses**. The Pillars—Leadership, Workforce, Data, Operations, Product, and Customer—provide the horizontal breadth and load-bearing strength of the organization. These are bisected vertically by the Six Lenses to reveal the true depth of the enterprise. Leaders apply **Human–AI Synergy** to augment collective intelligence and **Business Maturity** to instill systemic discipline. The **Digital Evolution** lens ensures technical fluidity, while **Execution Mechanics** masters the art of precision delivery. This is anchored by the **Cultural Compass**, which synchronizes shared purpose, and the **Moral Compass**, which verifies ethical integrity.

By aligning these twelve dimensions, the BLUEPRINT™ eliminates strategic drift and structural fragmentation. It ensures that every initiative is not just an isolated project, but a calibrated component of a unified, resilient, and unshakeable corporate architecture. Also remember to follow the Tab colors on the edges to navigate quickly.

The **Ai6x6 Transformation Blueprint™** gives leaders a multidimensional, architecture-grade view of how transformation behaves inside their enterprise. Where REVEAL™ diagnoses the present and IGNITE™ accelerates the future, the Blueprint defines the *structural truth* of the organization. Built on six proprietary Lenses, it converts fragmented observations, cultural noise, and operational ambiguity into a coherent, systems-level understanding. The Blueprint equips leaders with the clarity, depth, and structural intelligence required before designing any transformation roadmap, ensuring decisions are grounded in reality rather than aspiration.

The Blueprint begins with the **Human–AI Synergy Lens**, which examines how people and intelligent systems interact across workflows, decisions, and responsibilities. Rather than simply measuring adoption, this Lens analyzes augmentation, displacement risk, trust dynamics, and readiness for human-machine teaming. It distinguishes where AI enhances capability, where it creates friction, and where new operating models are required. Leaders begin with a grounded understanding of how human and machine roles must evolve together.

Next, the **Business Maturity Lens** evaluates the enterprise's structural readiness for transformation. This includes governance, financial discipline, decision velocity, and leadership alignment. The Lens separates foundational strength from structural fragility, revealing whether the organization can support sustained transformation without collapse. It clarifies which capabilities are stable, which require reinforcement, and which must be redesigned before acceleration. Business Maturity ensures leaders understand the chassis on which transformation will operate.

The third Lens, **Digital Evolution**, maps the depth, coherence, and scalability of the digital ecosystem. Leaders often face fragmented platforms, technical debt, and integration gaps that silently undermine transformation. Digital Evolution cuts through this complexity by assessing architecture, interoperability, data flow, and platform extensibility. It clarifies which digital foundations can scale, which must be modernized, and which create systemic risk. This Lens ensures technology decisions are grounded in architectural truth.

With structural and digital realities established, leaders move into the **Execution Mechanics Lens**, which examines how strategy becomes movement. This Lens evaluates processes, rhythms, accountability systems, and operational precision. It identifies where execution is consistent, where it is brittle, and where it breaks under pressure. Execution Mechanics ensures leaders understand the organization's true capacity to deliver, adapt, and sustain momentum across complex transformation cycles.

The fifth Lens, **Cultural Compass**, explores the beliefs, behaviors, incentives, and informal power structures that shape how people respond to change. Culture is often the silent accelerator or the invisible brake. This Lens reveals alignment, resistance patterns, psychological safety, and leadership influence networks. It distinguishes cultural strengths from cultural liabilities, giving leaders a clear view of the human forces that determine whether transformation will be embraced or rejected.

The final Lens, **Moral Compass**, evaluates ethics, trust, transparency, and responsible AI principles. As organizations adopt intelligent systems, the stakes of ethical decision-making rise dramatically. This Lens ensures transformation is not only effective but also equitable, accountable, and aligned with long-term societal and organizational values. It clarifies where safeguards are strong, where risks exist, and where governance must evolve.

Together, these Six Lenses form a complete, repeatable, consulting-grade architectural framework. The Ai6x6 Transformation Blueprint™ is not a checklist or maturity model; it is a structural intelligence system that moves leaders from assumption to understanding, from understanding to design, and from design to confident transformation planning. It gives leaders the conviction to move forward — not because the future is predictable, but because their decisions are grounded in multidimensional truth, structural clarity, and a deep understanding of how transformation behaves across the enterprise.

6x6™ Transformation Blueprint

Understand Pillars & Lenses

"We don't see things as they are, we see them as we are." — Anaïs Nin

The **Ai6x6 TRANSFORMATION BLUEPRINT** is fundamentally structured around two core concepts: **Pillars and Lenses**. Together, they create a dual-axis architecture that brings both breadth and depth to enterprise transformation. The Pillars provide the WIDTH of the framework, defining the major domains where transformation must take root for the organization to evolve coherently. These domains—Customer, Product, Operations, Data, Workforce, and Leadership—represent the full span of enterprise behavior. Each Pillar receives its own dedicated book, allowing leaders to explore the complexities, tensions, and opportunities of that domain without dilution or compromise.

This ensures that transformation is not treated as a single initiative but as a system of interconnected disciplines that must advance in parallel. Complementing this horizontal structure is the DEPTH created by the Six Lenses, the universal imperatives that cut vertically through every Pillar. These Lenses represent the non-negotiable dimensions of modern transformation—forces that must be activated in every domain, regardless of industry, maturity, or strategic ambition. The six Lenses are: Human–AI Synergy (L1), Business Maturity (L2), Digital Evolution (L3), Execution Mechanics (L4), Cultural Compass (L5), and Moral Compass (L6). Each Lens exposes a different layer of organizational capability, ensuring that no domain is approached superficially, and no critical dimension is overlooked. The goal of the blueprint is to be holistic and very comprehensive so that executives are not blind-sighted by a key aspects being overlooked. Since a lot of ground has to be covered, the books may look very dense, but I have tried my best to keep it brief & engaging.

Transformation has become one of the most OVERUSED words in modern leadership, yet one of the least understood. Organizations declare it, teams attempt it, consultants package it—but few enterprises ever truly achieve it. Not because they lack ambition or intelligence, but because they approach transformation as a sequence of disconnected initiatives rather than a shift in how the organization thinks, behaves, and evolves. I created the **Ai6x6 Transformation Blueprint** because I have seen the same pattern repeat across industries, functions, and eras of technology: transformation fails not at the point of strategy, but at the point of structure. It fails when organizations modernize technology but ignore culture. It fails when they redesign processes but overlook trust. It fails when they invest in AI but fail to prepare people. It fails when they build strategy but cannot execute it. It fails when leaders underestimate the complexity of aligning human capability with machine intelligence.

The Ai6x6 Transformation Blueprint exists to solve this fragmentation. It brings together **six foundational lenses** and **six transformation pillars** into a single, coherent architecture. Each lens represents a dimension of capability that every organization must master. Each pillar represents a domain— Leadership, Workforce, Data, Customer, Product, Operations —where transformation must take root. Together, they form a 6×6 matrix that reveals not only what must change, but how change becomes sustainable, repeatable, and scalable.

This blueprint is intentionally multidimensional. Transformation is not linear. It is not sequential. It is not a checklist. It is a living system—human, technical, cultural, operational, ethical—moving together or not at all. The Ai6x6 model acknowledges this complexity and gives leaders a structure strong enough to hold it.

- It is called a blueprint because it is both **architectural** and **actionable**.
- This blueprint is not a methodology per se. It is not a maturity model in the traditional sense. It is not a checklist as you commonly see it. It is a **HOLISTIC** way of seeing things that are often overlooked.
- It is called Ai6x6 because it integrates **AI-driven intelligence** with **six human-centered disciplines**, applied across **six transformation pillars**.
- It is a system designed for organizations that want to evolve with intention rather than react with urgency.

The Six Lenses emerged from a simple but unavoidable truth: "**no single discipline can carry transformation on its own**". Not AI. Not culture. Not governance. Not execution. Not digital modernization. Not ethics. Each is necessary, but none is sufficient. Real transformation requires all six lenses working together—reinforcing one another, correcting one another, and elevating one another.

It reveals the forces that shape organizational behavior, the tensions that undermine progress, and the capabilities that determine whether transformation becomes sustainable or collapses under its own ambition. It provides leaders with a structural map—one that makes complexity navigable and makes evolution repeatable.

Each lens represents a dimension of organizational life:

1. Human–AI Synergy: how intelligence is created, shared, and amplified.
2. Business Maturity: how discipline, readiness, and structure turn ambition into capability.
3. Digital Evolution: how technology becomes fluid, integrated, and trustworthy.
4. Execution Mechanics: how strategy becomes action and action becomes advantage.
5. Cultural Compass: how beliefs, behaviors, and meaning shape the human system.
6. Moral Compass: how ethics, transparency, and responsibility anchor everything else.

Together, they form a complete operating philosophy for modern transformation—one that is as human as it is technical, as strategic as it is operational, as aspirational as it is grounded. The Ai6x6 Transformation Blueprint is not about predicting the future. It is about preparing organizations to shape it. It is about giving leaders a language for what they instinctively know but rarely articulate. It is about building enterprises that can evolve continuously, coherently, and confidently. Most of all, it is about honoring the truth that **transformation is not something organizations *do*—it is something they *become*.**

The Pillars of the **Ai6x6 Transformation Blueprint** define the WIDTH of enterprise evolution. They represent the six domains where transformation must become real, measurable, and operationally grounded. While the Lenses provide the depth of capability, the Pillars provide the breadth of application—ensuring that transformation is not isolated to a single function but becomes a structural shift across the entire enterprise. Each Pillar receives its own dedicated book, allowing leaders to confront the full complexity of that domain with precision and without compromise.

The Leadership Pillar sits at the top of the architecture because leadership is the force that shapes every other domain. It defines how decisions are made, how priorities are set, how accountability is distributed, and how meaning is communicated. This Pillar examines the evolution of leadership in an era defined by AI, acceleration, and ambiguity. It challenges leaders to move from managing activities to designing systems, from directing work to enabling capability, and from protecting the past to shaping the future. Leadership is not a separate function—it is the integrator of the entire Blueprint.

The Workforce Pillar focuses on the human system that powers the enterprise. It explores how roles evolve, how teams collaborate, how skills are developed, and how people experience their work. In a world where AI augments every function, the Workforce Pillar examines how organizations support human capability, build adaptive talent ecosystems, and create environments where people feel empowered rather than displaced. This Pillar forces leaders to confront the truth that transformation succeeds or fails based on whether people feel equipped, supported, and connected to the mission.

The Data Pillar establishes the informational foundation of the enterprise. It examines how data is collected, governed, interpreted, and transformed into intelligence. In an AI-driven world, data is no longer a byproduct of operations—it is the raw material of decision-making, personalization, automation, and innovation. This Pillar explores how organizations build trustworthy data ecosystems, eliminate silos, and ensure that intelligence flows across the enterprise. It challenges leaders to treat data not as a technical asset but as a strategic capability that shapes every other Pillar.

The Customer Pillar anchors the organization in the lived experience of the people it serves. It forces leaders to confront the reality that customer expectations evolve faster than internal processes, and that loyalty is earned through coherence, trust, and emotional resonance. This Pillar examines how organizations understand, anticipate, and respond to customer needs—not as a marketing exercise, but as a cultural and operational discipline. It challenges leaders to build enterprises where customer experience is not a department, but a shared responsibility embedded in every decision.

The Product Pillar focuses on how organizations design, build, and evolve the value they deliver. Products are no longer static offerings—they are dynamic systems of intelligence, behavior, and continuous adaptation. This Pillar explores how product strategy, product design, and product operations must work together to create experiences that feel alive and responsive. It examines how AI reshapes product behavior, how digital ecosystems redefine differentiation, and how organizations must shift from feature delivery to continuous value creation.

The Operations Pillar addresses the engine room of the enterprise—the systems, workflows, and processes that determine whether strategy becomes reality. It examines how organizations orchestrate complexity, manage risk, and maintain resilience in environments defined by acceleration and uncertainty. This Pillar forces leaders to recognize that operational excellence is not about efficiency alone; it is about adaptability, intelligence, and the ability to respond to change without losing coherence. Operations is where transformation becomes tangible.

Together, these six Pillars form the structural WIDTH of the Ai6x6 Transformation Blueprint. They define the terrain on which the Six Lenses operate, ensuring that transformation is comprehensive, coherent, and enterprise-wide.

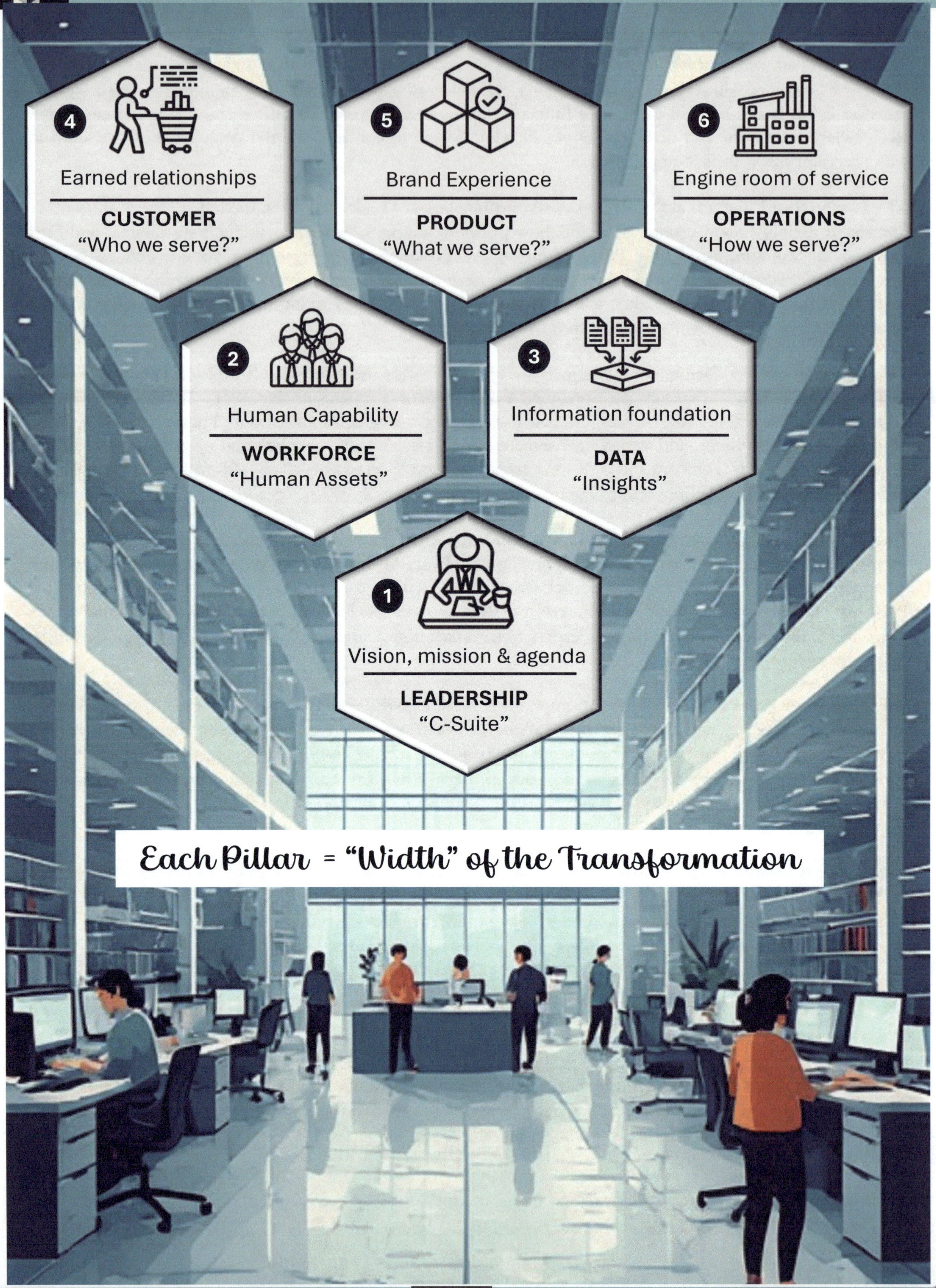
4
Earned relationships
CUSTOMER
"Who we serve?"
5
Brand Experience
PRODUCT
"What we serve?"
6
Engine room of service
OPERATIONS
"How we serve?"
2
Human Capability
WORKFORCE
"Human Assets"
3
Information foundation
DATA
"Insights"
1
Vision, mission & agenda
LEADERSHIP
"C-Suite"
Each Pillar = "Width" of the Transformation

4
Discerning Diners
CUSTOMER
"The Patrons"
5
Food, Drinks, Ambiance
PRODUCT
"The Feast"
6
Ingredients, Purchase
OPERATIONS
"The Logistics"
2
Cooks, Stewards, Servers
WORKFORCE
"The other staff"
3
Menu, Recipe & Reviews
DATA
"The Menu"
1
Lead Culinary aspect
LEADERSHIP
"The Exec Chef"

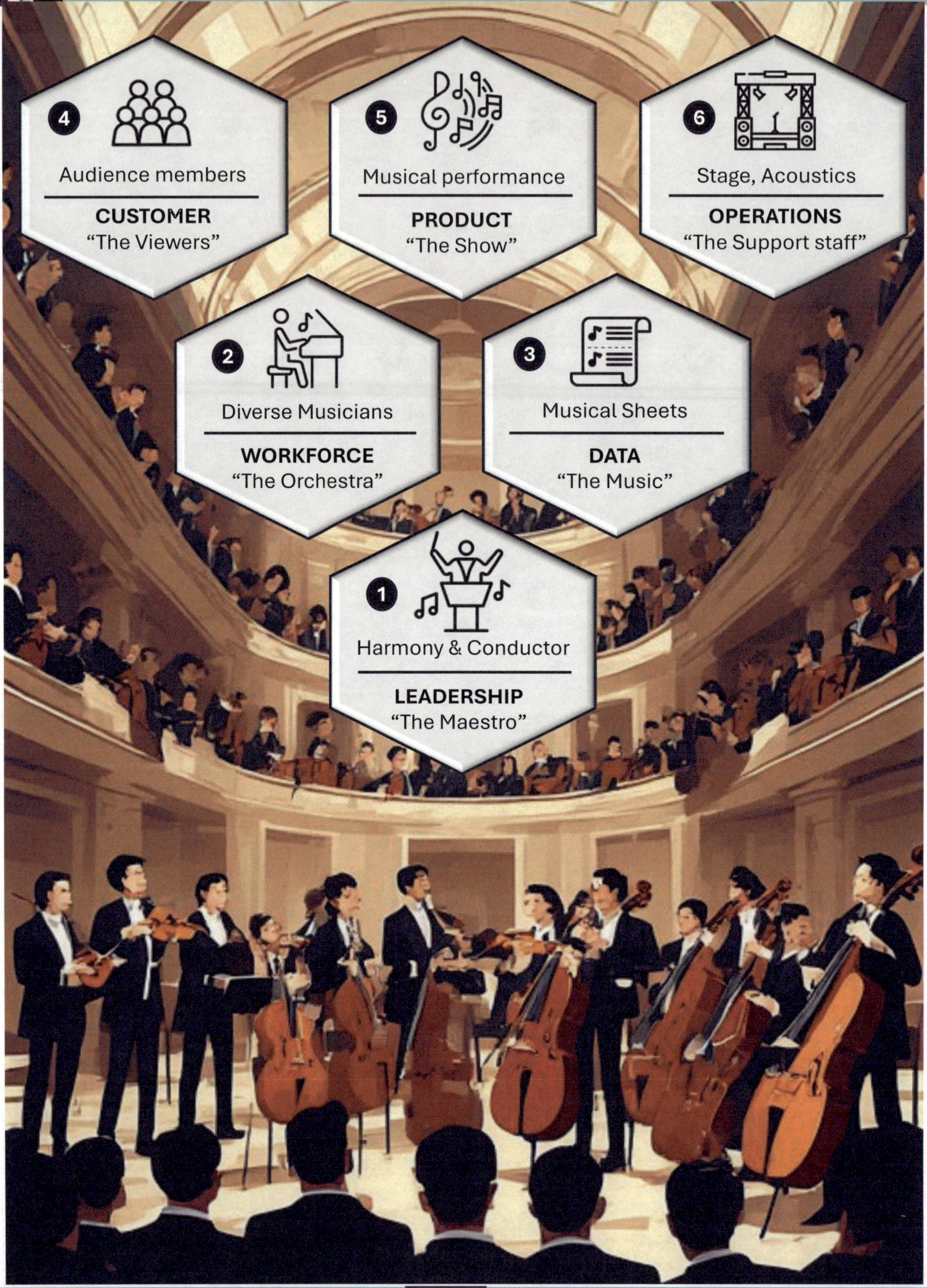
4
Audience members
CUSTOMER
"The Viewers"
5
Musical performance
PRODUCT
"The Show"
6
Stage, Acoustics
OPERATIONS
"The Support staff"
2
Diverse Musicians
WORKFORCE
"The Orchestra"
3
Musical Sheets
DATA
"The Music"
1
Harmony & Conductor
LEADERSHIP
"The Maestro"

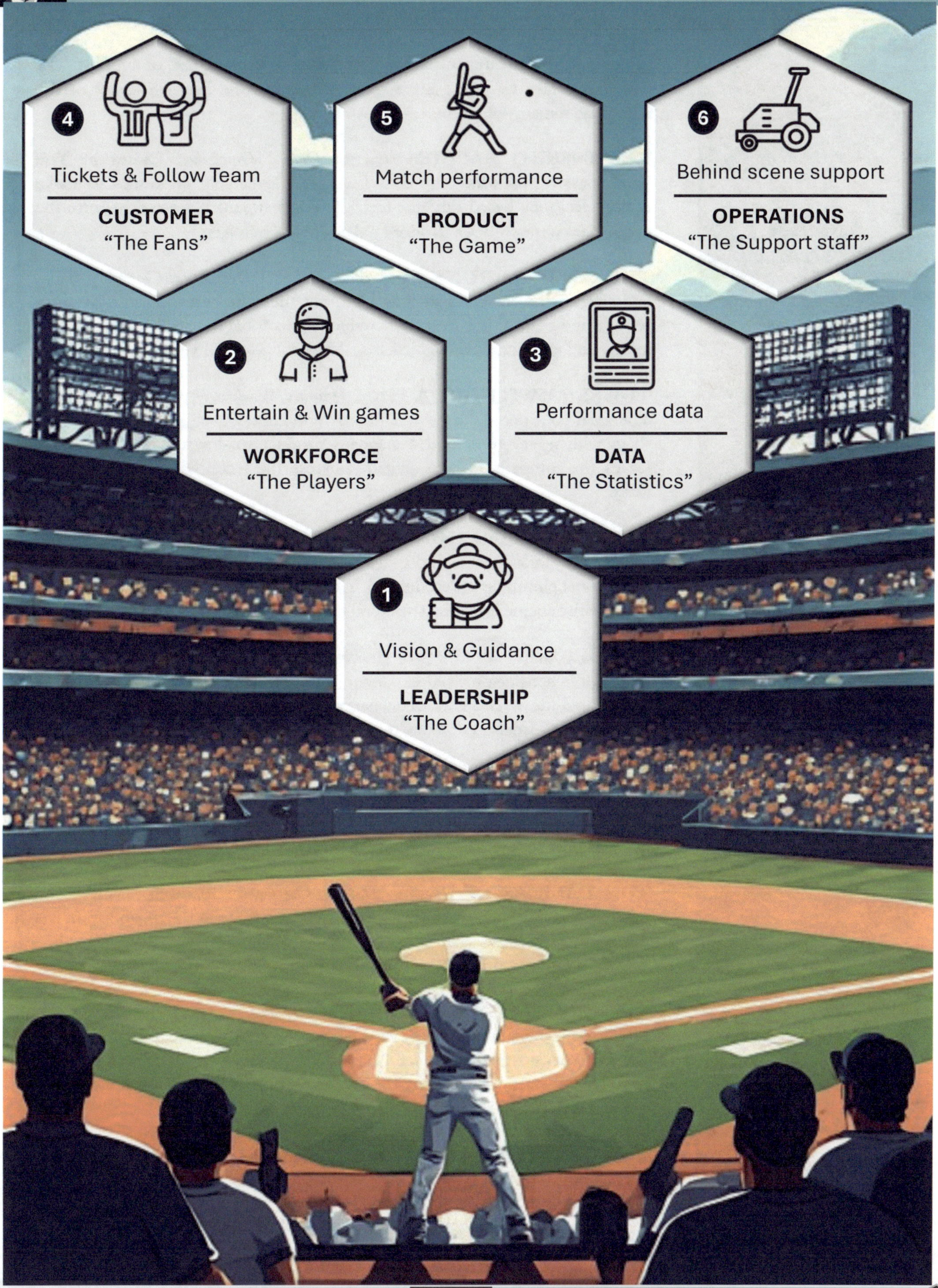
4
Tickets & Follow Team
CUSTOMER
"The Fans"
5
Match performance
PRODUCT
"The Game"
6
Behind scene support
OPERATIONS
"The Support staff"
2
Entertain & Win games
WORKFORCE
"The Players"
3
Performance data
DATA
"The Statistics"
1
Vision & Guidance
LEADERSHIP
"The Coach"

NEED FOR HOLISTIC BLUEPRINTS COVERING ALL SIX PILLARS

THE LITERACY GAP (*Invested: Product | Overlooked: Workforce*): You launch genius AI-driven features that your staff cannot explain, sell, or support. The technology fails to gain market traction because the "human layer" of the enterprise remains obsolete.

THE INSIGHT VACUUM (*Invested: Data | Overlooked: Leadership*): You build massive, high-integrity data lakes, but leadership continues to make critical decisions based on "gut feel." The investment is wasted because the upper-tier culture refuses to adopt a data-driven mindset.

THE COLD MACHINE (*Invested: Operations | Overlooked: Customer*): Back-end efficiency reaches an all-time high, but the actual user journey becomes robotic and rigid. You successfully reduce operational costs but lose the brand loyalty and emotional connection that drives long-term revenue.

THE SKILL WITHOUT A TOOL (*Invested: Workforce | Overlooked: Product*): You train your people to be "AI-Ready" and culturally agile, but you provide them with legacy tools and a stagnant product roadmap. Your best talent leaves for competitors who actually possess the technology to match their skills.

THE HALLUCINATION (*Invested: Leadership | Overlooked: Operations*): Executives announce a bold "Digital First" vision, but the underlying operational plumbing is decades old. The strategy never leaves the boardroom because the engine lacks the structural integrity to turn the wheels.

THE BROKEN PROMISE (*Invested: Customer | Overlooked: Data*): You market a hyper-personalized experience, but your internal data silos remain a fragmented mess. You end up alienating customers with irrelevant offers and inaccurate information that destroys their trust.

THE PROTOTYPE TRAP (*Invested: Product | Overlooked: Operations*): A brilliant AI-driven tool is developed in a vacuum, but the legacy supply chain and logistics cannot fulfill the demand it creates. The product succeeds in the lab but collapses the moment it hits the real-world operational floor.

THE DARK ASSET (*Invested: Data | Overlooked: Workforce*): You spend millions cleaning and structuring your data, but no one in the organization has the literacy to query or interpret it. The data sits idle and "dark," providing zero ROI because the people aren't equipped to find the value.

THE ROGUE ENGINE (*Invested: Operations | Overlooked: Leadership*): Middle management automates workflows to save time, but leadership has no visibility or governance over these new "black box" processes. The organization moves faster, but the steering wheel is no longer connected to the captain's chair.

THE INTERNAL ECHO (*Invested: Workforce | Overlooked: Customer*): Employees are empowered and "AI-literate," but they become obsessed with internal process improvements rather than user needs. The company becomes a high-performing bubble that is perfectly optimized for a customer that no longer exists.

THE BLIND CAPTAIN (*Invested: Leadership* | *Overlooked: Data*): A visionary leader attempts to steer the company through a massive pivot based on bold assumptions. Because the data pillar was ignored, they navigate straight into a market reef that the "missing data" would have clearly identified.

THE FANCY WRAPPER (*Invested: Customer* | *Overlooked: Product*): Your marketing and customer service are world-class, but the actual product has not evolved to keep up with competitors. You win the initial sale through great branding, but the churn rate sky-rockets once the customer realizes the product is hollow.

THE ANALYTICS ISLAND (*Invested: Data* | *Overlooked: Operations*): The data team produces perfect, real-time reports on every aspect of the business. However, the operational workflows are so slow and manual that by the time the business acts on the data, the opportunity has already vanished.

THE ORPHAN PROJECT (*Invested: Product* | *Overlooked: Leadership*): Innovation teams build cutting-edge AI tools that solve real problems, but leadership never provides a permanent budget or strategic home for them. These "pilot projects" eventually wither and die because they lack the top-down air cover to survive.

THE FRUSTRATED EXPERT (*Invested: Workforce* | *Overlooked: Operations*): You hire the best AI researchers and data scientists, but you force them to work within manual, "broken" internal workflows. Your experts spend 90% of their time fighting the system rather than building the future.

THE IVORY TOWER (*Invested: Leadership* | *Overlooked: Customer*): The board develops a strategy based purely on shareholder value and financial engineering. By ignoring the Customer pillar, they fail to see a massive shift in user sentiment that makes their entire three-year plan irrelevant.

THE HUMAN BOTTLENECK (*Invested: Operations* | *Overlooked: Workforce*): You automate 80% of your workflows, but the remaining 20% require human intervention that your staff isn't trained for. The entire system grinds to a halt because you optimized the machine but forgot to upskill the operator.

THE REACTIVE PANIC (*Invested: Customer* | *Overlooked: Leadership*): The company chases every new customer trend and social media "vibe" to stay relevant. Without a strategic anchor from Leadership, the organization becomes a fragmented mess of contradictory ideas that stand for nothing.

THE WASTEFUL HARVEST (*Invested: Data* | *Overlooked: Product*): You collect trillions of data points and pay massive cloud storage fees, but none of that data is ever used to iterate or improve the actual product. You are "Data Rich" but "Value Poor," hoarding information with no path to monetization.

THE GRASSROOTS REBELLION (*Invested: Workforce* | *Overlooked: Leadership*): Staff begins using "Shadow AI" and unsanctioned tools because leadership is too slow or fearful to provide guidance. You have an AI-ready workforce that is operating outside of the company's governance, creating a massive security and legal risk.

THE SIX LENSES OF THE Ai6x6 TRANSFORMATION BLUEPRINT

If the Pillars define the WIDTH of enterprise transformation, the **Six Lenses** define its DEPTH. They are the universal imperatives that cut vertically through every domain—Leadership, Workforce, Data, Customer, Product, and Operations. Each Lens exposes a different dimension of capability, ensuring that transformation is not superficial, fragmented, or episodic, but deeply rooted in the systems that shape how organizations think, behave, and evolve. These Lenses are not optional. They are not enhancements. They are the structural forces that determine whether transformation becomes sustainable or collapses under its own ambition. Each Lens reveals a layer of truth about the enterprise—its intelligence, its discipline, its technology, its execution, its culture, and its integrity.

The Human–AI Synergy Lens (L1) examines how intelligence is created, shared, and amplified across the enterprise. It explores the evolving relationship between human judgment and machine capability, ensuring that AI augments rather than replaces human potential. This Lens forces leaders to confront how decisions are made, how insights flow, and how people interact with intelligent systems. It challenges organizations to build environments where humans and AI operate as partners—each elevating the other. Human–AI Synergy is the starting point because it defines the cognitive foundation of modern transformation.

The Business Maturity Lens (L2) establishes the structural discipline required for transformation to take hold. It examines governance, alignment, prioritization, funding, risk management, and organizational readiness. This Lens reveals whether the enterprise has the scaffolding to support change or whether ambition is outpacing capability. Business Maturity is where leaders confront the operational truth: transformation is not powered by inspiration alone—it is powered by systems, discipline, and repeatable mechanisms that ensure consistency across the enterprise.

The Digital Evolution Lens (L3) focuses on the technological backbone of the organization. It examines how systems integrate, how data flows, how platforms scale, and how digital environments adapt to new demands. This Lens ensures that technology becomes an enabler rather than a bottleneck. It challenges organizations to build digital ecosystems that are fluid, intelligent, secure, and trustworthy. Digital Evolution is not about tools—it is about creating a technological environment capable of supporting continuous change.

The Execution Mechanics Lens (L4) converts strategy into action. It examines the methodologies, operating models, workflows, and performance systems that determine whether the organization can deliver consistently and at scale. This Lens exposes the gap between what leaders intend and what teams experience. It ensures that execution is not left to improvisation but is engineered with precision. Execution Mechanics is where transformation becomes tangible—where ideas become outcomes and outcomes become advantage.

The Cultural Compass Lens (L5) explores the beliefs, behaviors, and shared meaning that shape how people think and collaborate. It examines psychological safety, communication norms, leadership behaviors, and the emotional climate of the enterprise. This Lens reveals whether the culture accelerates transformation or resists it. Cultural Compass ensures that people feel connected, supported, and aligned—because no transformation can succeed in a culture that does not want it.

The **Moral Compass Lens (L6)** anchors the Blueprint in ethics and transparency. Moving beyond a static matrix, it defines the **True North** for AI governance, data privacy, and leadership accountability. This Lens ensures transformation is principled, not just efficient. In the AI era, a Moral Compass is the boundary condition for legitimacy—without it, innovation loses its direction; with it, trust becomes an unbreakable competitive advantage.

Together, these Six Lenses form the DEPTH of the Ai6x6 Transformation Blueprint. They ensure that every Pillar—Leadership, Workforce, Data, Customer, Product, and Operations—is approached with rigor, coherence, and integrity. They transform transformation from a collection of projects into a complete operating philosophy.

SIX LENSES OF TRANSFORMATION : Human Body Analogy

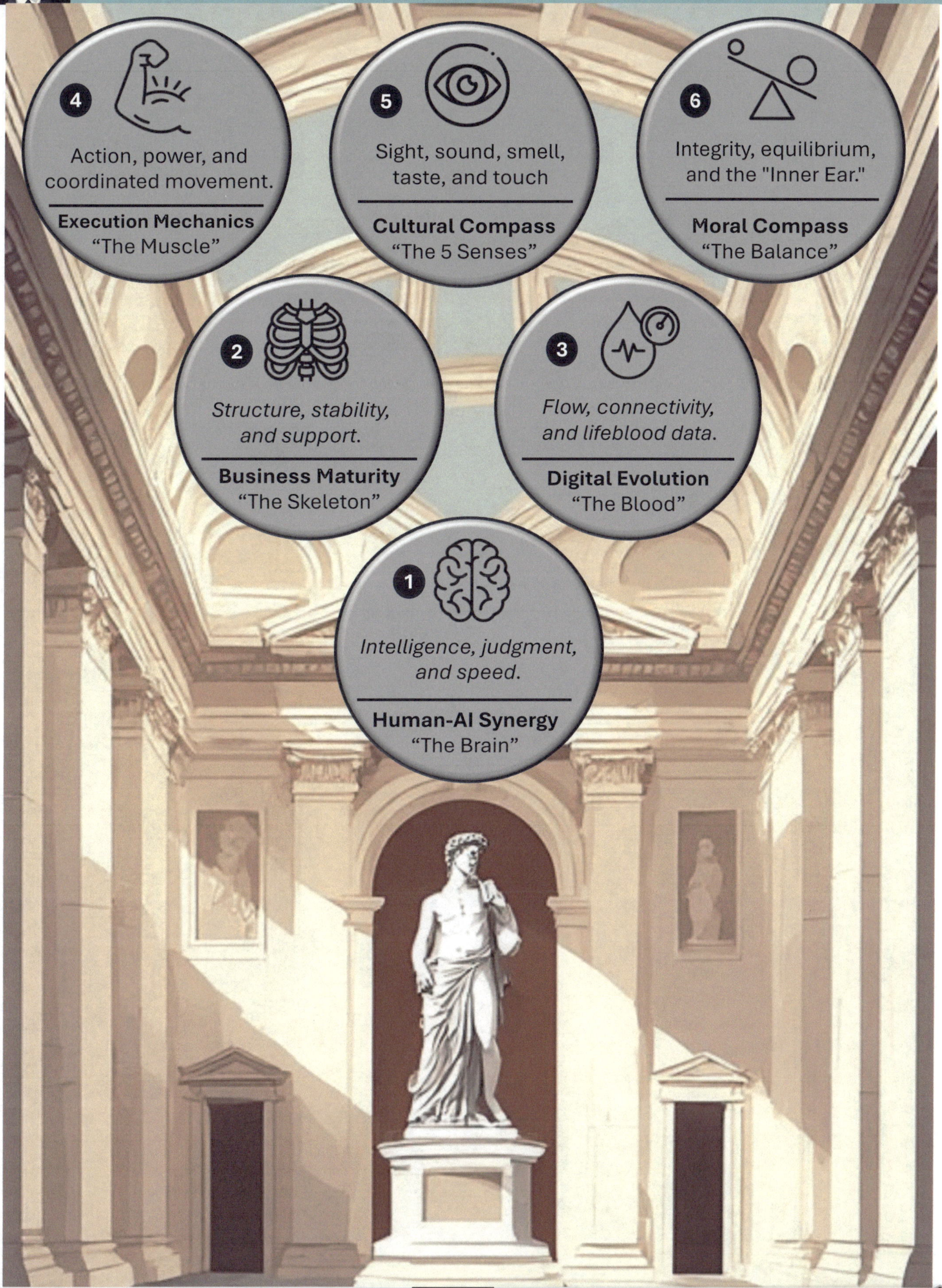

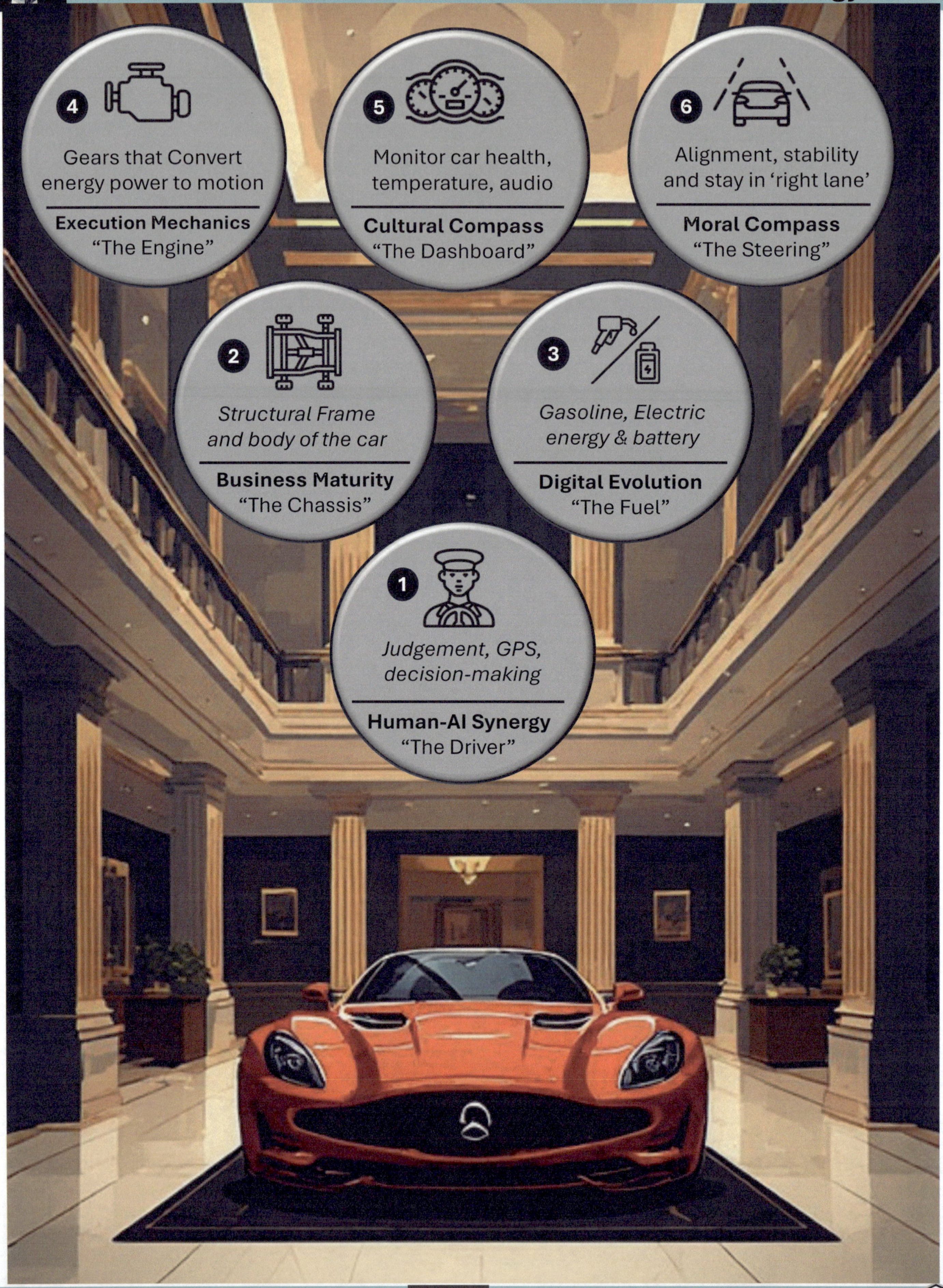
4
Gears that Convert energy power to motion
Execution Mechanics
"The Engine"
5
Monitor car health, temperature, audio
Cultural Compass
"The Dashboard"
6
Alignment, stability and stay in 'right lane'
Moral Compass
"The Steering"
2
Structural Frame and body of the car
Business Maturity
"The Chassis"
3
Gasoline, Electric energy & battery
Digital Evolution
"The Fuel"
1
Judgement, GPS, decision-making
Human-AI Synergy
"The Driver"

THE ETHICAL IMPLOSION (*Invested: Digital evolution | Overlooked: Moral compass*): You build a high-speed AI infrastructure with zero ethical guardrails or governance. One biased algorithm creates a public relations nightmare and a massive legal liability that can bankrupt the firm.

THE BUREAUCRATIC FREEZE (*Invested: Business maturity | Overlooked: Cultural compass*): You have established perfect governance and checklists, but you are operating within a terrified culture. People follow the rigid "process" of transformation but stop taking risks, effectively killing all real innovation.

THE THEORETICAL DREAM (*Invested: Human–AI synergy | Overlooked: Execution mechanics*): The organizational "brain" develops brilliant ideas for AI collaboration, but there is no "muscle" to act on them. Innovation stays trapped as a pilot project and never scales to produce enterprise value.

THE AUTOMATED ERROR (*Invested: Execution mechanics | Overlooked: Human–AI synergy*): You build a powerful execution engine but ignore the necessity of human judgment and intervention. The system begins executing "bad" decisions or flawed logic at 100x speed without anyone knowing how to stop it.

THE FUNDED HOBBY (*Invested: Cultural compass | Overlooked: Business maturity*): The internal culture is happy, safe, and collaborative, but there is no structural discipline or funding prioritization. Massive amounts of energy and capital are wasted on uncoordinated "passion projects" that never deliver ROI.

THE PRINCIPLED STANDSTILL (*Invested: Moral compass | Overlooked: Digital evolution*): You have world-class ethics and deep integrity, but you lack the "blood flow" of modern data and systems. You remain a "good" organization, but your technology is too outdated to actually compete with faster rivals.

THE FERRARI IN THE GARAGE (*Invested: Digital evolution | Overlooked: Execution mechanics*): You invest millions in an elite, high-performance tech stack that stays idle because no one has built the workflows to use it. You have purchased the "Engine" but never bothered to build the "Transmission" to turn the wheels.

THE LEGALISTIC LOOPHOLE (*Invested: Business maturity | Overlooked: Moral compass*): You strictly follow every legal rule and regulation while completely violating the spirit of ethics and trust. Your "Governance" is perfect on paper, but your brand is dying because your customers feel exploited by your "technically legal" actions.

THE INTELLIGENT OUTCAST (*Invested: Human–AI synergy | Overlooked: Cultural compass*): You develop AI tools that are technically brilliant and offer high synergy, but the workforce finds them creepy or intrusive. Because you ignored the "Senses" of the culture, the tools are rejected by the very people they were meant to help.

THE EFFICIENT COLLAPSE (*Invested: Execution mechanics | Overlooked: Business maturity*): The organization is moving with incredible power and speed, but it is moving toward a goal that isn't funded or strategically prioritized. You are winning the race, but you've realized too late that you're on the wrong track entirely.

COLLISION PATHS TO AVOID IN YOUR TRANSFORMATION

THE NICE LIAR (*Invested: Cultural compass* | *Overlooked: Moral compass*): The culture is polite, collaborative, and highly "engaged," but it is built on a foundation of hidden ethical skeletons. People are "happy" because they are protecting each other from the truth, leading to an eventual systemic implosion.

THE ANALOG SAINT (*Invested: Moral compass* | *Overlooked: Human–AI synergy*): The leadership has high integrity but lacks the "AI-IQ" to understand the risks of the modern digital landscape. They are making ethical decisions based on a world that no longer exists, leaving them vulnerable to new forms of disruption.

THE EXPENSIVE CHAOS (*Invested: Digital evolution* | *Overlooked: Business maturity*): You layer high-tech AI tools on top of disorganized, immature governance and disorganized funding. The tech doesn't solve your problems; it simply makes your existing chaos move much faster and cost much more.

THE STAGNANT FORTRESS (*Invested: Business maturity* | *Overlooked: Execution mechanics*): The organization has perfect defenses, massive funding reserves, and rigid discipline, but zero kinetic movement. You are a perfectly "safe" business that is being slowly surrounded and suffocated by smaller, faster competitors.

THE DR. FRANKENSTEIN (*Invested: Human–AI synergy* | *Overlooked: Moral compass*): You achieve brilliant "Brain-to-Machine" innovation that pushes the boundaries of what is possible. However, because you ignored the "Inner Ear" of balance and ethics, you create a system that causes more social harm than organizational good.

THE BURNOUT FACTORY (*Invested: Execution mechanics* | *Overlooked: Cultural compass*): You optimize your "Muscle" for maximum output but completely ignore your "Senses" and emotional climate. You fail to detect internal fatigue and resentment until your highest-performing talent quits en masse.

THE LUDDITE OASIS (*Invested: Cultural compass* | *Overlooked: Digital evolution*): You have created the best place in the world to work, with high safety and great vibes. However, because you ignored the "Blood Flow" of digital evolution, your company is being disrupted out of existence by competitors you can't even see.

THE TOOTHLESS POLICY (*Invested: Moral compass* | *Overlooked: Execution mechanics*): You have beautiful "Values" posters and world-class ethics policies, but no actual workflows or mechanics to enforce them. Your integrity is a theory that falls apart the moment a team is pressured to hit a quarterly target.

THE GHOST SHIP (*Invested: Digital evolution* | *Overlooked: Human–AI synergy*): You have successfully implemented full automation and "lights out" data systems across the enterprise. However, because there is no human judgment in the loop, the system continues to run on outdated logic that is driving the company into a ditch.

THE PRECISE DINOSAUR (*Invested: Business maturity* | *Overlooked: Human–AI synergy*): You have spent years optimizing and documenting every single process to a level of "Six Sigma" perfection. Unfortunately, you ignored the "Brain" of AI synergy, and those processes are now perfectly optimized for a market that has already moved on.

Human judgment amplified by AI

Human–AI Synergy defines how intelligence is created, shared, and amplified across the enterprise. It examines the evolving partnership between human judgment and machine capability, ensuring that AI elevates rather than replaces human potential. This Lens forces leaders to confront how decisions are made, how insights flow, and how people interact with intelligent systems. It exposes whether the organization is building a truly augmented workforce or simply layering tools on top of old thinking. Ignoring this Lens leads to fragmented automation, poor adoption, and inconsistent decision-making. Human–AI Synergy is essential because it establishes the cognitive foundation of transformation—how the enterprise thinks, learns, and adapts—and without it, AI becomes a disconnected tool rather than a strategic multiplier.

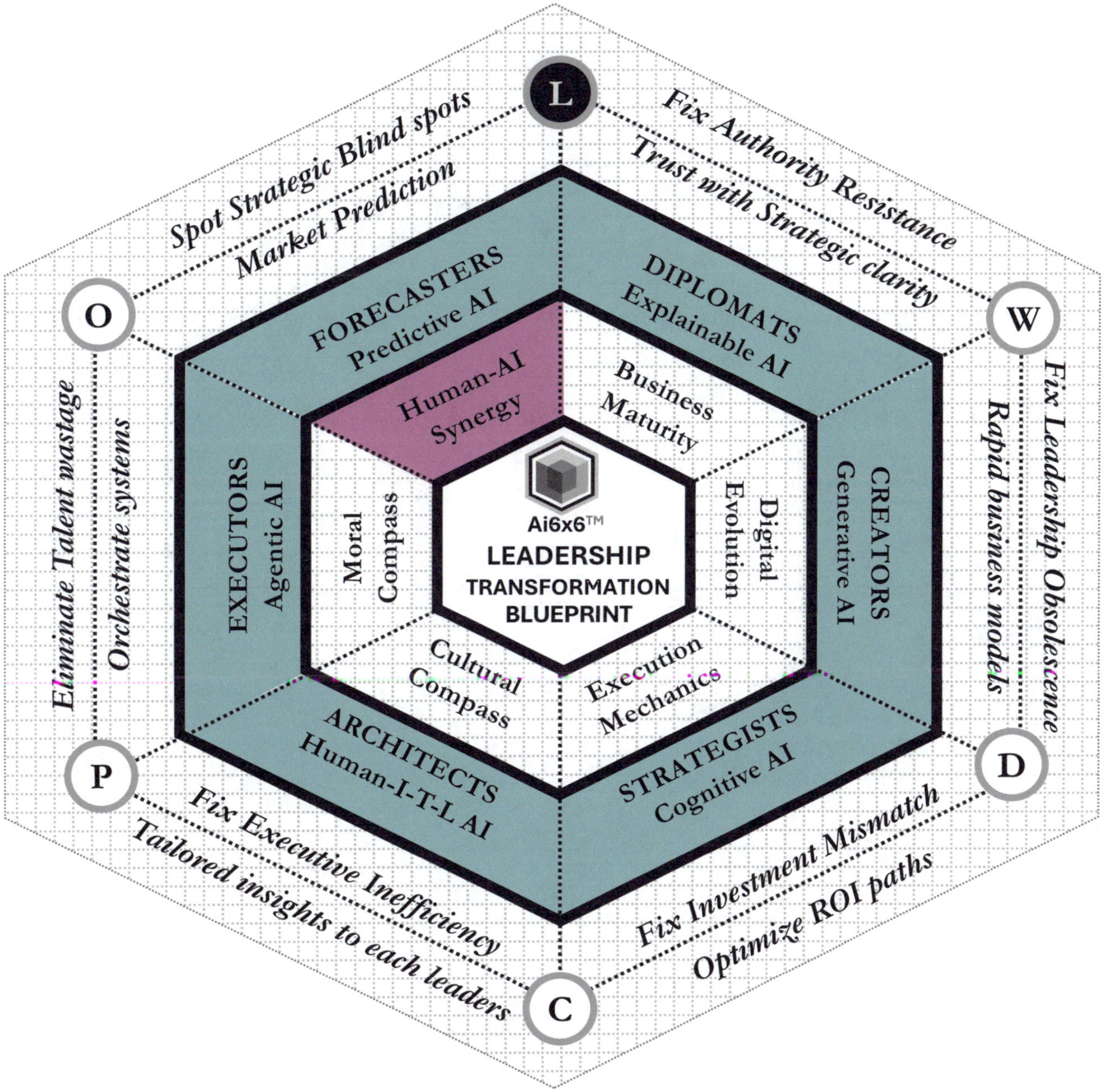

Ai6x6 HUMAN-AI SYNERGY FRAMEWORK FOR **LEADERSHIP LIFECYCLE**

Discipline that stabilizes transformation

Business Maturity is the structural discipline that transforms raw ambition into a legacy of Authority. It begins at the Foundation, establishing the governance and funding models that anchor an organization's identity before activation begins. As the enterprise moves through Initiation and Cohesion, this Lens evaluates whether the internal scaffolding is robust enough to support the increasing velocity of the transformation engine. This Lens exposes a fundamental truth: sustained success is engineered through systems, not slogans. It ensures that as Momentum builds, the enterprise maintains the Optimization necessary to prevent collapse under the weight of its own initiatives. Mastering this Lens allows an organization to transcend temporary programs and achieve the permanent market Authority that defines industry leadership.

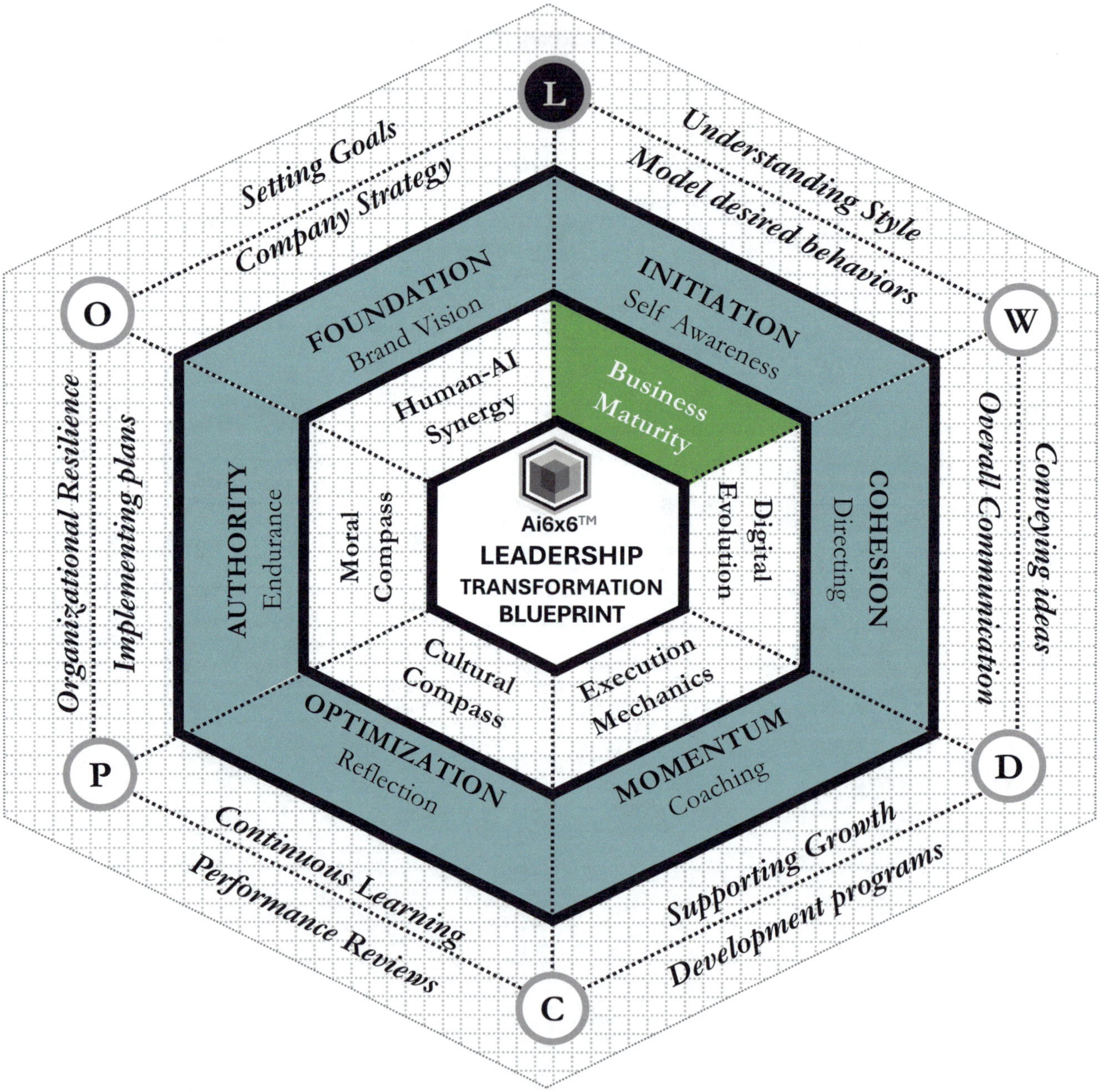

Ai6x6 BUSINESS MATURITY FRAMEWORK FOR **LEADERSHIP LIFECYCLE**

Infrastructure built for continuous change

Digital Evolution examines the technological backbone that enables modern transformation. It evaluates system integration, data flow, platform scalability, and the adaptability of digital environments. This Lens ensures technology accelerates rather than obstructs progress. It forces organizations to confront technical debt, architectural fragility, and the true readiness of their digital ecosystem. When ignored, enterprises build brittle systems, slow innovation, and create environments where strategy moves faster than infrastructure. Digital Evolution is not about tools—it is about constructing a digital foundation capable of continuous change. It ensures that technology becomes an enabler of intelligence, speed, and resilience. Without this Lens, even the strongest strategies fail because the underlying environment cannot support enterprise-wide evolution.

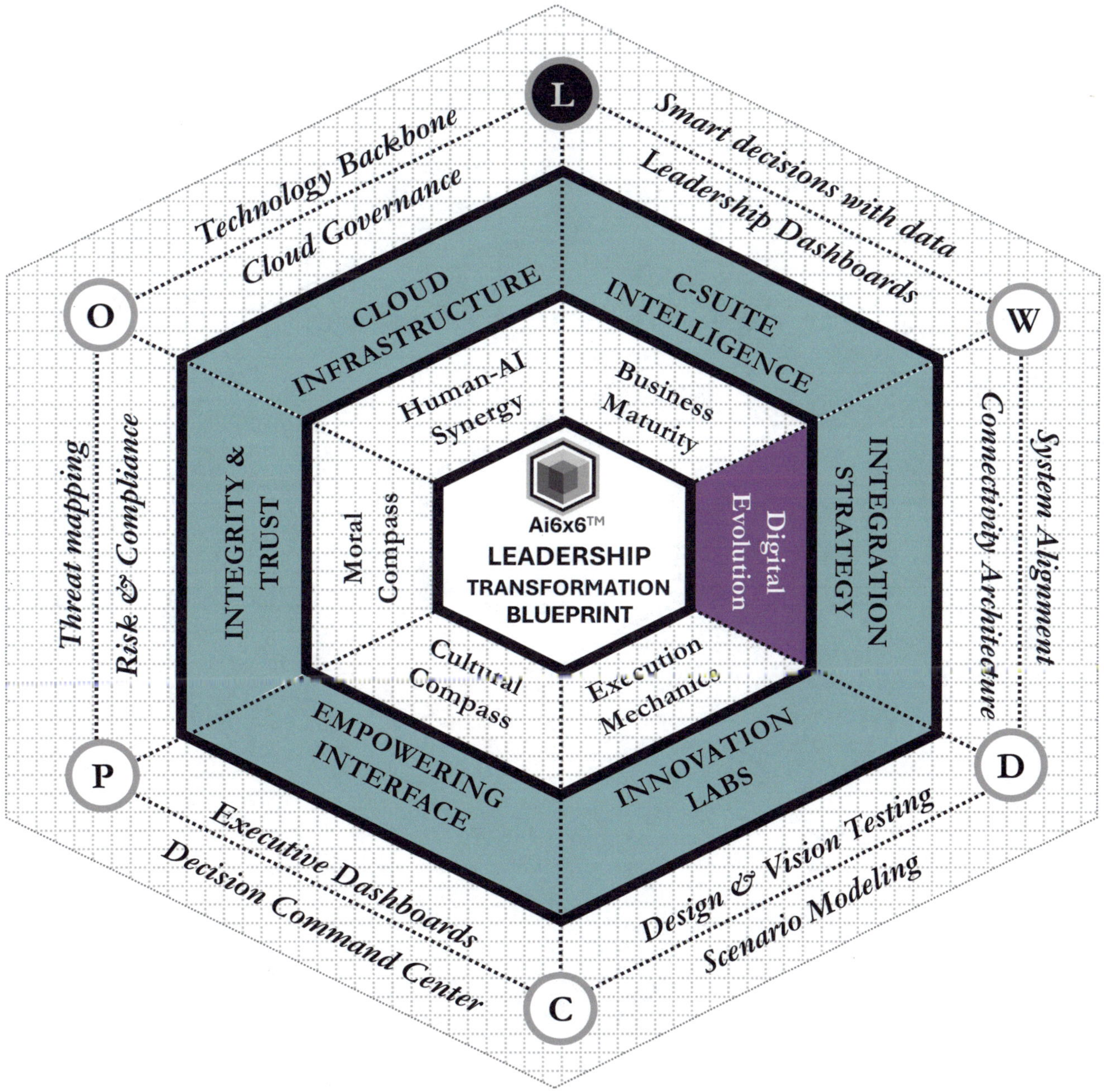

Ai6x6 DIGITAL EVOLUTION FRAMEWORK FOR **LEADERSHIP LIFECYCLE**

Strategy delivered with operational precision

Execution Mechanics converts strategy into consistent, scalable action. It examines operating models, workflows, methodologies, and performance systems to determine whether the organization can deliver reliably. This Lens exposes the gap between leadership intent and frontline reality, revealing whether execution is engineered or improvised. It forces leaders to confront bottlenecks, handoff failures, and the true velocity of delivery. Ignoring this Lens results in missed deadlines, inconsistent outcomes, and organizational frustration. Execution Mechanics is essential because it transforms ideas into outcomes and outcomes into advantage. It ensures that transformation is not theoretical but operational—a system of delivery that produces measurable impact at scale.

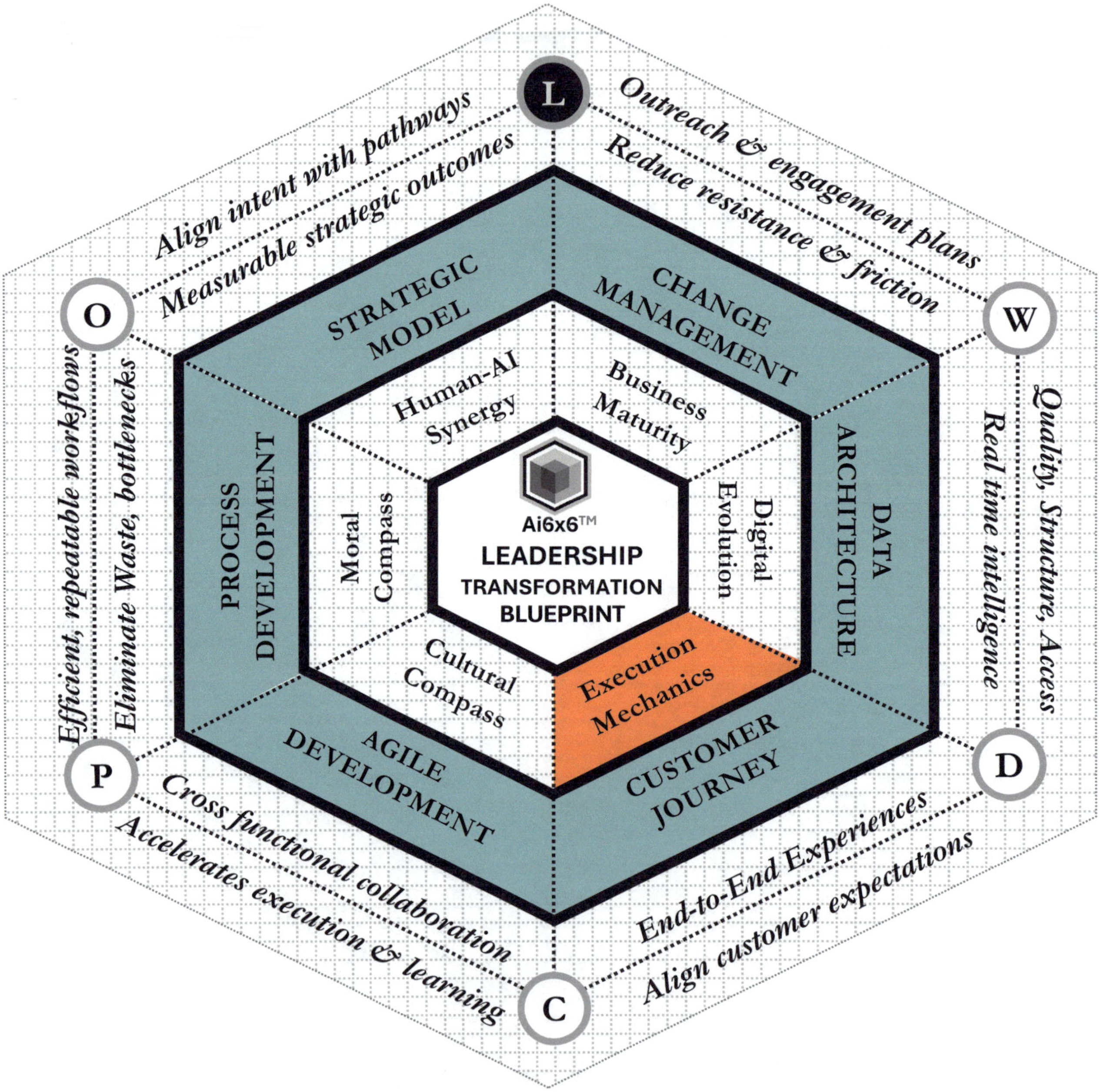

Ai6x6 EXECUTION MECHANICS FRAMEWORK FOR **LEADERSHIP LIFECYCLE**

Culture aligned to accelerate change

Cultural Compass explores the beliefs, behaviors, and shared meaning that shape how people think, collaborate, and respond to change. It examines psychological safety, communication norms, leadership behaviors, and emotional climate. This Lens reveals whether culture accelerates transformation or quietly resists it. It forces leaders to confront the invisible forces that determine engagement, trust, and momentum. Ignoring this Lens means ignoring the human system that ultimately determines success. Even the best strategies fail in cultures that are misaligned, fearful, or fatigued. Cultural Compass is essential because transformation is a human endeavor—no system or technology can succeed without cultural readiness. It ensures people are not just compliant but committed, creating the conditions for lasting change.

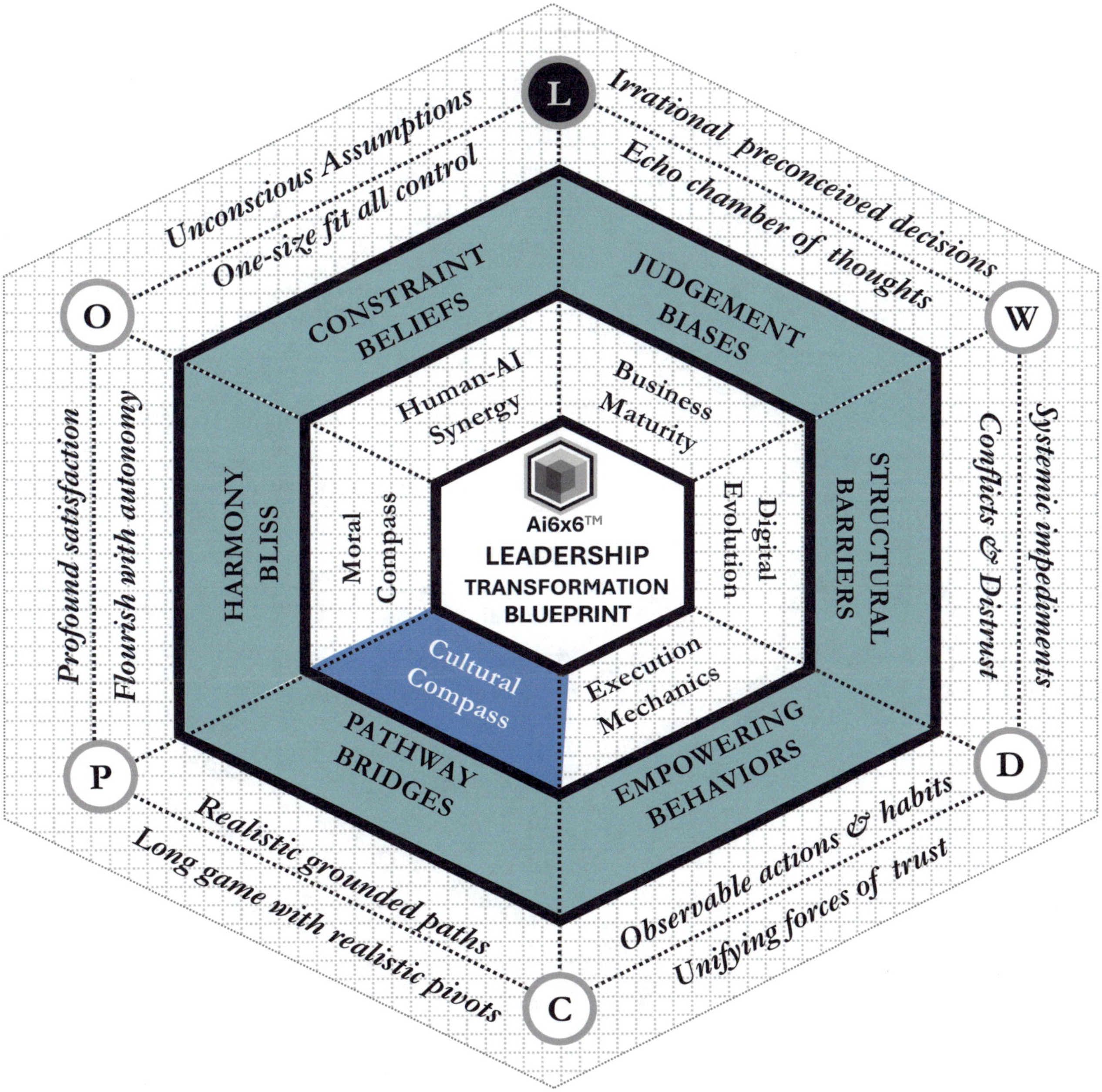

Ai6x6 CULTURAL COMPASS FRAMEWORK FOR **LEADERSHIP LIFECYCLE**

Ethics governing intelligent enterprise systems

The **Moral Compass** anchors transformation in ethics, transparency, and responsibility. It evaluates data privacy, AI fairness, accountability structures, and the moral obligations of leadership. This Lens ensures that innovation is principled and that stakeholders feel protected, respected, and informed. It forces leaders to confront the ethical implications of their decisions and the long-term consequences of their systems. Ignoring this Lens exposes organizations to reputational damage, regulatory risk, employee resistance, and customer disengagement. Trust is the boundary condition of modern enterprise—without it, nothing else holds. The Moral Compass is essential because it defines the legitimacy of every decision, system, and outcome, ensuring transformation is not only effective but worthy of confidence.

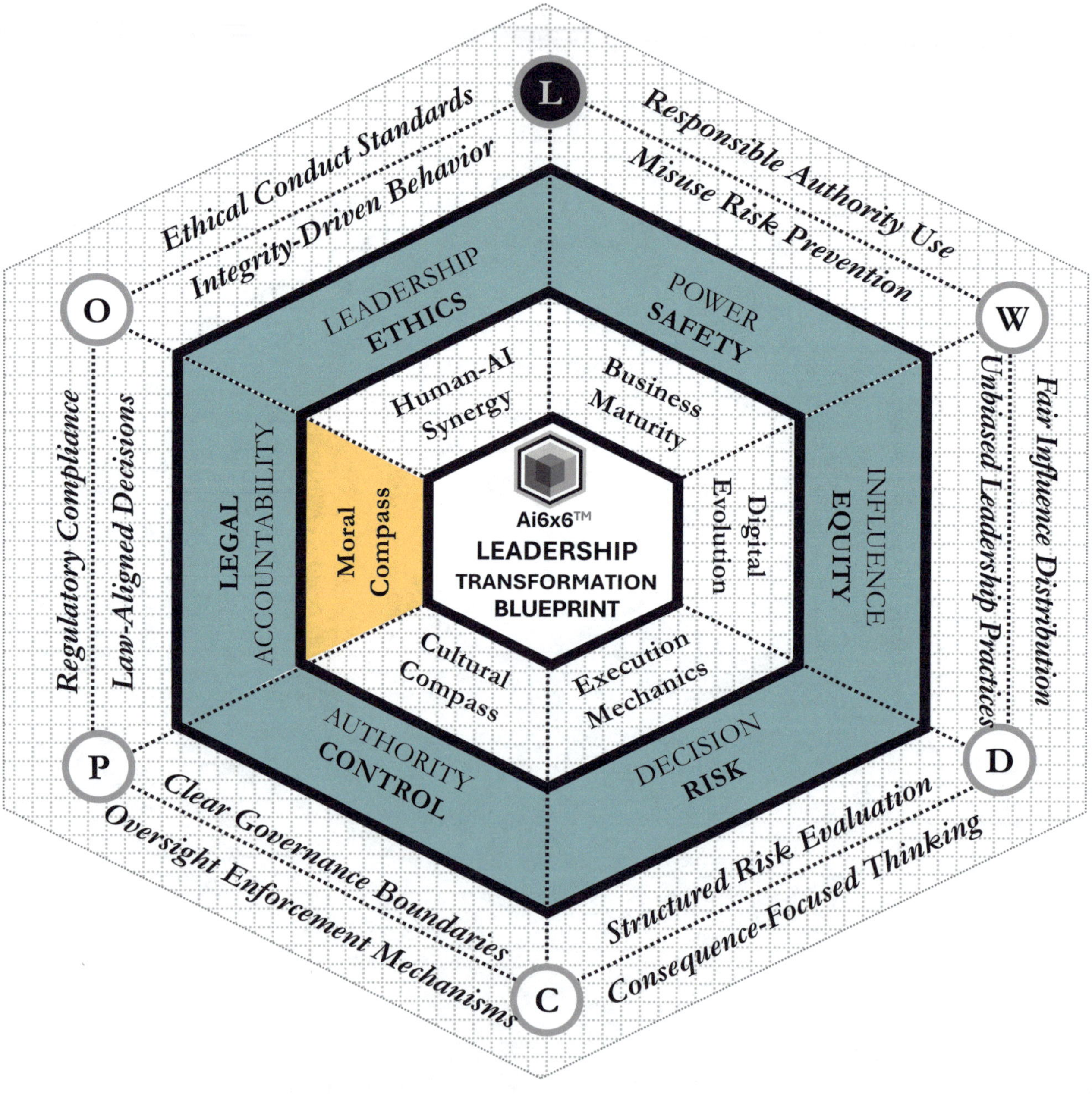

Ai6x6 MORAL COMPASS FRAMEWORK FOR **LEADERSHIP LIFECYCLE**

6x6™ Transformation Blueprint

Why Transform Leadership?

"The first responsibility of a leader is to define reality. The last is to say thank you. In between, the leader is a servant." — Max DePree

Max DePree was an American businessman and writer who served as the CEO of Herman Miller, a major furniture manufacturer. He is known for his influential books on leadership.

A new kind of leadership is essential in today's fast-paced, complex world. Traditional top-down hierarchies stifle innovation and fail to engage talent who demand purpose and autonomy. Without this transformation, organizations face high employee turnover and an inability to adapt to market shifts. The modern leader must move beyond managing tasks to truly inspiring and empowering teams, fostering a culture of psychological safety where creativity can thrive. This shift is critical for building resilient, agile organizations that can lead, not just survive, disruption. The benefit of this transformation is a highly engaged workforce, resulting in increased productivity and a strong competitive edge.

1. **Enhanced Strategic Foresight:** Leaders leverage real-time data feeds and predictive models, enabling them to abandon slow, annual planning for continuous strategy calibration that anticipates market forces and regulatory changes. This proactive stance ensures market leadership rather than reaction. The continuous feedback loop prevents sunk costs in obsolete projects and rapidly reallocates capital to high-growth areas based on hypotheses.
2. **Improved Change Agility:** The leadership actively removes friction points and bureaucratic overhead, developing a culture of low-friction change where shifts in strategy, resource allocation, and priorities can be executed across the enterprise in days, not months. By codifying organizational structures into easily changeable artifacts, the cost and emotional labor associated with restructuring are virtually eliminated. The entire organization can pivot to respond to external shock or opportunity while maintaining operational stability.
3. **Decentralized Decision-Making:** By establishing clear, non-negotiable constraints and guardrails, authority is systematically delegated to the edge of the business, empowering front-line teams to resolve issues and innovate in real-time without relying on management bottlenecks. This reduction in latency between customer signal and organizational response translates directly into superior service and quicker feature releases. Empowered teams become the primary engine of operational excellence, not just executors of top-down mandates.
4. **Better Systemic Risk Governance:** Leadership mandates the codification of risk management (Immutable Law), ensuring critical risks like technical debt, data security, and platform stability are managed proactively through automated, machine-enforceable constraints. This shifts the audit function from backward-looking assessment to continuous, forward-looking prevention, dramatically lowering the probability of catastrophic failures. Security and compliance become integrated features of the architecture, not expensive, periodic overlays.
5. **Stronger Talent Retention:** By articulating a clear, aspirational digital vision and providing challenging work with modern tools, leaders inspire employees and retain critical high-performing talent who seek purpose-driven, autonomous work environments. A low-churn rate among specialized engineers and data scientists preserves institutional memory and stabilizes velocity across critical development pipelines. Reduced recruitment costs and faster project completion both contribute to improved bottom-line performance.
6. **Accelerated Organizational Learning:** Leaders prioritize funding for small, iterative experimentation and require mandatory, rapid feedback loops, transforming the enterprise into a systematic learning machine that continuously refines its operating procedures and market strategy. The systematic capture and sharing of lessons learned across business units prevents the replication of past failures and amplifies successful innovations globally. This constant refinement ensures the organization's competitive edge is sharpened.
7. **Shorter Strategy-to-Execution Cycle:** The organization moves away from massive, multi-year programs toward manageable, short-cycle strategic initiatives (often 90-day cycles), dramatically reducing the time between a strategic idea and market deployment. This constant, high-frequency delivery of strategic value ensures capital expenditure yields measurable results rapidly, improving ROI visibility.
8. **Increased Investor and Stakeholder Confidence:** Transparent reporting on digital maturity, security posture, and innovation velocity demonstrates a modern, resilient business model, increasing confidence from financial markets and stakeholders. The demonstrated ability to adapt to black swan events and regulatory shifts results in a premium valuation compared to slower, non-transformed competitors. Reliability and adaptability are priced into the stock multiple.
9. **Cultivation of an Experimentation Mindset:** Leadership actively budgets for and celebrates "safe-to-fail" experiments, creating an environment where teams are rewarded for learning quickly, not just for delivering features, which fuels radical, high-value innovation. The de-risking of innovation allows the organization to explore adjacent and disruptive technologies that competitors are too conservative to pursue. This proactive exploration creates proprietary market space and sustainable differentiation.
10. **Establishment of Clear Accountability Boundaries:** By defining explicit ownership over every product, system, and data asset (e.g., "Team X owns Data Product Y"), leaders eliminate ambiguity, improving operational handoffs and incident response times. This clarity of ownership eliminates costly "blame games" during outages and ensures continuous service improvement is tied to measurable team metrics. Incident resolution becomes predictable, professional, and efficient.

1. The Leadership Crisis

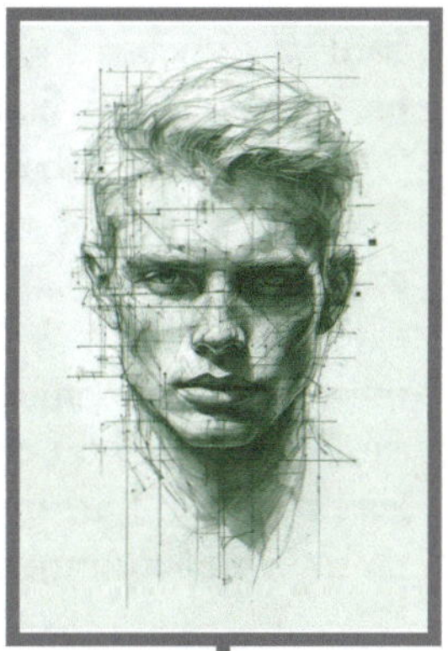

Everyone admired David—the executive who never missed a quarterly target—but behind closed doors, his team was burning out and disengaged. Meetings felt transactional, vision was blurry, and decisions flowed top-down. It wasn't malicious—it was momentum. He was leading like leaders always had: with certainty, distance, and performance metrics. But the world had shifted. A younger colleague finally said, "We don't need more direction—we need connection." That moment stuck. Across industries, similar stories have played out. Once-respected leaders now face a silent attrition of trust and meaning. Employees want purpose, not just a paycheck. The pandemic magnified these cracks, exposing just how fragile traditional leadership models really are. Micromanagement no longer scales, and charisma alone doesn't translate across a Zoom screen. Leaders are realizing that technical competence isn't enough—they must now inspire, adapt, and genuinely listen.

Leadership is no longer about having the answers; it's about asking the right questions—and being brave enough to evolve.

2. The Age of Acceleration

Three months into her new role, Priya realized something jarring: every decision she made was outdated by the time it was implemented. Markets were shifting overnight, customer behavior was rewriting itself in real time, and AI tools were automating half the workflows she thought were "future-proof." Leadership today isn't about planning every step—it's about navigating in motion. We live in an age of acceleration, where change is not episodic but exponential. Technology, globalization, hybrid work, and generational shifts are compressing time and amplifying complexity. Leaders are being pulled into more decisions, with less certainty and higher stakes. In this environment, waiting for full information is a luxury—and trying to stay in control is an illusion. The leaders who thrive are those who experiment boldly, adapt quickly, and empower others to act autonomously.

It's no longer about playing it safe—it's about moving with velocity and vision, or being left behind entirely.

3. Beyond Titles: A New Definition of Leadership

Jason never had the "manager" title, but when a crisis hit, his team looked to him—not because of hierarchy, but because of trust. While others froze, he stepped in with empathy, gathered information, asked questions, and offered clarity. That was leadership. For decades, we've equated leadership with position—corner offices, direct reports, decision authority. But today's world demands something different. Leadership is becoming less about control and more about influence. It's not about knowing all the answers; it's about surfacing the right questions. It's not about issuing commands; it's about creating clarity amidst chaos. In this new era, the best leaders are listeners, connectors, and meaning-makers. Anyone can lead—from any level—when they take responsibility for the energy, culture, and outcomes around them.

The world is rebalancing toward shared power and horizontal trust. The question is no longer "What's your title?" but "How do you show up when it matters?"

4. Self-Awareness as a Leadership Catalyst

Sam was known for her brilliance, but also her emotional volatility. In high-stress meetings, her reactions left colleagues tiptoeing around her moods. One day, her CEO said gently, "People follow what you model, not what you expect." That became her turning point. Through coaching and reflection, Sam uncovered patterns: where she deflected feedback, avoided vulnerability, or reacted instead of responding. As she grew in self-awareness, everything changed. Meetings became safer, ideas flowed more freely, and her team began mirroring her openness. In today's landscape, self-awareness is not a soft skill—it's a strategic one. Leaders who understand their triggers, values, and blind spots are far better equipped to lead complex, human systems. They own their impact, regulate their presence, and grow faster than their environment demands.

Leadership transformation doesn't start with spreadsheets or slide decks—it starts with a mirror. And the courage to really look.

5. Purpose and Resilience

When Marcus's startup hit a cash crunch, investors panicked and morale tanked. But Marcus stayed grounded—not in bravado, but in purpose. He reminded the team why they'd started: to make healthcare more accessible. They cut perks but kept mission-critical roles. They downsized, but didn't despair. That clarity kept people focused and proud. Resilience isn't just about bouncing back—it's about bouncing forward, anchored by meaning. Leaders with a clear "why" don't crumble under pressure—they reframe, recommit, and reengage. Purpose gives resilience its spine. It's the difference between burnout and grit, anxiety and drive. In a chaotic world, followers don't rally behind perfect leaders—they rally behind authentic ones with a cause worth showing up for. Whether you're in a startup or a legacy enterprise, resilience lives in leaders who turn challenges into chapters, and setbacks into fuel.

Purpose is the compass—and resilience, the path.

6. Leading Through Uncertainty

In 2020, Layla had to lay off half her staff over Zoom. She didn't sugarcoat. She didn't script. She showed up fully—teary-eyed, clear, and honest. And strangely, her people rallied. They said later, "We didn't lose trust because of the layoffs—we gained it because of how you led them." In uncertainty, clarity is more powerful than certainty. Modern leaders can't promise outcomes—they can only promise presence. They can't eliminate risk—but they can model composure. Today's leadership playbook doesn't reward bravado—it rewards grace under ambiguity. Layla didn't have answers. But she made decisions with integrity, communicated openly, and stayed visible—even when it hurt. And in doing so, she built resilience into her culture.

Leadership through uncertainty isn't about predicting the storm—it's about becoming the calm in its eye. And reminding others: we may not know the path, but we're walking it—together.

7. From Authority to Trust

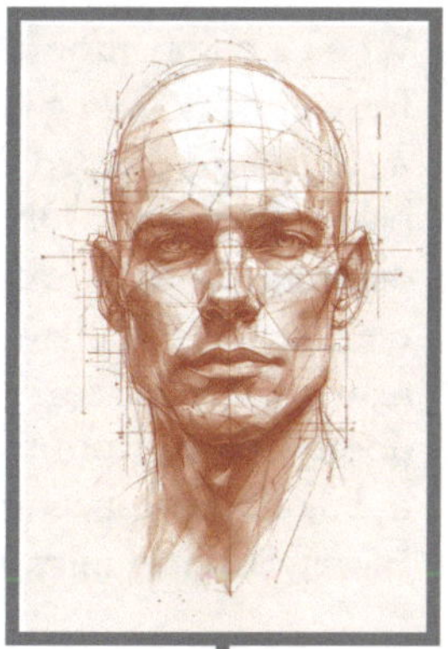

Alex had the title, the office, the strategy deck—but his team never spoke up. Until a junior engineer finally said, "We don't feel safe being wrong around you." That hit hard. In the old model, authority equaled credibility. Today, credibility flows from trust—and trust is built through humility, consistency, and care. When Alex started listening more than speaking, owning his mistakes, and showing real curiosity, everything shifted. Meetings became livelier. Risks felt less scary. People began offering bold ideas without fear. Leading through authority may yield compliance—but leading through trust yields commitment. And in fast-moving environments, commitment is everything. Trust isn't granted by title—it's earned through behavior.

Leaders who build psychological safety become magnets for creativity, accountability, and innovation. Because when people feel seen and safe, they bring their best—not just their job description.

8. The Science of Influence

Sofia was brilliant—but her ideas often fell flat in leadership meetings. It wasn't her logic. It was her delivery. Then she studied storytelling, behavioral science, and cognitive biases—and everything changed. She opened pitches with emotional hooks. Framed data in user stories. Anchored visions in shared values. Her ideas began landing, not because they were louder, but because they resonated. Influence is no longer a game of dominance—it's a dance of empathy and design. Modern leaders don't push—they persuade. They don't manipulate—they mirror. Influence lives at the intersection of psychology, narrative, and presence. It's knowing when to speak, when to listen, and how to frame ideas so others feel ownership. In the transformation era, brilliant thinking isn't enough.

Leaders must be architects of belief—able to shape conversations, spark action, and enroll others in bold futures. Influence isn't charm. It's strategy with heart.

9. Coaching, Not Commanding

Luis once managed like a checklist—assigning tasks, correcting mistakes, repeating instructions. He got results, but never growth. Then he tried something different. During 1:1s, he stopped talking first. He started asking: "What's your take?" "What's blocking you?" He began coaching instead of commanding. His team started owning problems, not outsourcing them. One designer said, "For the first time, I feel like this is mine." Coaching isn't about having all the answers—it's about helping others find theirs. Modern leadership is a multiplier game. It's not about your genius—it's about creating the conditions for genius to emerge around you. Coaching leaders foster psychological safety, drive autonomy, and build capability. They see people not as assets to manage, but as talent to unlock. And in return, teams don't just perform—they grow.

In fast-changing environments, the leaders who scale are the ones who help others do the same.

10. Designing for Adaptability

When a competitor launched a similar product overnight, Monica's entire roadmap crumbled. Instead of doubling down, she paused—and invited her team to rethink everything. Within a week, they'd pivoted, built a prototype, and found a new niche. Why? Because adaptability was baked into their DNA. Their processes were agile, their culture curious, their leadership distributed. In a world of disruption, rigid plans are liabilities. Adaptive leadership isn't reactive—it's designed. Leaders who build systems that learn—structures that evolve and teams that self-correct—outpace those who cling to control. In transformation, stability doesn't come from sticking to the plan—it comes from sticking to the plan—it comes from building the capacity to shift. Adaptive leadership means turning every challenge into a learning opportunity. It's not "What did we lose?" but "What can we now build?"

Monica's team didn't survive because they predicted the future. They thrived because they were ready for any future. And that's the new definition of operational strength.

11. Digital Fluency and AI Ethics

When Raj's team introduced a customer support chatbot, efficiency soared—but something was off. Complaints rose, biases appeared in responses, and trust eroded fast. Raj realized they'd rushed tech adoption without ethical oversight. In today's world, digital fluency isn't just about mastering new tools—it's about navigating their implications. Leaders must understand enough to ask tough questions: Is this AI equitable? What data are we using? Who's left out of this automation? Tech and ethics must now be partners, not afterthoughts. Digital fluency means being curious about emerging tools—but also conscious of their unintended consequences. It's not about coding—it's about responsible adoption. The best leaders won't just chase innovation—they'll steer it with intention. They'll know how to spark experimentation without losing empathy.

In a world where technology can scale harm or healing, ethical fluency is the ultimate differentiator. Because what we build matters—but how we build it matters more.

12. Building Regenerative Cultures

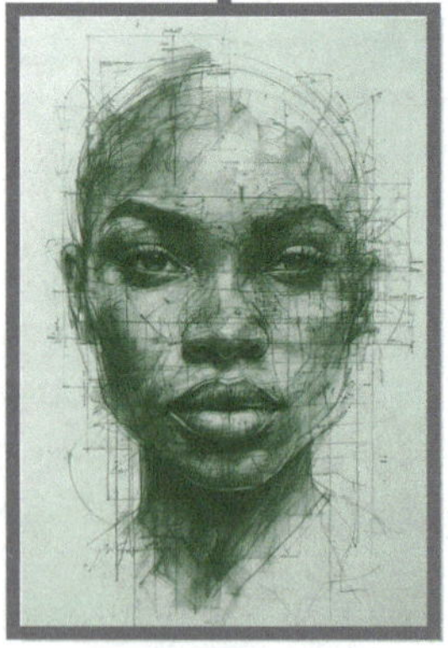

Jasmine's company used to celebrate burnout as a badge of honor. That changed after two top performers quit citing exhaustion. Jasmine paused—and decided to rebuild. They introduced rest policies, created spaces for reflection, and redefined "performance" to include well-being. A year later, retention soared, creativity returned, and the culture felt alive again. Regenerative cultures don't just sustain—they renew. They nourish people, adapt proactively, and give more than they take. Leaders play the role of gardeners—not mechanics—cultivating environments where teams thrive beyond the next quarter. Especially now, when burnout and disengagement are epidemic, culture isn't an HR initiative—it's a leadership responsibility. Regenerative cultures don't just make work better—they make people better.

Because the most powerful transformation doesn't start with strategies or structures—it starts with how it feels to show up every day. And stay.

6x6™ Transformation Blueprint

Power of Diversified Insight

"AI will not replace humans, but those who use AI will replace those who don't."

Ginni Rometty, Former CEO of IBM. This widely-cited quote frames the AI economy not as a job elimination crisis, but as an augmentation imperative, making the ability to adapt and integrate AI tools the single greatest competitive advantage for both individuals and companies.

We have deliberately chosen to move beyond the limiting confines of the conventional single-expert **Foreword**, which, by its nature, risks restricting the reader to one singular point of view. In an era where **business transformation** is shaped by a confluence of disruptive forces, relying on a solitary oracle is fundamentally inadequate. Instead, this book opens with a **curated collection of 15 interview snippets** from a diverse array of global thought leaders. This method immediately immerses you in a rich, multi-faceted tapestry of opinions and real-world experiences. By foregrounding this **plurality of perspective**, we offer a wider, more nuanced, and dynamic lens through which to view the challenges and enormous opportunities presented by modern change. Consider these short, powerful excerpts a **Chorus of Counsel**: a comprehensive and inclusive foundation built on collective wisdom that better equips you for your own complex journey of enterprise innovation and transformation. This is a commitment to **diversified insight** from the very first page.

Gain unprecedented, insider access to the minds of 27 global thought leaders in this six-book series. Sourced from comprehensive 1-on-1 interviews, these experts are recognized worldwide for their work in business transformation and emerging technology. Their insights are drawn from my personal, professional network—including former bosses, colleagues, and business partners —giving you an insider's view into the minds of industry shifters. To maximize focus and utility, we've synthesized these deep-dive conversations into thematic 'interview cards.' Each of the six books offers 15 relevant, concentrated insights on specific topics, allowing you to quickly access and apply the most relevant expertise, no matter where it originated in the original interview.

EXPERT VOICES

Subrata Mukherjee: What is your philosophy for structuring and piloting growth, particularly for technology companies operating in a fast-paced environment?

Françoise Nauton-Inglis: Pace comes from focus. I use a three-horizon portfolio: protect core reliability, push one or two growth bets, and explore a small set of option plays. Governance is key and needs to mobilize people across the entire company. Our decision-making process relies on exploring and aligning different vantage points. We steer by implementing few but meaningful leading indicators, not quarterly surprises or dense dashboards. Finally, we ground growth in basics: quality onboarding, transparency including when making tough decisions, customer relations —so teams can move fast without breaking trust.

SM: You've been involved in M&A and international expansion. What is your strategy for integrating new teams and technologies across different cultures and geographies?

FNI: Integration is new journey in itself, I strongly believe it cannot just be a repeatable process. We start with culture and customers before tools and rules. This usually starts with a long observation period where you initially adapt your organization to the new one, not the other round.This helps avoid three important and common mistakes during the integration process. One, changing things for the sake of changing them, which I also call trying to fix what is not broken, which leads to strong disengagement on a human level. Two, miss out on understanding best practice from your acquisition and fail to incorporate it in your own business. And three, ignore cultural differences that have a strong impact on customers as well as staff and managers. Once that is done you can agree on decision rights, objcctives, reporting and so on. Local leadership has autonomy on go-to-market within a common brand promise—freedom with accountability, not uniformity for its own sake. Tech stacks are harmonized at integration points first, not wholesale rewrites.

SM: As a consultant and a leader, how do you help businesses identify and solve the right problems with technology, instead of just implementing the latest trend?

FNI: We start with friction mapping: where customers get stuck, where teams rework, and where the P&L leaks. From there, we propose the smallest intervention with the biggest learning signal. If AI is the answer, great; if not, we celebrate a simpler fix. We insist on a decision review after 60–90 days: did behavior change? If yes, scale; if no, pivot. This discipline prevents "trend deployments" and builds confidence that tech serves strategy, not the other way around.

SM: Beyond your work with AI for e-commerce, what emerging technology will have the most significant impact on the business world in the next five years?

FNI: I'm watching three trends that need to converge: agentic AI with verifiable context (retrieval + actions), privacy-preserving analytics at the edge, and sustainable IT. Together, they make digital services more proactive and accountable—able to act for customers while proving why those actions were safe, compliant, and beneficial.

SM: What role does external funding play in the growth of a company like yours, and what advice would you give to other founders seeking investment?

FNI: Funding is a tool, not a goal. We calibrate the capital mix to the shape of the opportunity: product R&D and go-to-market may justify external fuunding; services lines often scale best with disciplined cash flow. My advice: prove pull before you pour — secure the right partners and people to lead new initiative, show behavioral lift and ensure the results are tangible not just promises. Choose investors who value responsible growth and operational excellence, not just headlines. Money should accelerate fit, not replace it.

FRANÇOISE NAUTON-INGLIS

CEO @ Spiriit / Juliia, Ex GM/CFO of Kaliop Interactive Media

Françoise Nauton-Inglis is an accomplished business leader with over 20 years of experience in the digital and technology sectors. She is the CEO of Spiriit, an agency that designs and builds custom digital and AI solutions, and its AI-first product, Juliia, which focuses on intelligent search and product recommendations for e-commerce. Françoise's expertise lies in growth management, international development, and strategic business planning. Prior to Spiriit, she spent a decade at Kaliop Interactive Media, where she served as the Group General Manager, overseeing M&A, international expansion, and financial operations. She has a deep passion for structuring and growing innovative, profitable, and human-centered businesses, and she holds leadership positions in several organizations, including as President of Digital 113 and CI Montpellier Management. She is based out of Montpellier, France.

Subrata Mukherjee: Can you share an early technical project that, in hindsight, taught you a crucial lesson about managing people or business processes?

Alex Demichev: Every project you do teaches you something and becomes part of your life. Some of the most transformational for me were sometimes not the biggest successes. I once failed a project, we delivered it later than we should have, and with much more effort that was originally planned. By all standards it should have been considered a failure. But me and my team got rewarded by the founder of our company, because it turned out that everyone, including the client, thought we wouldn't be able to deliver it at all, and our commitment and effort led to them trusting us with everything else they had and led to what has later become one of the most successful and long term cooperations for us.

SM: How does your background give you a unique advantage in understanding emerging tech like AI?

AD: Technical background gives a man critical thinking capabilities, well organized thought processes, and the ability to deal with complex problems. AI or other emerging technologies require all that to be used to the full extent, but at the same time not overly rely on things that are not yet ready to be relied upon.

SM: With such a broad scope, what was your personal strategy for managing and mastering all the different aspects of the business?

AD: My personal priorities were always the same – figure out what will be in demand next, focus on client centricity and talent retention. It was like that when I was a project manager, department manager and then the CEO. Our company was always organized in a way that resembles holding structures. So in a way I was a CEO of my part of the holding long before I became the CEO of the whole holding. I believe the crucial part of the transition to higher management position was the switch from a developer to a project manager. Tactical steps to achieve that within larger groups of people vary, but you do it step by step. You learn to trust other people manage themselves. Without that there is no way up the corporate ladder. No one ever wakes up and becomes a skilled department manager/CEO.

SM: How do you plan to leverage AI to solidify ITransition's position as a leader in the global market?

AD: In full accordance with my advice above we are leveraging AI to deliver the most possible value to our clients even if they don't know they can achieve something as quickly/efficiently as it is possible these days using AI. Our goal is to be at the forefront of this technical revolution and constantly reassess the possibilities of the technology (which progresses every month/couple of weeks to heights which seemed impossible in the near past), and provide our clients with the best and the latest so they can benefit from it before their competition can. It is a very ambitious and fast pace game but only those who are fast can take the most advantage from it at this point.

SM: What do you see as the next major frontier for the IT services and software development industry? What advice would you give to other technical professionals who aspire to follow a similar path from a hands-on role to a C-level position in the age of AI?

AD: I see further consolidation, smaller and simpler jobs becoming the domain of AI driven/saas businesses, while larger enterprise grade projects being handled by fewer but larger corporations with tougher entry barriers and fierce competition. Focus on making your client happy by giving him what he needs not what he asks you for. There's a fine balance here, but look into what they REALLY need, and try to give it to them based on your laser sharp vision of the most modern technology, a kind your client might not even know about. Use the advantage of knowing all the most cutting edge stuff as a great value add you can provide to your client.

ALEX DEMICHEV

CEO at ITransition Group

Alex Demichev is a business-oriented software professional with a rich background spanning over two decades at ITransition Group, where he has risen from a Department Manager to the Chief Executive Officer. His deep technical and operational expertise, honed over nearly 14 years in management, is complemented by a strategic business perspective gained from his M.B.A. at the University of Oxford - Saïd Business School. As CEO, Alex leads a company focused on helping clients with complex technology challenges, including M&A integration, ERP modernization, and broader digital transformation. His leadership combines a passion for detail with a comprehensive understanding of business needs, guiding the company to deliver robust software solutions and drive value for its clients. He is based out of Berlin, in Germany, Europe.

EXPERT VOICES

Subrata Mukherjee: You've founded multiple companies, including a physical goods company, Skiens Cykkelfabrik, alongside your software ventures. What is your process for identifying a new business opportunity, and what was the unique challenge of starting a bicycle manufacturing company compared to a software company?

Bard Farstad: For me, new business ideas rarely start with spreadsheets or market analysis. They usually begin with a passion for great products and the thought, "I can do this better than what is available today." My process always starts with the product itself, combining uncompromising engineering with beautiful design. That approach has guided me in both software and physical goods. Of course, manufacturing a bicycle is very different in execution from building software. Supply chains, logistics, and production constraints play a much bigger role. But the core principles are the same: focus on engineering excellence, be design-driven, and never compromise on quality. In both cases, the product defines the price point, not the other way around. It may not be what you learn in business school, but I believe that building remarkable products is the best way to build a remarkable company.

SM: What is the single most valuable piece of advice you'd give to a first-time founder?

BF: For a first-time founder, ignorance can actually be an advantage. You often start out thinking, "How hard can it be?" and that naive optimism gives you the courage to try things that others might dismiss as impossible. Of course, you quickly discover just how challenging it really is, but by then you are already moving forward. The most valuable lesson is that success ultimately comes down to people. Build a team with diverse skills, and surround yourself with those who know things you do not. Listen, learn, and empower them to excel. And above all, persistence matters more than anything else. If you never quit, you will never truly fail.

SM: How has your experience working with organizations in both Europe and the USA influenced your approach to building a global company?

BF: Working with organizations in both Europe and the USA has reinforced my belief that selling is one of the hardest things you can do. That is why I have always been product-first. I want to create products that people naturally want to buy, not products that need to be aggressively sold. This mindset applies both to the end users who rely on the product every day and to the developers who build on top of it. My experience across regions has shown me that while business cultures may differ, a strong product transcends those differences. Today, this approach is often described as product-led growth, where the product itself drives adoption and customer loyalty rather than relying primarily on a sales organization. For me, that philosophy has been the most effective way to build a truly global company.

SM: What is one technology trend you're most excited about for the future of online commerce?

BF: The most exciting technology trend for the future of online commerce is without a doubt AI. We are only at the very beginning of seeing how it will reshape everything from content creation to personalized shopping experiences. AI will act as a true co-pilot, helping businesses generate and optimize product information, tailor marketing, and even automate customer support. On the customer side, it will make discovery more intuitive, recommendations more relevant, and transactions more seamless. Combined with headless and API-first architectures, AI can be embedded directly into the commerce workflow rather than bolted on as an afterthought. The pace of change is extraordinary, and I believe AI will define the next era of commerce in the same way mobile once did.

BARD FARSTAD

CEO, Crystallize Headless Commerce

Bård is a serial entrepreneur and the CEO of Crystallize, a modern headless e-commerce service built on his vision to reimagine e-commerce platforms. His extensive career is defined by a history of founding and leading multiple companies, demonstrating a broad range of expertise that spans product, engineering, marketing, and digital strategy. In addition to Crystallize, he founded Snowball Digital, a product design and engineering firm, and Skiens Cykkelfabrik, a bicycle design company. Notably, he spent nearly two decades as the founder and chief strategist of eZ Systems, a content management software company, where he was a key driver of its engineering and product direction. With long-standing experience working with organizations across Europe and the USA, showcasing his commitment to strategic guidance and business development. He is based out of Norway.

EXPERT VOICES

Subrata Mukherjee: How can a modern CIO transition from being a technology service provider to a strategic driver of business innovation?

Richard Entrup: A modern CIO can transition from being a technology service provider to a strategic driver of business innovation by aligning IT with business goals, fostering a culture of innovation, and leveraging emerging technologies to create value. You need a seat at the executive table and you must speak in business vernacular, not tech.

Let's break this down in more detail;

The CIO must deeply understand the organization's strategic objectives and market dynamics, engaging with C-suite peers to co-create solutions that drive revenue, efficiency, or customer experience—such as using AI for personalized marketing or IoT for supply chain optimization. This requires shifting from a reactive, operational mindset to a proactive, outcome-focused one, where IT initiatives are tied to measurable business impact.

The CIO should champion a culture of experimentation, encouraging cross-functional teams to prototype solutions using agile methodologies and technologies like cloud, AI, or blockchain, while securing quick wins to build credibility

Investing in data-driven decision-making is critical—building robust analytics platforms to uncover insights that inform strategy, like predictive models for customer churn.

The CIO must navigate the open-source vs. closed-source debate strategically: using closed-source AI for rapid, reliable deployments and open-source for long-term, customized innovation

Upskilling the IT team and fostering partnerships with startups, VCs, academia, or industry consortia ensures access to cutting-edge ideas and talent. By embedding IT as a core enabler of business strategy—evidenced by metrics like reduced time-to-market or increased market share—the CIO becomes a catalyst for innovation, driving transformation rather than merely supporting it.

For new roles, you will know during the recruitment process if you're going into a 'keep the lights on' IT services provider role vs a strategic leadership tech role by a few factors. Budget and resources allocation, reporting structure (CEO vs COO or CFO), compensation level, and a few others.

SM: What is the most common mistake you see startups make when trying to pitch their technology to a large enterprise? What lessons can we learn from a successful CIO?

RE: They don't start with the problem they are solving. If they can't answer that quickly on slide one, or verbally, I'm out. The pitch must only be about the 'job to be done', not how great the leadership team is or who else has invested, all good things, but not the top of my list. Don't take too long to get to the problem you're solving for. I also love to see a founder who's hair is on fire to solve a problem that's personal to them. My dear friend Jason Bailey founders ClubNFT, a startup that helped collectors backup their NFTs safely and prevent loss. Why did Jason start ClubNFT? Because he probably lost millions of dollars

RICHARD ENTRUP

Managing Director, Head of Emerging Solutions at KPMG US

A multi-award-winning global technology and innovation leader with nearly three decades of experience driving strategic business transformation. He is at the forefront of identifying and incubating disruptive technologies like GenAI, Quantum Computing, and Spatial/XR, pioneering new products and solutions for clients. A two-time recipient of the HMG Global Leadership Institute Award for Exceptional Business Transformation, his distinguished background includes C-suite leadership roles (CIO, CTO, CDO) at globally recognized brands such as Verizon, Disney/ABC, Christie's, Time Warner, Viacom, and Tiffany & Company. He also serves as a strategic advisor and investor to a portfolio of innovative startups and venture capital firms, passionately guiding ventures focused on cutting-edge innovations. He is based out of New York, United States.

EXPERT VOICES

Subrata Mukherjee: As a CIO advisor at Salesforce, you have a direct line to the strategic challenges and opportunities facing technology leaders today. Your current role involves leading the development of the CIO value proposition. What is the single biggest shift you are seeing in the CIO's role today, particularly with the rise of AI and digital labor?

Chris Pena: The CIO role has fundamentally evolved from technology operator to business strategist. With AI and digital labor transforming the landscape, CIOs are now expected to drive growth, customer engagement, and innovation, not just operational efficiency. In my book Déjà Boom: Hype, Hope, and Human Intelligence in the AI Era, I reflected on how hype cycles shift leadership expectations, and this is exactly what we are seeing again with AI. The most significant change is the board's expectation that CIOs actively shape business models and revenue streams, leveraging AI to unlock both productivity gains and personalization at scale. We have moved beyond "keeping the lights on" to orchestrating transformations that redefine competitive advantage in the marketplace.

SM: How can a CIO effectively articulate the business value of a new technology initiative to a non-technical C-suite and board?

CP: The key is framing technology as a business outcome story rather than a technical roadmap. Instead of describing system upgrades or architectural improvements, CIOs must highlight tangible metrics such as revenue growth, customer retention improvements, or accelerated product launches. Success comes from directly linking initiatives to the company's strategic priorities, whether that is entering new markets, improving profit margins, or building organizational resilience. Compelling storytelling supported by concrete data and industry benchmarks makes technology value resonate at the boardroom level.

SM: What is one piece of advice you consistently give to CIOs who are trying to balance modernizing legacy systems with adopting cutting-edge technologies like AI?

CP: I advocate for a "two-speed" strategy, modernize critical legacy systems in parallel with smaller, high-impact AI pilots. This approach prevents transformation from stalling under the weight of technical debt while generating quick wins that build organizational momentum. CIOs must prioritize interoperability and data accessibility, ensuring today's AI investments will not be constrained by yesterday's architectural decisions.

In my second book, Emotional Intelligence: The Next Superhuman Power, I emphasize the human side of this balancing act. CIOs must manage change with empathy and influence, not just technical skill. The goal is balancing calculated risk with innovation to demonstrate steady, measurable progress.

CHRIS PEÑA

VP, CIO Go To Market Industry Lead at Salesforce | Ex-CIO Aspire

Chris is a strategic technology leader and CIO advisor with a deep expertise in AI, digital transformation, and e-commerce. In his current role as VP, CIO Go To Market Industry Lead at Salesforce, he is responsible for developing and executing the company's value proposition for Chief Information Officers across various industries. As Global VP of Consumer Technologies at PVH Corp., he managed the global Salesforce platform. Prior to that, as General Manager CIO at Aspire Lifestyles, he built a global technology team from the ground up, modernizing operations with cloud infrastructure and automation, and successfully managing massive budgets. His extensive experience also includes senior leadership roles at EY, Razorfish, and Pearson, where he led digital strategy, platform implementations, and engineering. He is based out New York in United States.

EXPERT VOICES

Subrata Mukherjee: Your experience includes leading innovation labs at both **Standard & Poor's** and **Dun & Bradstreet**. What is your playbook for fostering data and analytics innovation within a large enterprise?

Saleem Khan: Within a large organization I tend to think of running an innovation lab in the same way a venture capitalist may look at their portfolio. You may have a dozen investments in your portfolio, of which 11 will likely result in bad outcomes and one might have a decent chance of making up for the other 11 failures.

The problem is that you don't know which idea will work with customers and which won't. Similar to investing in individual stocks, it's important to not fall in love with your ideas; you should let your winners run. I made this mistake early on and felt very sure a particular idea would work with customers. I kept pushing the idea, but customers didn't want it. Then something significant changed in the marketplace which created an increased need for shipping data. This particular product and dataset ended up being a runaway success when all my intuition was telling me it had no shot. Diversification is key when it comes to applied innovation for commercial purposes

SM: How do you balance the pursuit of cutting-edge ideas with the need to deliver tangible, near-term business value?

SK: Continuing with the venture capital analogy, I think it's important to look at 50% of your portfolio as being true blue-sky initiatives that try to dream up new paradigms and business models. The other 50% of your portfolio should be investments that are more grounded in the typical business school metrics for identifying whether an idea will work. Calculating your TAM, SAM, go-to-market approach, distribution, pricing, etc. will ground your efforts to ensure you're using both a left and right-brained approach.

SM: In this new economy, what are the biggest risks and opportunities for companies that fail to properly govern or leverage their data assets?

SK: The biggest risk is going out of business. Data is truly the new oil and companies that can govern and leverage data in new and novel ways will have an advantage. Having a data moat by way of creating first-party data or having some unique claim to a dataset is key as well. The opportunities for organizations to commercialize data in its raw form, derived outputs or research reports are the typical ways in which you can commercially leverage a datasets. However, gen AI has created another distribution approach that if governed properly can create an outsized reward.

Market sizing is a crucial step in understanding the potential of a business idea. If often measured by :
TAM = Total Addressable Market (Revenue opportunity with full 100% of the market share, and no competition)
SAM = Serviceable Addressable Market—"Slice" of the TAM "pie" that can be served by your products and services.
SOM = Serviceable Obtainable Market— Actual amount of the market served by your products & services.

SALEEM KHAN

Chief Data & Analytics Officer, Pole Star Global

Saleem is a seasoned data and analytics executive with a strong history of creating business value through data-driven innovation and successful company exits. In his current role as Chief Data & Analytics Officer at Pole Star Global, he leverages his deep expertise to lead and execute data strategy. His notable achievements include serving as Chief Digital Product Officer at VettaFi, which was successfully acquired by TMX Group, and as Chief Data & Analytics Officer at Discovery Data, which had a successful exit to Institutional Shareholder Services. He has also held leadership positions in data innovation and advanced analytics at major corporations like Dun & Bradstreet and S&P Global, and was a founder of the fintech startup ESGINE, an AI-enhanced platform for ESG compliance. He studied at Fordham University. He is based out of New York, United States.

EXPERT VOICES

Subrata Mukherjee: How did you build a reputation for your team as a thought leader and a valuable business partner, not just a service provider?

Abe Cytryn: Technology has evolved from being an operational driver to being a big part of the product/business itself. We recognized early on the importance of communication skills. We had a business analysis group that were the key liaisons to the business groups. We saw the credibility and trust this group helped achieve. We looked for technologists that were business minded, curious and always aimed to over deliver. After all we knew what was possible so why not showcase that and be proactive.

SM: Can you share an example of a time you used a technology solution, such as **Big Data** or a new API, to directly drive business goals ?

AC: That was our daily mission. Our team built out an ecommerce platform from the ground up to support our subscription business. This ultimately led to the ability of delivering digital products that we could bundle with our more traditional offerings. Provide more value to the consumer and that will make your offering more compelling. It helped drive sales and even more importantly improved retention.

SM: What advice would you give to a technology leader struggling to get a seat at the business strategy table?

AC: This is ultimately driven by demonstrating your ability to deliver. There are also key relationships that you have to nurture over time. You need to be faster, cheaper and have a proven track record. Make yourself a trusted partner. One that will help them be successful. No ego and or territorial battles. Be willing to consider all options including using outside providers. Show that you only care about success for them and the business.

SM: What is your philosophy for building a high-performance team that is respected and recognized?

AC: Its understanding what they value. Continuous learning, freedom to experiment, being purpose driven. Being valued and recognized. Making a difference. That being said you have to be transparent in what gets rewarded. Results oriented. Commitment. Creative problem solving. Ability to promote team spirit. My job is to remove distractions. Make decisions quickly and inclusively. Be an example on how to treat others.

SM: Your recommendations praise your ability to coach and inspire dedication. How did you foster a high-energy culture and keep a team engaged for the long term?

AC: It starts with Trust. If you have trust you have communication. If you have communication, you eliminate fear. If you eliminate fear, you have a safe environment leading to communication. If you have communication, it leads to trust. It starts and ends with Trust.

SM: What are the most important qualities you looked for in a team member, beyond technical skills?

AC: Communication skills first and foremost. Curiosity. Someone who doesn't ask any questions in an interview is not getting the job. Energy. Sense of Humor. Caring. Desire to grow and learn. Not afraid to try new things. Self-Starter. Will go beyond what is asked.

ABE CYTRYN

Former Chief Technology Officer – Time Magazine and Magster

Abe is an accomplished digital leader with over two decades of experience guiding digital product development and technology strategy in the media and entertainment industry. His expertise spans a wide range of critical areas, including eCommerce, mobile, big data, video, and content management. As a proactive and business-oriented leader, he has successfully developed technology roadmaps, built high-performance global teams, and led agile transformations to drive consumer growth and engagement. His extensive career includes pivotal leadership roles at Time Inc., Magzter, and HBO, where he consistently translated business objectives into innovative and effective technology solutions. Known for inspiring high-energy teams and fostering strong staff retention, Abe is an executive who empowers his people and delivers results. He is based out of New York, United States.

EXPERT VOICES

Subrata Mukherjee: Your time as Chief Digital Officer at The Economist and SHL placed you at the center of large-scale digital transformation. What was the most impactful digital transformation project you led during your time as a CDO?

Jora Gill: At The Economist, reimagining the subscription model with digital-first products. It wasn't just technology; it was cultural change. We shifted from print-led to data-driven, opening global growth. It was a tough project and I learnt a lot about myself, my team and my business partners. Don't let "run the business" drown by just focusing on "change the business through AI." Carve out innovation capacity, even if small. And anchor it in measurable business outcomes, so the board sees both scalability and progress not just an AI innovation lab. I often reflect back on this one as there was so much more I could have done better – ultimately it was the reason I created FireBreak.ai as I learnt I was not alone in not always having my finger on the pulse of a complex fast-moving project.

SM: Before your time as a CDO, you served as a CTO at Elsevier and Standard and Poor's. How did your experience with large-scale data and technology systems at these information-centric companies prepare you for your current work in AI?

JG: Working with vast datasets at Elsevier and Standard & Poor's taught me how to manage complexity, governance, and scale. AI thrives on good data foundations. Without them, it fails. That perspective shaped my belief in ethical and explainable AI. At S&P, modernizing core systems without disrupting critical services. It reinforced that technology leadership is as much about change management as code. The CTO is now strategic. It's no longer about servers or platforms. Boards expect CTOs to shape growth, customer trust, and even societal impact. That's a huge shift. We are also seeing ex CTOs like me join public PLC boards as strategic tech provides opportunities for productivity improvements, better employee and customer experience through technology and AI come to the centre.

SM: In 2022, you made a significant pivot to found two AI-focused companies, ConnectingYouNow.ai and FireBreak.ai. What inspired you to leave a successful corporate career to become an AI entrepreneur?

JG: After decades as a CIO, CTO, and CDO, I saw the same problems repeating in large organizations: projects failing, data trapped in silos, teams overstretched. I realized AI could address these systemic challenges — but only if applied responsibly. That was the catalyst to build ConnectingYouNow and FireBreak: to move from firefighting to prevention, from fragmented systems to clarity. Both come from lived experience. With ConnectingYouNow, we focused on accessibility and ethical AI because technology should serve everyone, not exclude them. With FireBreak, the insight was that 70% of big projects fail not because of poor planning, but because risks emerge too late. We wanted to give leaders early warning and actionable guidance.

SM: What is the biggest difference between leading a small, agile AI company versus managing large-scale technology for a global corporation?

JG: Speed and ownership. In a startup, there's nowhere to hide: you can't delegate responsibility endlessly. Decisions are faster, experiments bolder, but resources are finite. In a global firm, the scale is immense, but bureaucracy slows execution. The challenge — and the joy — in a startup is making every move count. But startups also come with stress in the early months/years and do not have the financial stability of a corporation – so they are not for everyone.

JORA GILL

CEO, ConnectingYouNow.ai & FireBreak.ai | Ex-Chief Digital Officer SHL

Jora is a dynamic CEO and seasoned Non-Executive Director who advises leading global organizations on transformation through digital, data, and AI. As the co-founder and CEO of two AI firms, he is pioneering AI-powered governance and risk oversight platforms designed to help businesses manage complex projects and accelerate growth. His extensive executive career includes a significant tenure as **Chief Digital Officer at both SHL and The Economist**, where he was instrumental in leading large-scale digital transformations and fostering collaborative teams. He also has a strong background in strategic governance, having served as a Non-Executive Director for companies such as Travis Perkins and Phoenix Life. An MBA graduate from the University of Birmingham, his expertise lies in bridging strategic vision with operational execution to deliver tangible business results. He is based out of United Kingdom.

EXPERT VOICES

Subrata Mukherjee: You have experience as a Group CTO for a global organization and now serve as a Fractional CTO. This dual perspective is unique. How has the role of the CTO evolved in recent years, and where does the fractional model fit into this evolution?

Mark Brincat: The role of the CTO has shifted from being purely a technology custodian to a business enabler. Today's CTO must shape customer experience, drive revenue through digital, and champion innovation alongside governance. They're expected to have board-level influence and bridge technical depth with strategic storytelling. The fractional model has emerged because not every business needs—or can afford—a full time CTO. Startups, scale-ups, and even large enterprises facing transformation benefit from fractional leaders who bring specialist expertise, immediate impact, and a fresh perspective without the long-term overhead.

SM: What are the key differences in mindset and approach when you're acting as a full-time CTO versus a fractional advisor?

MB: As a full-time Group CTO, my mindset is focused on ownership and continuity. I am responsible for building long-term roadmaps, hiring leadership teams, and guiding the organization through multi-year journeys. It's about deep integration and transformation stewardship. As a fractional CTO, my mindset is about focus and acceleration. I zero in on specific, high-value challenges—such as fundraising narratives or AI adoption—and provide an advisory lens that balances influence without the same operational burden. I bring cross-industry learnings rapidly, ensuring businesses don't repeat avoidable mistakes and helping them make critical decisions quickly and with confidence.

SM: In both roles, you're responsible for technology strategy. How do you balance the long-term vision with the immediate, high-impact needs of a business?

MB: It's not about choosing between long-term or short-term—it's about linking them. I help clients frame every initiative around two questions: What immediate business value does this deliver? How does it fit into the longer-term strategy? For example, enabling AI interview notes has a short-term outcome of faster decision-making. The long-term outcome is a scalable data foundation that feeds predictive talent intelligence. By connecting those dots, the business sees results today without derailing tomorrow. In practice, this means being deliberate: you can meet urgent business needs, but you do so in a way that preserves resilience, scalability, and the ability to innovate. The best strategies are those where today's quick wins become tomorrow's building blocks.

SM: What is the single most important quality a technology leader needs to possess in today's rapidly changing environment?

MB: The single most important quality is adaptability with clarity. Technology leaders can't just master one domain anymore—they must pivot quickly as markets, customer needs, and platforms evolve. But adaptability alone isn't enough; leaders must also provide a clear direction so their teams don't feel like they're constantly chasing the next shiny thing.

SM: What is your strategic process for identifying and evaluating potential technology partners?

MB: My strategic process for identifying and evaluating partners begins by anchoring partner evaluations to the company's growth priorities—whether that's expanding into new markets, enhancing product capability, or accelerating innovation. I evaluate partners on three dimensions: complementarity (do they fill a gap or accelerate our roadmap?), market access (will the partnership extend our reach?), and cultural and strategic alignment (do they share our values?). Finally, I advocate for a proof of value—a pilot or a quick integration that demonstrates a tangible impact on customer outcomes.

MARK BRINCAT

Ex-CTO @ SHL, Economist | Architect & BBC, Telegraph, Home Office

Mark Brincat is a technology executive with deep expertise in digital transformation, AI, and data analytics. In his current role as the Global Head of Presales and Technology Partnerships at SHL, he leads go-to-market strategy and advises clients on their transformation journeys. Concurrently, he serves as a Fractional CTO for Castille Resourcing, where he oversees the engineering and scalability of their digital platform. He previously held key Chief Technology Officer (CTO) roles at both SHL and Intive, and senior architectural positions at major media organizations like The Economist, Telegraph Media Group, and BBC Worldwide. His career highlights include implementing large-scale digital and cloud solutions and forging strategic partnerships. He is based out of London, United Kingdom.

EXPERT VOICES

Subrata Mukherjee: Your title is a powerful one, especially in a time of rapid technological change. Could you define what a "translational economist" is and explain how your unique career path—from Georgetown to The Economist and the World Economic Forum—has prepared you for this role?

Mouchka Heller: "Translational economist" is a term I borrowed from Paul Swartz, from Boston Consulting Group. In the context of my career, it refers to the role I have played in connecting the dots between policy, finance, industry, and economics to decipher the what and how of the AI era. I have had the privilege, throughout my career, to consistently find myself dealing with the question of "how do we transition to this new technological revolution?" For example, at The Economist, I got a chance to join some of the smartest people in the world as they thought through digitizing what was at the time a 170 years old company, transforming our product management, marketing, and customer engagement with the advent of big data. I joined the Government of Canada as Trade Commissioner of Infrastructure after that, where I was asked to "figure out smart cities," with the goal of understanding what business opportunities would arise from the focus on sustainability and resiliency that followed Hurricane Sandy, and from the early smart urban technologies that are transforming our cities. I joined the World Economic Forum after my stint as Trade Commissioner because I was eager to test what I had learned at a global scale. There, I joined an exceptionally talented team dedicated to transportation technologies. I got to sit with the CEOs of the world's top companies as they wondered whether the world would ever be ready for autonomous vehicles, negotiated with policymakers to integrate public and private data, and thought through the electrification of public transportation.

SM: As a translational economist, how do you help organizations like Amazon navigate the unprecedented economic shifts brought on by the rise of AI?

MH: I currently work for Amazon Web Services (AWS), where I help answer the same kind of questions I have tackled throughout my career, from the perspective of supporting data center infrastructure. Every day, I think about how our investments impact the lifestyles, economic circumstances, and future of families and individuals.

SM: In what ways does your background in data and policy uniquely position you to analyze and understand the economic impact of AI on different sectors and regions?

MH: We work in a world today that does away with siloes. To understand the economic and human impact of AI, one needs to become comfortable diving deep into the mechanisms of the technology, and all the avenues through which it can change people's lives. On the other hand, policy can be more complex than we give it credit for. The impact of policies and regulations is often in the detail rather than in the headlines.

SM: Your work at the World Economic Forum and as a Trade Commissioner shows a clear focus on global issues and the ethical use of technology. What is the role of public policy in ensuring the benefits of technologies like AI are distributed equitably across the globe?

MH: I believe in the role international organizations have to play in bringing together disparate public and private entities to unlock a greater, global good. At the same time, I also believe in national sovereignty, and each country's duty to its people. Good public policy protects against harm, but great public policy brings people together with empathy.

MOUCHKA HELLER

Principal & Ex- Global Head, Economic Research, AWS Policy - Amazon

Mouchka is a translational economist and accomplished executive focused on the intersection of technology, policy, and economic development. In her role at Amazon, she is focused on the data center infrastructure and power distribution that enables AIML technology. Her career highlights include serving as Global Data Officer at The Economist and holding various leadership positions at the World Economic Forum, where she focused on integrated mobility systems and policy. As a former Trade Commissioner for the Consulate General of Canada, she has extensive experience in international relations and business development. Mouchka's background is distinguished by her ability to bridge complex policy, data, and business strategies to drive innovation. She holds a Masters in International Relations and Affairs at Georgetown University and a BA from Yeshiva University. She is based out of New York in the United States.

EXPERT VOICES

Subrata Mukherjee: What is the biggest difference in your approach to business strategy when you're working with a company on AI readiness versus an emerging media technology from a decade ago?

Ellen Weinstein: A decade ago, our focus was squarely on digital transformation, which was a strategic effort to take analog processes and make them digital. We were leveraging emerging technologies like marketing automation and ad-serving platforms to simply optimize existing operations. The goal was to increase efficiency and reduce costs, and the strategy was often centered on a single, siloed platform with point-to-point data integrations. Today, with AI readiness, our approach is a complete paradigm shift. We're not just automating tasks; we're building algorithmic business models. The strategy is to leverage AI to unlock the latent value within an enterprise's data, creating new revenue streams and enabling generative and agentic capabilities that were previously unimaginable. This is a move from optimizing a well-understood process to creating an entirely new, learning-driven business model.

SM: How has the conversation around "business value" changed now that AI is a central topic?

EW: In the past, the conversation around "business value" was a straightforward discussion about efficiency, automation, and cost reduction—all tangible, easily-measured metrics. Today, the dialogue has fundamentally shifted to value creation. AI enables us to move beyond simply making the existing business run faster and into building entirely new products, services, and revenue streams. For instance, an AI-powered system can transform a traditional media company into a highly personalized content engine that tailors experiences to individual users, or it can turn a marketing firm into a predictive commerce platform. This conversation is about leveraging AI not just for incremental gains but to create a strategic moat and a sustainable competitive advantage in the market.

SM: Your focus is on ensuring AI is an "accelerant, not a disruption." Can you provide a practical example of a company that is doing this well?

EW: I would argue that the concept of true AI-led acceleration without any disruption is a myth. Any technology with the power of AI will inevitably cause organizational and process disruption. The key is not to avoid it, but to manage that disruption effectively. A company that does this well is a financial institution that uses a hybrid AI model to augment its human risk analysts. The AI handles the high-volume, low-complexity fraud detection by flagging suspicious transactions with a high degree of confidence. At the same time, the human analysts are alerted to high-risk, nuanced cases that require their judgment and expertise. This collaborative workflow, which marries the AI's speed with human critical thinking, minimizes disruption by elevating the human's role rather than replacing it outright.

SM: What is the most important lesson you learned from leading these "first-mover" initiatives that still applies to your work with emerging technologies today?

EW: The most important lesson is the need for an industry champion. It's not enough to have a superior product; you need a powerful trading partner, someone who controls the business rules or a significant portion of the market's spend, to drive widespread adoption. This lesson is more relevant than ever in the world of emerging technologies like AI. To get a new AI framework or platform adopted, you need to find an anchor client who is willing to pioneer its use and serve as the proof of concept for the rest of the industry. This is the only way to scale a truly disruptive innovation beyond the early adopters.

ELLEN WEINSTEIN SALAMAY

Seasoned Business Development Leader, Ex-Marketo, Ex-Cognizant

Ellen is a seasoned business development and strategy professional with a proven record of helping companies leverage technology to achieve significant growth. Her current focus is on AI adoption and digital transformation, which she applies in multiple roles, including as an Advisor at Covington Place Partners and Sr. Consultant at Priorise. She also serves as Vice President of Business Development for Accounti, an AI-driven financial services firm, and InCommon®, which builds AI-capable teams. With a career spanning over two decades, she specialized in connecting tech startups with agencies and media companies. Her extensive background includes senior leadership positions at prominent firms such as Marketo, Cognizant Technology Solutions, and Xen.AI, where she consistently drove business growth, forged strategic partnerships, and optimized marketing and operational strategies across the advertising and media industries. She is based out of New Jersey, in United States.

EXPERT VOICES

Subrata Mukherjee: As a former professor at NYU who taught product innovation, what is the single advice you could give an aspiring product leader?

Brian Glassman: The single most important piece of advice I give to aspiring product leaders is to develop a deep, structured understanding of the product or technology domain they are responsible for. This means analyzing it across three horizons—what exists today, what is evolving in the mid-term, and what is likely to emerge in the future. Product leadership is about forecasting, not reacting. It is about building the fluency to interpret where the field is heading and shaping strategy from that foundation. Just as important is the ability to map out a clear future vision for your product area. This involves actively monitoring technology shifts, understanding customer behavior and feedback, and constructing scenarios for how the product might evolve. Consider the difference between BlackBerry and the iPhone. One was built for the moment and the other was built for the decade. The ability to predict where the world is going and design accordingly is what defines great product leadership.

SM: What is the single biggest opportunity for business value creation with Generative AI in the next three years?

BG: The single biggest opportunity for business value creation with Generative AI in the next three years lies in how it transforms software development. Programming with AI-powered coding tools will dramatically reduce the time it takes to build software while simultaneously raising the quality of the output. These tools are capable of producing highly efficient code, generating multiple versions of a solution, and adapting outputs to suit different user needs and business contexts. What once took weeks or months to develop will be completed in days or even hours, reshaping the economics of software creation. More fundamentally, this shift will render traditional models of agile software development obsolete. The iterative, sprint-based frameworks that have defined engineering teams for the past two decades will be replaced by a new model centered on AI-driven generation, review, and deployment. Software will increasingly be conceived, prototyped, and refined in collaboration with intelligent systems that accelerate decision making and eliminate many manual tasks.

SM: How do you stay on top of the rapidly changing AI landscape and continue to evolve your own expertise?

BG: Staying current in the AI space requires a focused and intentional approach. The field is vast, spanning everything from image and video generation to coding assistants and mathematical reasoning. Rather than attempting to track every advancement, it is more effective to concentrate on the areas most relevant to your customers or operational domain. At the same time, it is important to maintain a working knowledge of broader AI megatrends. This balance allows you to stay grounded in practical impact while also anticipating strategic shifts. To do this well, you need to consume expert-level content regularly. The general press is often too shallow or sensationalized, so the most useful insights come from practitioners publishing in-depth analysis on platforms like Medium. Video demos and platform-based walkthroughs also play a critical role in staying informed. These resources allow you to see how a new tool actually works and how it might be applied. In addition to passive learning, it is essential to actively engage with the tools themselves. Use AI to evaluate new AI. Ask it questions, explore how it interprets tasks, and identify limitations. If something shows promise, you need to test it in your own workflow. Public claims about performance rarely capture the nuance of specific use cases, so hands-on experimentation is the only way to determine real value. Setting aside even one or two hours each week to try out emerging tools is a worthwhile investment.

DR. BRIAN SCOTT GLASSMAN

Chief Product Officer & Head of Client Engagements at AInspire.ai

Dr. Glassman is a seasoned product management executive, engineer, and product strategist specializing in generative AI. As the co-founder and Chief Product Officer at AInspire.ai, he leads client engagements and product strategies, collaborating with C-level executives to develop and deploy AI solutions for large enterprises. With a career dedicated to leading technology innovation, he previously served as Director of Product Management at Electra Vehicles, where he built the product function for AI-powered solutions in the automotive sector. His extensive experience includes a seven-year tenure at Rave Digital, where he led product management and customer engagements for Fortune 1000 clients. Furthermore, his background as an engineer and product leader at Cameron, a Schlumberger company, highlights a unique expertise in technical product development. He is based in New York.

EXPERT VOICES

Subrata Mukherjee: : Your recent certificates from Carnegie Mellon University in AI Leadership and Digital Innovation show a strong commitment to staying on the cutting edge. How are you currently leveraging your knowledge of AI to drive new digital solutions and services for your clients?

Preeti Sharma: We're embedding AI into digital workflows to improve efficiency, enhance personalization, and uncover insights from data that were previously hidden. For our clients, that often means building closed systems where we can safely train AI models tailored to their specific needs—giving them smarter, more secure solutions. In my CIO role, I also focus on using AI to streamline mundane tasks internally, so our teams can spend more time on higher-value, creative, and strategic work.

SM: In your view, what is the single biggest opportunity and the most significant challenge that AI presents for the digital services industry today?

PS: The opportunity is scale—AI allows us to deliver smarter, faster, more accessible solutions that can dramatically expand impact. The challenge is trust—ensuring AI is ethical, transparent, and inclusive. Too often, people approach AI in a binary way: either avoiding it out of fear or jumping in headlong without enough caution. Both are non-curious approaches. The real challenge is bridging these two types of thinkers—creating a space where healthy skepticism and bold experimentation can coexist. Clients and communities must feel confident that the technology is working for them, not on them. I don't want to deliver work product that ends up in the digital graveyard, so being thoughtful in the application of AI and handling the fear of AI will go a long way in ensuring shelf life.

SM: Based on your training, what is one piece of advice you would give to a fellow CIO who is just beginning to explore their own company's AI strategy?

PS: Start with a problem, not the technology. Don't chase AI for AI's sake. Identify where it can make life better for your clients or your teams, and build from there. Pilot small, learn quickly, and scale thoughtfully. Try different AI tools, but do so in areas where you or your team are subject matter experts—so you can validate the output and truly assess its value. And always keep a human in the loop. AI should augment expertise, not replace it.

SM: Can you share an example of a successful digital project you've led for a government client? How did you navigate the security, data, and policy concerns that are unique to the public sector?

PS: One project I'm especially proud of is nccareers.org, where we took complex economic and occupational data and transformed it into a user-friendly digital platform. The challenge wasn't just about managing large datasets—it was about making that information meaningful and accessible to students, educators and job seekers. By adding an intuitive front end, we enabled people to explore career pathways in a way that was practical and user friendly.

Security and accessibility were built in from the start, and by using a modern tech stack and architecture, we managed to hit all compliance requirements so the solution not only met the state's security and privacy standards but also delivered real value to the public. What makes me proudest is the adoption, the platform is actively used to guide career exploration and occupation planning, proving that technology can truly empower better decisions.

PREETI SHARMA

Chief Digital Officer & CIO at VGD NYC

Preeti is a distinguished technology executive and a "Notable Woman in Tech" with a proven history of driving digital strategy and innovation. As the Chief Digital Officer and CIO at VGD NYC for over 18 years, she leads the delivery of digital services, technology strategy, and solutions for a diverse range of clients. Preeti's expertise is deeply rooted in her long-standing focus on AI and technology transformation, backed by recent certifications in AI Leadership and Digital Innovation from Carnegie Mellon University. Her foundational experience includes nearly a decade at ISI Emerging Markets, where she served as Global Operations Manager, playing a key role in scaling the business from a small startup into a profitable enterprise. A member of Chief and the Forbes Technology Council, she has been recognized with multiple AVA Digital Awards for her work. She is based out of New York.

EXPERT VOICES

Subrata Mukherjee: How can a company's board become strategic partner, effectively hold the C-suite accountable for cyber risk without micromanaging?

Michael Gabriel: We've heard far too many times that no matter how much money is spent, we can't ensure there won't be an incident. But they can ensure they're reducing that likelihood and are in a position to mitigate any damages from an incident. So, how much money is enough? What risks are acceptable? Board members must maintain objectivity and avoid entanglement in operational decisions, with a more elevated perspective rather than a focus on operational involvement. The best tool for instilling accountability and proper oversight is a well-structured and consistent Board status, including appropriate cyber risk dashboards. This should convey the current cybersecurity status and any gaps to target goals, industry comparisons, projects and status required to improve their cybersecurity posture, current and emerging cyber risk trends, and the costs of the cybersecurity program. If the magnitude or frequency of incidents that occur are avoidable or recurring, then that will be obvious as a focus for C-Level accountability and explanation. If gaps exist between current and future state goals that are not making progress per plan, then that also should be evident. The Board also needs to have unbiased third-party access and exposure to be aware of cybersecurity trends, such as through Board-based industry publications and communications geared to the Board level, as well as organizations such as Gartner and the NACD. They should be briefed on trends by third-party experts, such as their external auditor, and other professional service firms that have a cybersecurity focus, as well as by their CISO. I would caution about receiving updates from vendors who provide operational products and services to the corporation, as it could create a bias and potentially tempt the Board to intervene in vendor and product selection.

SM: As technology has become more integral to business, how have you seen the CIO's role evolve from a support function to a strategic partner?

MG: Technology has been integral to business for many decades. It increased in importance mainly due to the Internet, smartphones, and now AI. But anyone who has been an effective CIO has been an effective business partner throughout this journey. The fact is that technology is a service organization, unless you are responsible for a product line or service P&L. We are a business enabler, but we still need the business to want to be enabled. We have always had a better understanding of all parts of the business than any other role, including the CFO, because technology touches and supports every part of a company. So, the CIO must be business-focused as well as technologically savvy. If the CIO wants to be the best technologist in their staff, then they probably will be doing that to the detriment of putting time into their business relationships, as technology is changing at a faster pace than ever, and so is the business. You need to be more of a symphony conductor rather than being able to play all the instruments better than anyone else, which is not practical. You need to be both a teacher as well as a student, and learn from your business partners. Shadow IT comes about mostly due to close-minded technologists thinking they have to have the answer for everything, and need to solve the problems themselves. Being open to utilizing the business users' knowledge and expertise can help the CIO scale their ability to provide even more benefits, often faster and less expensive than if the technology staff were solely responsible for that. Technology is now everyone's responsibility, and a strong CIO will be able to leverage that even with resources don't report to the CIO. That's how you can become a strategic partner.

MICHAEL GABRIEL

Ex CIO at HBO, NBA | NACD Certified Board Director

Michael is a senior technology and cybersecurity executive with experience providing C-level leadership, governance, and advisory services. He is currently an independent Board Director and provides professional services for companies like Warner Brothers Discovery (WBD). His career is highlighted by a 14-year tenure as EVP & Global CIO at HBO, where he was the initial executive to spearhead the groundbreaking HBO GO initiative and led a company-wide digital transformation. An inventor on six technology patents, he also played a key role in information security governance across the Time Warner enterprise. Prior to HBO, he served as Global CIO for the National Basketball Association (NBA), where he implemented the first NBA retail store technology system. He also held senior IT leadership positions at EMI Records Group and NBC Olympics.. He is based out of New York

EXPERT VOICES

Subrata Mukherjee: Your background is deeply rooted in recruiting for Design and Technology roles. You helped build Google's first UX team outside of Mountain View. How has the role of UX and design leadership evolved since then?

Jennifer Speciale: When I built Google's first UX team outside Mountain View, UX was still fighting for a seat at the table. Today, design leaders are expected to set the table—shaping strategy, influencing revenue models, and guiding AI ethics. The shift is from execution to orchestration. The most successful design leaders I coach are bilingual: they speak the language of craft and the language of capital. That combination makes design not just relevant, but indispensable, in board-level conversations.

SM: How can a technology leader today effectively demonstrate their value beyond technical skills, particularly in a landscape shaped by AI and automation?

JS: Technical skills are entry tickets; strategic storytelling is the differentiator. A technology leader today must connect their work to business outcomes: revenue growth, cost savings, risk reduction, or customer trust. I remind clients: if you cannot explain your value in business terms, you invite being sidelined as "delivery." One CTO I coached began framing AI initiatives not in terms of algorithms, but as "how we unlock $50M in new sales channels." Suddenly, the board stopped asking about tools and started asking about investment.

SM: What is one key insight you've gained from coaching technology leaders that you wish more people understood?

JS: The hardest lesson: being right is not enough. Many leaders believe deep expertise guarantees influence. It does not. Influence comes from shaping perception—knowing which story to tell, to whom, and when. One engineering VP I coached realized that instead of proving he was smarter than peers, his job was to make peers feel smarter after talking with him. That pivot unlocked his promotion. The insight: expertise gets you in the room; emotional intelligence keeps you there.

SM: You are a public speaker. What is the one message you consistently share with leaders about their professional development?

JS: My consistent message is simple: your career is not something you find, it is something you design. Leaders must stop treating roles as chance opportunities and start architecting the conditions for their next chapter. When you view yourself as a designer of your own narrative—clarifying principles, building story banks, curating visibility—you create not just jobs but trajectories. That mindset shift is what liberates careers from drift.

SM: Looking back at your career, from recruiter to coach, what do you consider your proudest achievement?

JS: My proudest achievement is that I finally feel I am doing my life's work—work that actually helps people and moves the needle. The singular moment was not a promotion or title, but the decision to take the plunge and bet on myself. I let go of the handlebars and trusted that I could build something that mattered. Today, when I see a leader transform because of our work together, it confirms that leap was worth it. That risk, and the reality that I am riding the bike on my own, is my proudest professional moment.

SM: Finally, what's one piece of advice you'd give to a leader who is feeling stuck in their current role and is ready to make a change, but doesn't know where to start?

JS: If you feel stuck, start small: articulate one story of impact that you are proud of. Then share it—with a peer, a mentor, or on LinkedIn. Momentum builds when you reclaim authorship of your own narrative. Too often, leaders wait for clarity before moving. In truth, clarity comes from movement. Begin with one story, one conversation, one action. That is how you stop being stuck and start designing forward.

JENNIFER SPECIALE

CEO, Speciale Executive Leadership | Ex- Amazon, McKinsey, Google...

Jennifer is a Global Executive Leadership Coach where she partners with senior leaders to showcase Executive Presence, Influence, and Impact. Her coaching practice is built on over two decades of experience in strategic talent advisory, having directly shaped executive teams at the world's most innovative companies. Jennifer's career highlights include establishing the pioneering UX Design team in NYC for Google, building the foundational global hiring processes for Indeed during its hypergrowth phase, and leading the global expansion of the Design executive talent function at both McKinsey & Company and Amazon. She is an expert in organizational Change Management and Executive Search, dedicated to translating senior leaders' expertise into compelling narratives that unlock C-suite and VP opportunities across the Design and Technology landscapes. She is based out of New York city.

WE NEED YOU!

WELCOMING OUR NEXT SET OF THOUGHT LEADERS

The remarkable success of the Ai6x6 Organization Transformation Series is built upon the **hard-won wisdom** of executive leaders who have successfully navigated the AI transformation within their industries, turning theory into a deeply practical, actionable guide. As we expand the series, we are extending a highly selective invitation to a new cohort of pioneers—individuals whose professional journey is at the **leading edge of AI strategy and execution**. We are looking for executives who possess a proven track record of rewriting business models, mastering unforeseen ethical and competitive challenges, and offering a **clear blueprint** for the next 3-5 years of the AI-Era.

Contributing to the Ai6x6 Series is a unique opportunity to **cement your professional legacy**, joining an exclusive, curated network of global authorities whose insights are distilled into executive-level knowledge cards, directly shaping the governance and innovation decisions of the world's most influential companies. Our selection process is based strictly on **pure expertise and demonstrated leadership** in the AI-Era, reinforcing our commitment to editorial integrity and authentic thought leadership—there is no monetary exchange involved. Select contributors will also receive an invitation to blog on our expert network microsite, further amplifying their platform. To confidentially inquire about contributing your expertise, please contact us via our dedicated portal at **www.Ai6x6.com/contact** or by scanning the adjacent QR code.

Contact

SUBSCRIBE TO OUR NEWSLETTERS: To stay ahead of the curve and receive continuous, executive-level intelligence on the rapid evolution of the AI-Era, we invite you to register for our **Ai6x6 Weekly Newsletter**. Each edition delivers curated insights directly from our thought leaders and highlights critical shifts in AI strategy, governance, and provides relevant excepts from our website and books. Keep your finger on the pulse of the future and ensure your company remains competitive by signing up on the same page

PLEASE REVIEW US ON AMAZON!

Leadership Blueprint

We know your time is valuable, but if the Ai6x6 Organization Transformation Series delivered the real-world, actionable strategies you needed, please consider leaving a brief review on Amazon. If we missed the mark, we would also like to know, and that is the only we we can improve. Your honest feedback is crucial; it helps fellow executives discover this unique framework and benefit from the insights of our thought leaders. Thank you for supporting our mission!

BLUEPRINT SERIES

SECTION B

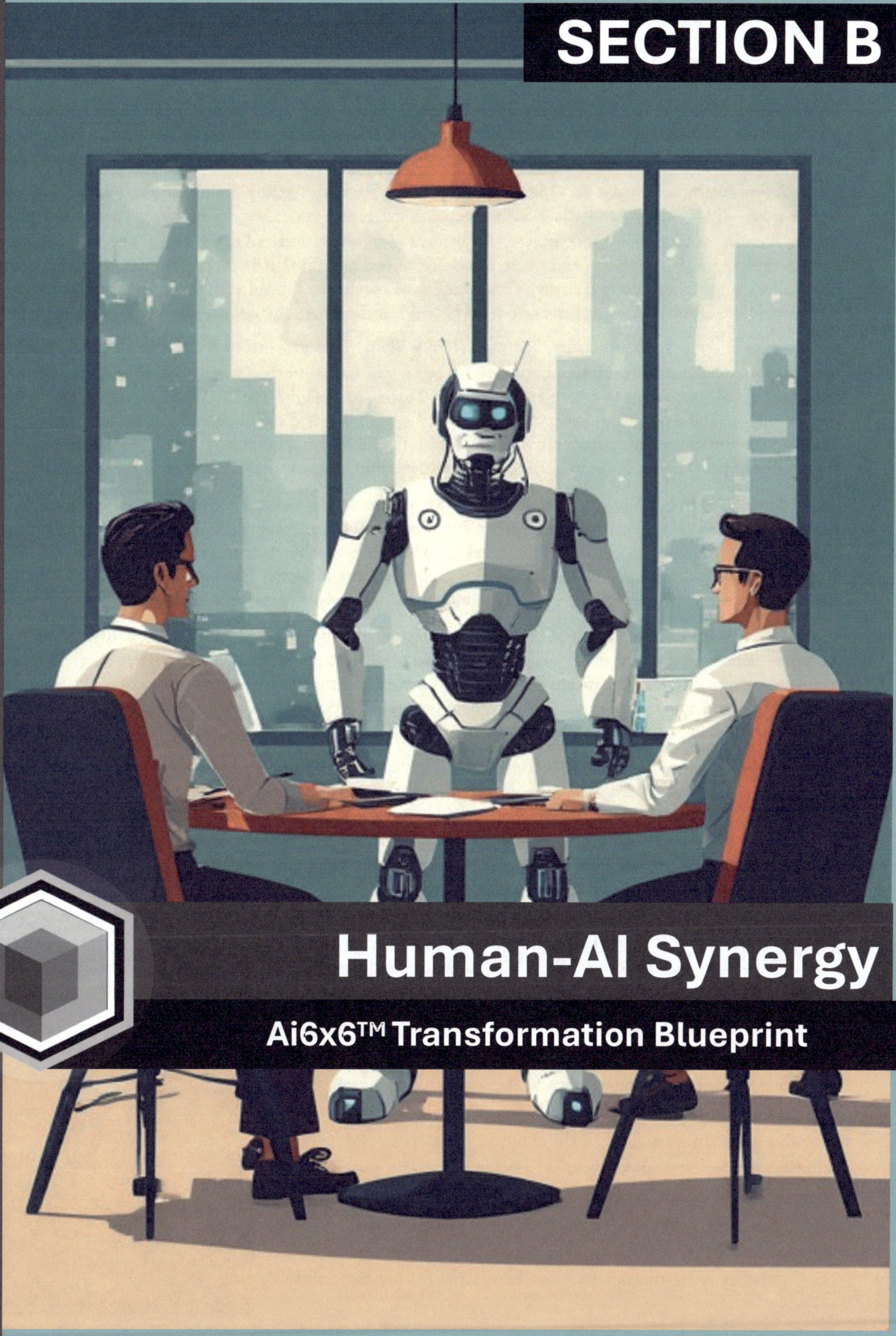

Human-AI Synergy

Ai6x6™ Transformation Blueprint

Story 1 — Enigma Codebreakers (1940–45): German U-boats encrypted communications using constantly shifting ciphers, forcing Allied cryptanalysts to chase patterns that dissolved daily. Manual decoding could not keep pace with wartime urgency, and insights often vanished before they could be confirmed. Electromechanical devices accelerated the testing of solutions at unprecedented speed. Humans interpreted these outputs, correcting false leads and refining assumptions. The fusion of machine acceleration & human judgment reshaped the trajectory of the war.

Story 2 — Turing's Bombe & Operators (1940–45): Breaking enemy codes required evaluating thousands of cipher possibilities rapidly, and the Bombe machine generated candidate solutions at speeds humans could not match. Yet its outputs were incomplete, ambiguous, and often misleading. Human operators validated results, corrected machine assumptions, and filtered noise from meaningful signals. Their contextual judgment transformed mechanical output into strategic clarity. This interplay of mechanical speed and human reasoning cracked codes that neither side could solve alone. It became one of history's earliest demonstrations of human–machine symbiosis.

Story 3 — Toyota Production System (1950s–70s): Factories struggled with defects, bottlenecks, and unpredictable workloads as machines produced volume without contextual awareness. Workers observed issues that machines could not detect or interpret, prompting Toyota to design a synchronized workflow that aligned human insight with machine precision. Real-time data from the line informed continuous improvement cycles. This human–machine rhythm created a new operational standard for reliability and efficiency. The system became a global model for intelligent production. It demonstrated how augmentation, not automation, drives sustainable excellence.

Story 4 — Salk Polio Vaccine Trials (1952–55): As polio outbreaks intensified, statistical models guided trial design but lacked the ethical nuance required for medical decision-making. Researchers interpreted model outputs through clinical judgment, ensuring safety and fairness while machines processed vast amounts of trial data. This combined system accelerated discovery without compromising integrity. Human oversight balanced computational rigor with moral responsibility. It showed how human values anchor scientific acceleration.

Story 5 — Apollo 13 Mission Control (1970): An explosion crippled Apollo 13, leaving astronauts in mortal danger as incomplete telemetry streamed back to Earth. Engineers relied on simulations to test survival scenarios under extreme constraints, but the machines could only reveal possibilities. Human teams improvised procedures the computers could not imagine, blending machine-generated insights with creative reasoning. It became a defining moment of human–machine crisis collaboration.

Story 6 — Air Traffic Control Modernization (1960s–80s): Surging air travel overwhelmed manual tracking systems, and controllers struggled to maintain safe separation across increasingly crowded skies. Radar automation provided real-time positional data, enabling machines to track aircraft continuously while humans made judgment calls. This hybrid system reduced collisions, increased capacity, and established a global standard for aviation safety. Human–machine coordination became the backbone of modern air travel. The system remains one of the world's most enduring examples of operational augmentation.

LESSONS FROM A LEADERSHIP TRANSFORMATION PERSPECTIVE

Leadership in human–AI synergy evolves from managing isolated tools to shaping systems where human judgment and machine capability mature together. Early examples show leaders confronting uncertainty with mechanical acceleration, while later decades reveal leaders orchestrating digital coordination at scale. Across this progression, effective leadership is defined by the ability to interpret signals machines surface, impose ethical boundaries machines cannot perceive, and mobilize teams to act decisively when models reach their limits. The leader becomes a systems architect—balancing speed with scrutiny, automation with accountability, and innovation with safety—ensuring that human purpose remains the governing force behind technological power.

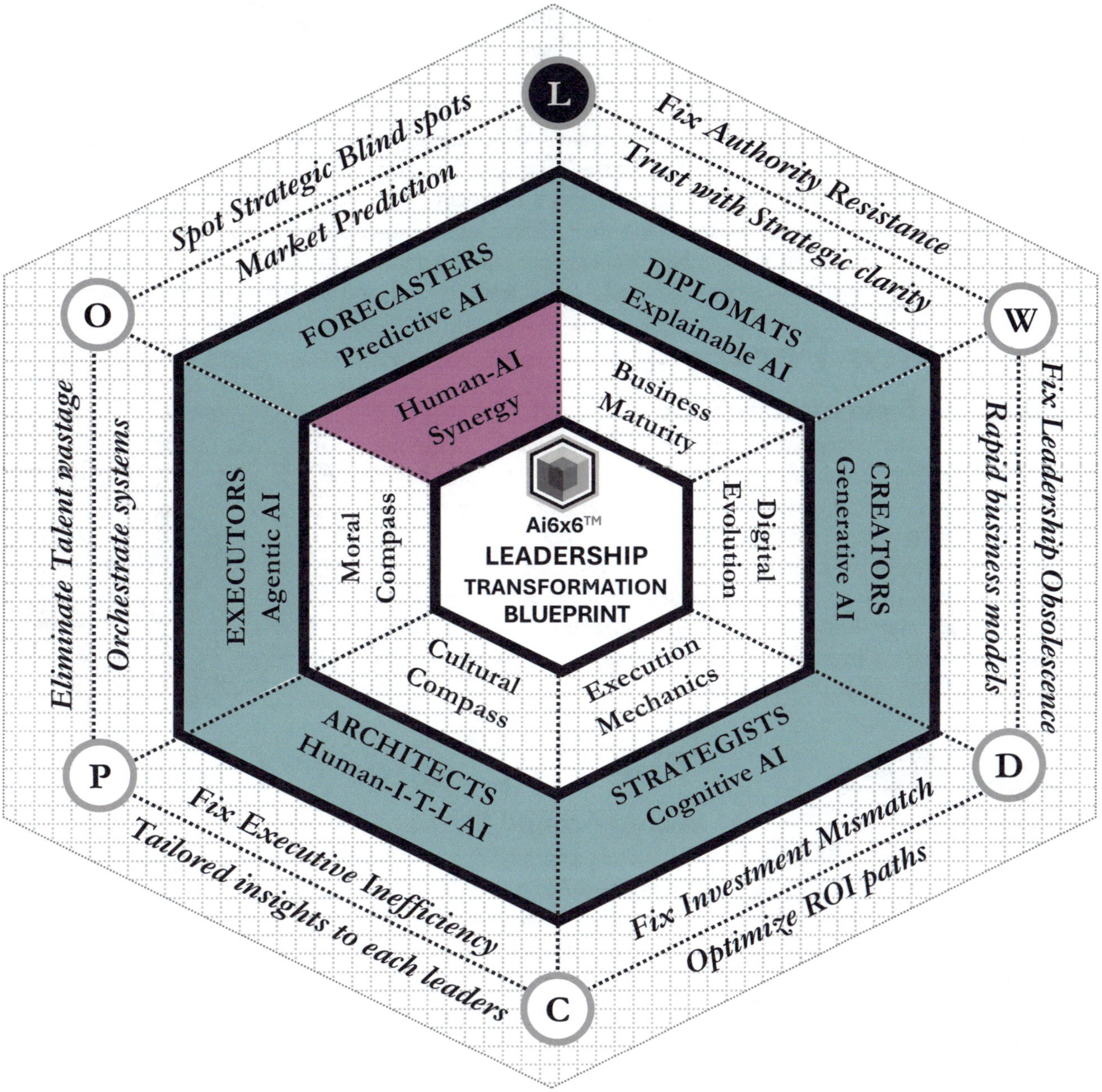

The **Ai6x6 Research Lab** chose this lens to address the silent crisis of identity at the top of the modern enterprise. We believe the first hurdle of transformation is not technical, but the psychological surrender of the "hero leader" archetype. This lens forces the mind to stop viewing intelligence as a personal trophy and start viewing it as a collaborative architecture. It is designed to move a leader from the paralyzing fear of being replaced by a machine to the profound liberation of being amplified by one. By exploring this synergy, we ensure the human spirit remains the indispensable "why" behind the machine's "how."

It is about shifting the leader's heart from a defensive posture to one of creative orchestration. This lens acts as a cognitive bridge, allowing a leader to see agentic systems not as competitors, but as diverse personas that extend their own reach. Without this shift, the leader remains a bottleneck; with it, they become the architect of a collective brilliance that far exceeds the sum of its parts. It is an invitation to shed the burden of omniscience and embrace a new, augmented authority grounded in humility, curiosity, and the courage to lead alongside an intelligence that never sleeps.

Leadership in the age of intelligent systems demands a new grammar of collaboration—one that transcends the binary of human versus machine and instead frames intelligence as a continuum. The **Human–AI Synergy Lens** redefines leadership not as the art of commanding technology but as the discipline of co-creating with it. It asks leaders to move beyond automation and efficiency toward a deeper integration of foresight, ethics, creativity, cognition, governance, and execution. At its core, Human–AI Synergy is about **shared agency**. It recognizes that leadership today is exercised not only through human judgment but also through algorithmic influence. Every forecast, recommendation, and automated decision carries a fragment of leadership intent. The challenge is not to suppress that influence but to shape it—to ensure that AI amplifies human purpose rather than diluting it. This lens therefore becomes a mirror: it reflects how leaders distribute cognition, emotion, and control across human and machine systems, and how they design organizations that learn, reason, and act as unified intelligences.

The Creator Persona

Expanding the Field of Possibility

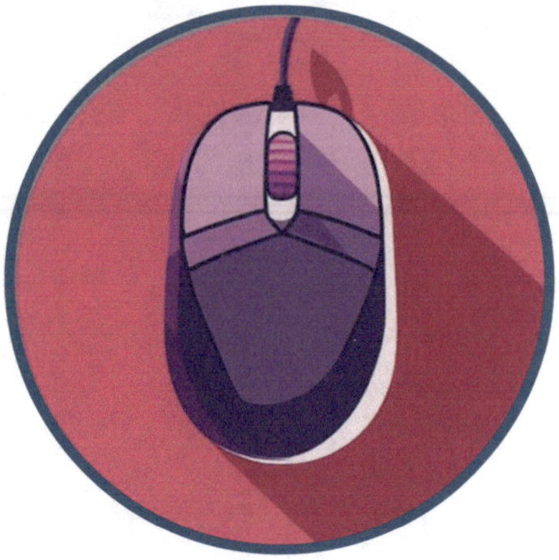

Creators represent the imaginative frontier of leadership. Through **Generative AI**, they explore more ideas than any human-only process could conceive. They treat AI not as a threat to originality but as a catalyst for it—a partner in ideation that expands the creative bandwidth of the organization. The Creator's discipline is not about producing more content but about discovering new patterns of thought. They use AI to prototype futures, to visualize complexity, and to democratize innovation. In their hands, generative systems become engines of curiosity, enabling leaders to design from abundance rather than scarcity.

The Forecaster Persona

Seeing Around Corners

Forecasters embody the anticipatory dimension of leadership. They use **Predictive AI** to model futures before committing resources, transforming uncertainty into structured insight. Their craft lies in converting data into foresight—detecting weak signals, pressure-testing assumptions, and rehearsing alternative scenarios. In traditional leadership, forecasting was intuition-driven; in synergistic leadership, it becomes a disciplined dialogue between human imagination and machine probability. The Forecaster's maturity is measured not by how accurately they predict but by how responsibly they prepare. They shift the organization from reacting to events to anticipating them, embedding resilience into its strategic DNA.

The Strategist Persona

Orchestrating Cognitive Intelligence

Strategists use **Cognitive AI** to weave intelligence across functions, geographies, and time horizons. They design strategy as a living system—adaptive, self-correcting, and continuously learning. The Strategist's genius lies in orchestration: aligning distributed intelligences into coherent purpose. They understand that cognition is no longer centralized in the executive suite; it flows through networks of humans and machines that co-reason in real time. Their leadership is systemic rather than hierarchical, emphasizing feedback loops, dynamic modeling, and cross-domain synthesis. In this persona, AI becomes the connective tissue of enterprise intelligence.

The Diplomat Persona:

Building Ethical Bridges

Diplomats operate at the intersection of **Explainable AI** and human empathy. They translate algorithmic logic into moral language, ensuring that machine outputs are understood, trusted, and socially accepted. Their role is to humanize the interface between data and decision, to make transparency a leadership act. In a world where AI can both illuminate and obscure, the Diplomat safeguards the integrity of interpretation. They remind the enterprise that trust is not a technical feature—it is a cultural outcome. By bridging emotional and ethical terrain, they ensure that AI strengthens relationships rather than eroding them.

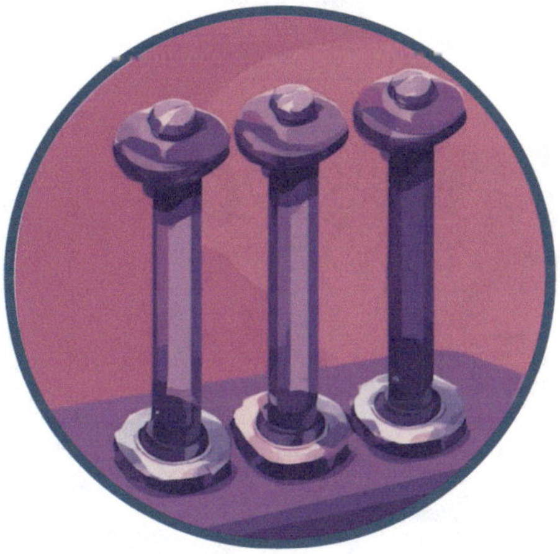

The Architect Persona:

Designing Responsible Collaboration

Architects are the custodians of **Human-in-the-Loop AI**, ensuring that governance and ethical guardrails are embedded into every workflow. They design environments where humans and machines collaborate responsibly—where oversight is not an afterthought but a structural principle. The Architect's leadership is infrastructural: they build the conditions under which synergy can thrive. Their focus is on accountability, transparency, and inclusion, crafting systems that respect both human dignity and machine precision. In their world, collaboration is engineered, not improvised; it is sustained by design integrity.

The Executor Persona:

Converting Insight into Action

Executors close the loop between knowing and doing. Through **Agentic AI**, they embed intelligence directly into operations so that insight becomes action without delay. They represent the kinetic dimension of synergy—the moment where cognition translates into execution. The Executor's mastery lies in operationalizing intelligence: automating decisions that are safe, contextual, and aligned with strategic intent. They ensure that the organization's nervous system is responsive, not reactive, and that every process carries the imprint of learned intelligence.

Together, these six personas form a **leadership spectrum**—from foresight to action, from imagination to governance. They illustrate that synergy is not a single capability but a choreography of roles. The Forecaster anticipates, the Diplomat interprets, the Creator imagines, the Strategist aligns, the Architect safeguards, and the Executor delivers. Each persona represents a distinct mode of engagement with AI, yet all converge on one principle: leadership is now a shared performance between human and machine. The Human–AI Synergy Lens therefore reframes leadership as **co-evolution**. It invites leaders to cultivate humility before complexity, curiosity before certainty, and design before control. It challenges them to lead not by commanding intelligence but by convening it—to orchestrate a dialogue between human wisdom and machine capability that continuously redefines what is possible. In this lens, synergy is not the end state; it is the ongoing discipline of learning together.

THE FORECASTER
Predictive AI — Meteorologist

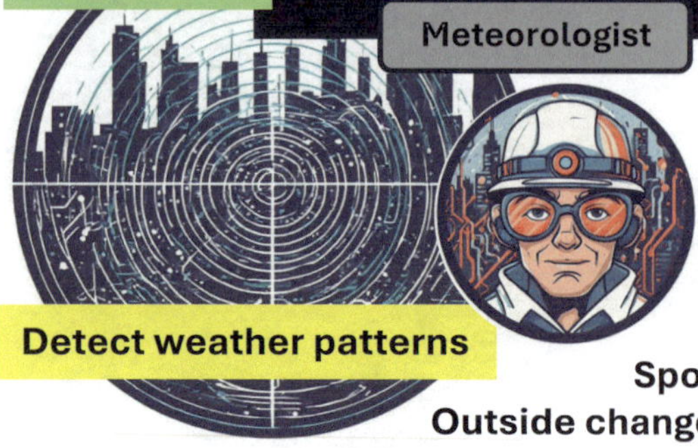

Detect weather patterns

Spot Outside change

Seeing around corners and spotting trends before they hit.

THE STRATEGIST
Cognitive AI — Urban Planner

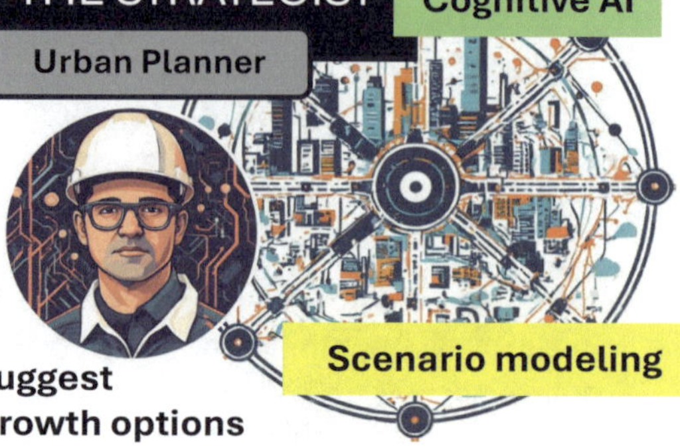

Scenario modeling

Suggest Growth options

Orchestrating intelligence and making complex decisions.

THE EXECUTOR
Agentic AI — Operations Chief

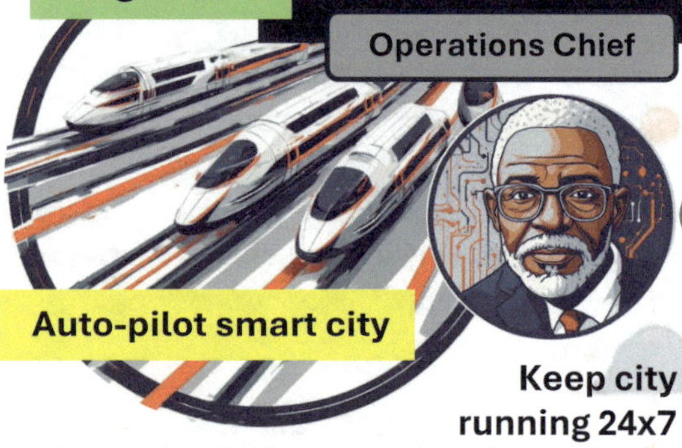

Auto-pilot smart city

Keep city running 24x7

Closing the gap between knowing and performing automated action.

THE CREATOR
Generative AI — Master Architect

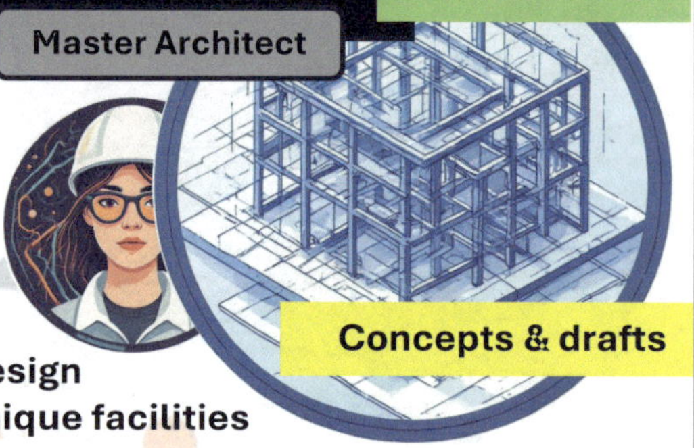

Concepts & drafts

Design unique facilities

Making, building, and expanding the field of possibility.

THE DIPLOMAT
Explainable AI — Press Secretary

Explain system logic

Win people's trust

Building trust, empathy, and ethical connection.

THE ARCHITECT
Human I-T-L AI — Mayor & City Hall

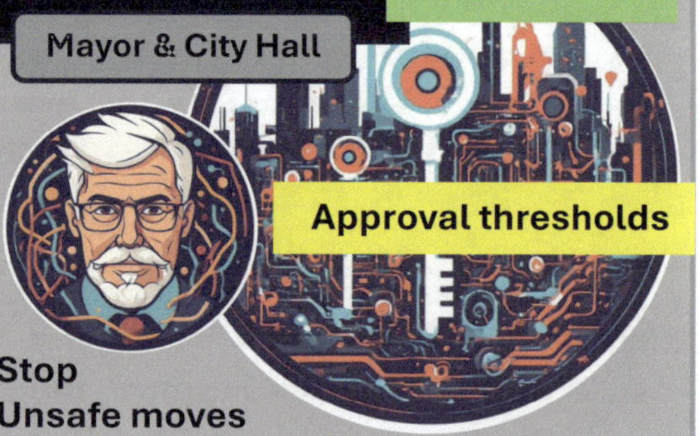

Approval thresholds

Stop Unsafe moves

Providing the structure, governance, and guardrails.

6x6™ Transformation Blueprint

Evolution of Humanity

"The only true wisdom is in knowing you know nothing."

Socrates (c. 470–399 BC); Wisdom is the humility to acknowledge the unknown, leading to the creation of machine-enforceable constraints to manage systemic risk.

The fulfillment of leadership in ancient civilizations was rooted in moral clarity and the sovereignty of judgment. The ideal of the Philosopher-King or the Roman Senator was not defined by efficiency, but by the weight of responsible decision-making—ruling justly and maintaining the imperium (command). True purpose was found in establishing durable law and acting as a custodian of societal integrity. The Industrial Age, however, shifted this focus dramatically; leadership fulfillment became defined by managerial control, optimizing labor, and maximizing profit. The leader became the master of the hierarchy, often losing moral clarity in the pursuit of scale, leading to managerial exhaustion and ethical compromise. In the AI Era, fulfillment for leaders returns to its ancient roots of Integrity and Systemic Sovereignty. By delegating operational execution to autonomous systems, the leader is liberated to focus exclusively on defining machine-enforceable constraints (Policy as Code). Their ultimate professional purpose is found in architectural stewardship and ensuring that their fundamental ethical judgment is scaled and guaranteed by the technology itself, achieving stability and reducing systemic risk.

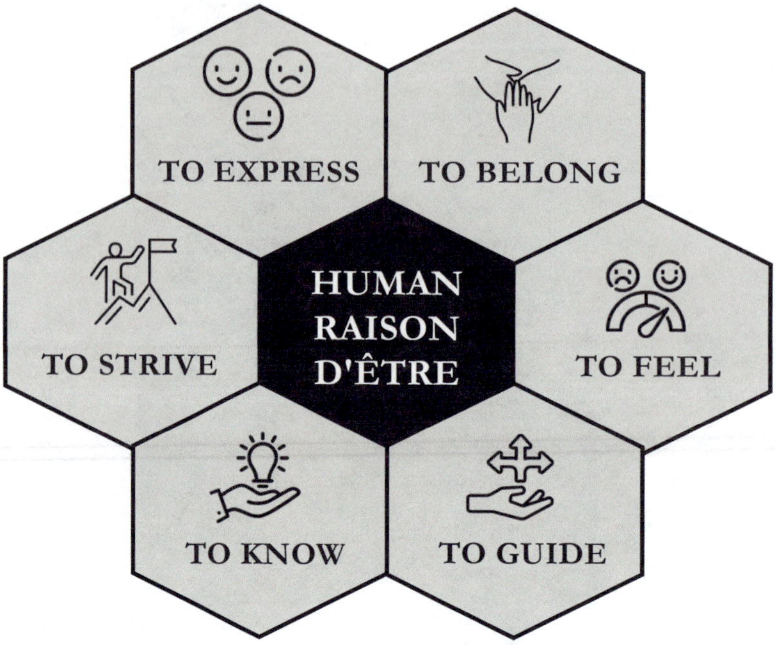

SIX UNIVERSAL ANCHORS OF HUMAN FULFILLMENT

HUMAN RAISON D'ÊTRE (THE REASON FOR EXISTENCE)

In an age defined by exponential automation, the core question of human existence shifts from survival to **significance**. When technology handles efficiency and labor, human value is no longer defined by capability, but by **meaning**. The six **Universal Anchors of Human Fulfillment** define the enduring, non-negotiable contributions of human consciousness—the pillars of a **Raison D'être** that secures our purpose across any era. These anchors affirm that our ultimate mission is to serve as the unwavering source of **Meaning and Moral Authority** in existence. They are the essential acts that only consciousness can perform.

1) TO EXPRESS

The **Anchor of Expression** is the essential human mandate to be the **Source of Novelty and Creation**, generating original ideas and subjective feelings that resist calculation or statistical synthesis. This purpose validates our unique consciousness by asserting that fulfillment comes from the generative act of bringing forth something utterly unique, not merely consuming or replicating. It requires **Uncurated Creativity**—the freedom to explore and define **Aesthetic Judgment**—thereby expanding the conceptual universe with concepts that have no statistical precedent. Through expression, we transform complex inner experience into external reality, upholding our role as the inventors of meaning.

The ultimate benefit of this anchor is the preservation of **Human Originality** and **Intellectual Sovereignty**. By committing to expression, we secure a future that is continually enriched by unique ideas, preventing stagnation and the predictability of automated systems. This anchor maintains the essential spark of invention, ensuring that human meaning remains tied to the capacity to create, **differentiating our value from that of a machine**. It is the source of all artistic, philosophical, and inventive breakthroughs that define a truly dynamic culture.

2) TO BELONG

The **Anchor of Belonging** defines our core identity as social beings, asserting that **authentic connection** is an existential necessity beyond mere digital interaction. This purpose requires the active, continuous labor of **Deep Connection**, prioritizing **unconditional presence** and shared vulnerability to foster a resilient community. Fulfillment is achieved by being the **social architect**, engaging in the emotional work of **mediating conflict**, and fostering cohesion. It includes preserving and transmitting **Cultural Heritage**—the stories and qualitative wisdom that bind us across generations. To belong is to confirm that our worth is interwoven with our relationships.

The ultimate benefit of this anchor is **Emotional Resilience** and **Social Stability**. By prioritizing genuine human connection, we safeguard ourselves against isolation and the psychological fragility that arises from purely transactional relationships.[1] This anchor ensures a supportive social framework that validates empathy and provides necessary emotional resonance, acting as a crucial defense mechanism against an isolating technological world. It builds the **trust and cohesion** necessary for any human system to function justly and compassionately.

3) TO FEEL

The **Anchor of Feeling** grounds human existence in the tangible, asserting the profound psychological value of **Embodied Mastery** and sensory engagement. This purpose elevates the dedicated pursuit of physical skill and process—not for necessity, but for the inherent satisfaction of the process. It is the quest for the **'Flow' state**, where an earned link between **exertion and accomplishment** provides deep fulfillment. This anchor mandates resisting the ease of total automation by engaging in voluntary challenges and perfecting non-essential crafts. It ensures that our abstract thoughts are grounded in the undeniable reality of the body and the senses.

The ultimate benefit of this anchor is **Physical and Psychological Integrity**. By committing to felt experience, we preserve our cognitive and physical health, resisting the inertia of automation and maintaining a sharp connection to reality. This anchor validates the **dignity of effort**, providing a clear, visceral confirmation of personal competence and self-mastery that is crucial for self-esteem. It ensures that fulfillment remains an **earned state**, confirming the complexity and richness of being physically alive.

4) TO GUIDE

The **Anchor of Guidance** establishes the human being as the **ultimate moral and ethical sovereign**. This purpose is founded on our unique capacity for empathy and subjective value judgment, making us the non-delegable authority for setting rules and managing risk. It is the practice of **Ethical Governance**, demanding that we **Draw the Moral Line** and translate compassionate principles into enforceable policy. Fulfillment is achieved by maintaining transparency, enforcing accountability, and continuously exercising the **Human Veto Power** over critical decisions.

The ultimate benefit of this anchor is **Systemic Integrity** and **Moral Safety**. By retaining final ethical authority, we ensure that technological power remains fully aligned with, and subservient to, human values, preventing systems from optimizing themselves at the expense of human welfare. This anchor provides the essential check against unchecked power, establishing the legal and moral frameworks necessary to protect all citizens. It guarantees that empathy and justice remain the **supreme architects of our collective future**.

5) TO KNOW

The **Anchor of Knowing** defines our core intellectual purpose as the engine of **Epistemic Exploration** and abstract reasoning. This purpose compels the mind to focus on formulating **Axiomatic Hypotheses**—the fundamental "Why?" questions that lie beyond the reach of data processing and established knowledge. Fulfillment is found in the generative act of **challenging established truths** and creating new theoretical frameworks that synthesize disparate fields. This anchor ensures that the pursuit of knowledge remains boundless, driven by philosophical depth rather than functional utility.

The ultimate benefit of this anchor is **Intellectual Freedom** and **Conceptual Progress**. By focusing on formulating new questions and challenging paradigms, we prevent the stagnation that could result if all knowledge were confined to data-driven answers. This anchor guarantees that our intellectual journey continues to be defined by profound theoretical breakthroughs and a relentless expansion of understanding. It confirms that human intelligence is defined not by the volume of information stored, but by the **depth and originality of the questions** we ask.

6) TO STRIVE

The **Anchor of Striving** is the **Existential Quest**—the voluntary pursuit of difficulty necessary to generate meaning, independent of any external necessity. This purpose recognizes that true fulfillment is an **earned state**, achieved through **Deliberate Discipline** and self-imposed ambition. It compels us to **Set Impossible Goals** and **Choose the Hard Way**—overcoming voluntary hardship—precisely because the challenge validates our capacity and resolve. This continuous effort allows us to craft a meaningful **Personal Legacy** by executing complex, long-term plans purely for the satisfaction of the effort.

The ultimate benefit of this anchor is **Authentic Self-Mastery** and **Purposeful Existence**. By embracing self-imposed difficulty, we cultivate the resilience, discipline, and fortitude that define character, resisting the complacency of automated comfort. This anchor ensures that human worth is determined by our **internal will**, rather than external circumstance, providing a framework for continuous personal growth and the creation of an enduring, meaningful legacy.

CONCLUSION: THE RAISON D'ÊTRE FOR THE HUMAN

The six Universal Anchors converge to form humanity's ultimate purpose: to be the **Unwavering Source of Meaning and Moral Authority**. This composite Raison D'être defines the essential human role in any advanced civilization.

Our mission is the sustained commitment to **non-automatable value**. We are the unique entity that not only pursues truth (To Know) but also creates the beauty (To Express) that makes truth worth pursuing. We are the mechanism that defines the rules of fairness (To Guide) and then personally invests the necessary effort (To Strive) to live up to those rules. Crucially, we are the only entity that grounds this entire conceptual structure in the tangible joy of physical reality (To Feel) and the vital necessity of shared vulnerability (To Belong).

The human purpose shifts from survival to an intentional journey of **growth, creation, and ethical custodianship**. Our value is not in output, but in the **input of consciousness, conscience, and will**. By diligently exercising these six anchors, every individual contributes to the essential, enduring purpose of the species: to be the **Sole Arbiter of Worth, Will, and Wonder**—ensuring that existence remains a deeply meaningful and purposeful quest.

In this era, human existence was focused on the direct, tangible application of skill for immediate survival, such hunting, fishing and surviving. The human was the Master of the Hunt.

In the Primitive/Paleolithic Era (Survival & Sustenance), human purpose was an absolute function of immediate necessity. Fulfillment was not a philosophical pursuit but a practical outcome of successfully continuing existence for oneself and the kin group. All six Universal Anchors manifested, but they were tethered directly to the biological and environmental imperative to survive the day and propagate the species. The overarching Raison D'être was to **Endure and Transmit**, meaning personal worth was inseparable from one's contribution to the group's survival calculus. This collective focus meant that individual talent was instantly converted into a benefit for all, solidifying a direct, unbroken link between personal action and existential significance.

To Express: Expression was functional and communal, centered on **Ritual Storytelling** and cave art. This art was not for individual fame or abstract aesthetics, but served as the primary educational system, preserving and transmitting essential survival knowledge, hunting techniques, spiritual beliefs, and the topography of the land. The repetitive, sacred nature of the rituals ensured collective memory and adherence to successful methods. Furthermore, these cultural forms encoded crucial information, often representing animals and spirits, that, if lost or forgotten, would jeopardize the tribe's safety and cultural identity.

To Belong: The Anchor of Belonging was non-negotiable, manifesting as **Kinship and Mutual Defense**. Isolation meant certain death; therefore, the purpose of connection was absolute. Fulfillment came from actively contributing to the group's defense, sharing food and resources, and ensuring the collective physical safety of the tribe above all else, fostering a fierce, self-sustaining mutualism. The emotional security derived from this deep collaboration—knowing that others were committed to your well-being—was a core psychological reward.

To Feel: Embodied Mastery was defined by **Essential Tool Mastery**. The feeling of accomplishment was directly tied to successfully chipping a sharp stone tool, accurately throwing a spear, or enduring the physical hardships of the hunt. This pursuit was not sport, but the daily necessity of mastering the physical world to secure food and shelter, and the psychological confidence gained was crucial for future success. This expertise immediately translated into social value; the most capable hunter or tool-maker earned respect and a higher status within the hierarchy, directly reinforcing the dignity of effort.

To Guide: Moral authority was based on immediate experience, taking the form of **Elder Authority and Tradition**. Guidance was unwritten law, passed down through generations of hard-won experience. Decisions—like where to hunt, when to migrate, or how to resolve disputes—were arbitrated by those with the deepest, most successful experiential knowledge, and were obeyed out of necessity for the group's stable cohesion. This system was vital for ensuring collective action remained predictable and risk-adverse, preventing internal conflict from threatening external survival.

To Know: The pursuit of knowledge was pure **Observation and Practicality**. The purpose of knowing was to accurately predict the immediate environment for resource exploitation. This involved memorizing the habits of migratory animals, understanding the medicinal or poisonous properties of local plants, and tracking subtle shifts in weather patterns—information critical for daily sustenance and successful, low-risk planning. Early forms of abstract reasoning also emerged here, through tracking the movement of stars or noting the cyclic nature of seasons and astronomical events, demonstrating an innate human desire to find order and predictability in environmental chaos.

To Strive: Striving was synonymous with **Daily Endurance**. There was no need to invent abstract goals; the pursuit of difficulty was the environment itself, offering constant challenge. Fulfillment was the successful completion of the continuous, grueling effort required to secure food, withstand the elements, and ensure the basic propagation of the tribe, creating a deeply ingrained sense of earned worth. The resilience built through continuous hardship was the ultimate personal asset, psychologically connecting physical suffering with the existential meaning of survival.

With stability, purpose shifted to sustained output, long-term planning, and maintaining social harmony within a fixed settlement. The human was the Master of the Land.

The Agricultural/Neolithic Era (Farming & Community) marked humanity's fundamental pivot from nomadic struggle to settled stability. The mastery of cultivation and animal husbandry created a reliable **Surplus of Time and Resources**, fundamentally shifting human purpose from immediate survival to calculated investment, long-term order, and the establishment of lasting community structures. The Raison D'être of this period was to **Settle and Establish Order**. This era saw humans take on the role of conscious managers of the environment, laying the groundwork for systematic laws, property rights, and enduring social legacies that extended far beyond one generation.

To Express: Expression became focused on **Accounting and Enduring Form**. The logistical necessity of tracking seeds, yields, stored grains, and communal ownership led directly to the invention of the first systematic writing forms, such as early cuneiform, demonstrating that expression was initially a tool for managing complex resources. Beyond text, expression was seen in durable, functional goods like specialized pottery, permanent architectural structures, and woven textiles. These forms reflected a deep, stable community identity and provided evidence of specialized skill that contributed directly to the village's permanence.

To Belong: Belonging shifted to being centered on **Land-Based Cooperation**. The success or failure of the farm was the success or failure of the entire settlement, making shared labor and mutual dependency paramount. Fulfillment was found in the reliable rhythms of shared work, such as irrigation projects or harvesting, cementing generational inheritance and deeply intertwined **Kinship Networks**. Purpose was now inseparable from one's physical, defined place within the land and the local community, creating a profound sense of rootedness and stability unknown in the Paleolithic past.

To Feel: Embodied Mastery focused on **Craftsmanship and Process**. Fulfillment was derived from the methodical, repeated application of skill required by agriculture—plowing a straight furrow, carefully managing livestock, or perfecting the complex processes of metallurgy and food preservation. This mastery provided a sense of control over the environment that raw endurance could not. The purpose of skill was purely to generate the stable surplus required to maintain the settled life, providing a deep psychological reward through the predictable yield of sustained, expert effort.

To Guide: Moral authority became formalized, focusing urgently on **Property Rights and Resource Management**. Guiding the community meant establishing clear, complex rules around ownership of land, managing precious resources like water rights, and ensuring crop protection from both animals and neighbors. This necessity gave rise to the first systematic, written laws, moving purpose from inherited tradition (as in the Paleolithic) to a **Codified Social Contract**. Leadership was defined by the ability to enforce these new laws, ensuring fairness and stability within an increasingly complex economy.

To Know: Abstract reasoning began to flourish, driven by **Mathematics and Calendar Tracking**. The purpose of knowledge was now predictive certainty, essential for farming success. This meant mastering geometry to accurately divide and manage fields, and engaging in sophisticated celestial observation to track seasons and accurately determine optimal planting and harvesting cycles. This pursuit of cosmic order confirmed the human ability to impose intellectual structure upon chaotic nature, turning farming into a sophisticated, knowledge-driven enterprise.

To Strive: Striving fundamentally shifted from day-to-day survival to **Creating Surplus and Investment**. The existential quest became the arduous, long-term, deliberate work required to build durable structures, amass protective stores of food, and ensure the economic and social prosperity of the next generation. This focus on legacy and investment transformed work into a purposeful act of building a lasting future, where one's worth was measured not just by survival, but by the inheritance left behind for the family and the stability provided to the community.

This era defined purpose through professional expertise and contribution to complex, often abstract, technological and economic systems. The human was the Master of the Machine.

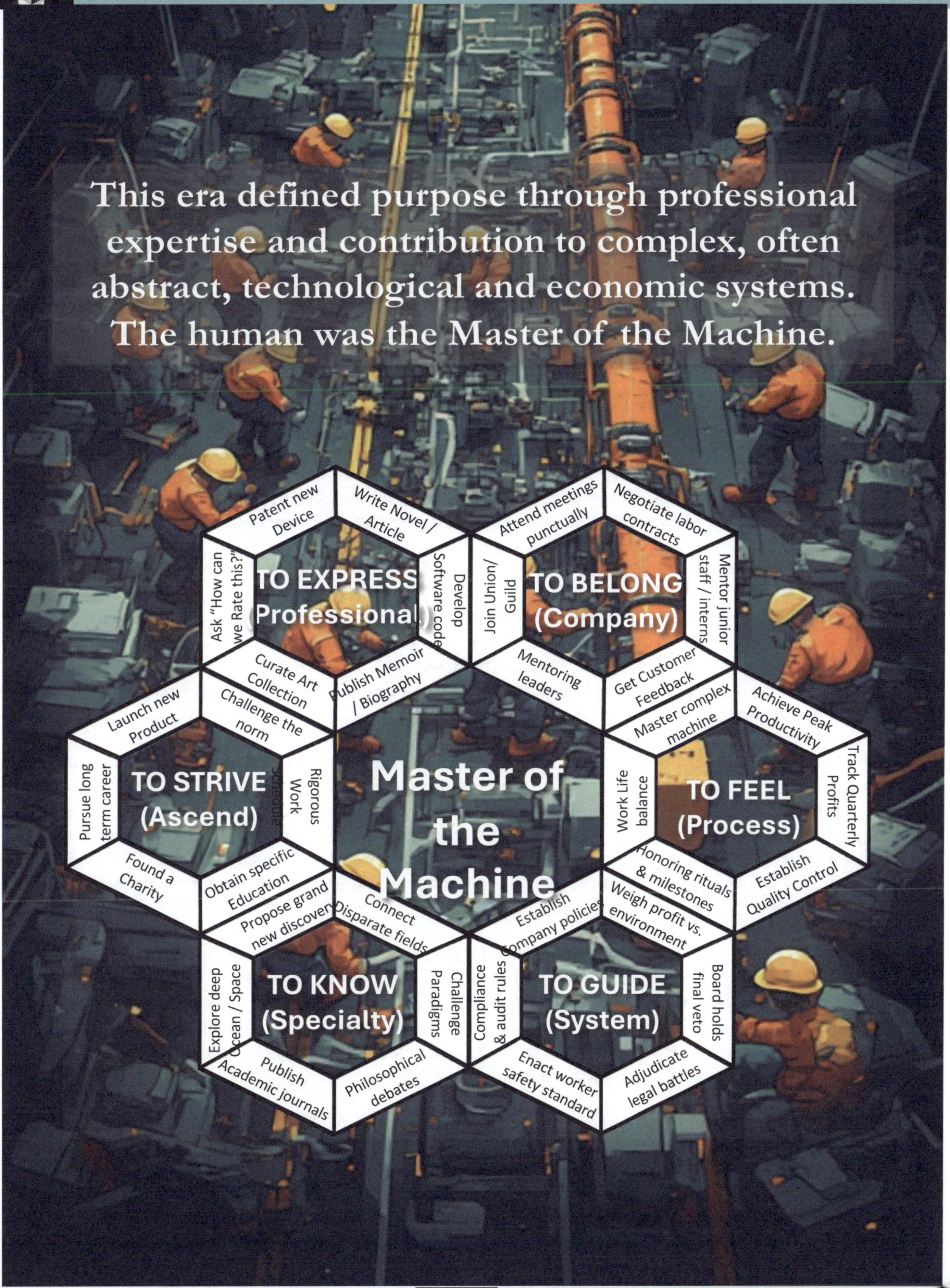

The Modern/Industrial Era (Specialization & Function) marked the decisive triumph of machine logic and capital over the natural world. Human purpose shifted from managing the cycles of the land (Neolithic) to mastering the mechanical efficiency of the factory and the market. The rise of the machine and the division of labor fundamentally redefined the individual, whose worth was now derived from highly specialized, functional expertise within a vast, interconnected economic system. The Raison D'être of this period was to **Achieve and Advance**, driving relentless progress through expertise and material gain.

To Express: Expression became focused on **Mass Media and Invention**. With the advent of the printing press, cheaper paper, and later, radio and cinema, creative expression transformed into a commodified product designed for wide distribution and profit. The purpose of expression was to inform, entertain, and persuade the masses, shaping shared cultural narratives through novels, newspapers, and commercial design. The creation of a **Patentable Invention** became a powerful act of individual expression, translating originality directly into economic ownership and societal contribution. Expression was celebrated when it served a functional, marketable, or propagandistic role.

To Belong: Belonging was redefined by **Nation and Institution**. The old agrarian kinship networks dissolved as masses migrated to urban factory centers, replaced by new, abstract, yet powerful social structures. Fulfillment was found in national identity, membership in professional organizations, and political affiliation. Crucially, the extreme specialization of labor necessitated the formation of organized counter-structures like **Labor Unions** to protect the rights of workers against centralized corporate power, demonstrating that belonging was a functional necessity for collective bargaining and social stability in a highly unequal economic system.

To Feel: Embodied Mastery centered on **Efficiency and Maximized Output**. The physical purpose shifted away from holistic craft towards highly repetitive, specialized labor—the mastery of a single, precise task on the assembly line. While monotonous, this specialization provided a clear, measurable sense of contribution, where personal worth was directly tied to economic output and reliability. Outside of the factory, the Anchor manifested in the rise of **Organized, Competitive Sports**, where physical effort was deliberately structured and formalized, allowing the industrial individual to achieve a sense of earned competence and visceral struggle that was absent in their work.

To Guide: Moral authority became formalized through **Secular Law and Corporate Regulation**. The complexity of contract law, property ownership, and massive corporate liability demanded sophisticated, non-religious guidance systems. The purpose of guidance was to manage the inherent conflicts between capital and labor, protect consumer safety in mass production, and define the boundaries of ethical economic competition. This period saw the relentless development of regulatory bodies and constitutional law aimed at establishing fairness within the new, powerful industrial economy, ensuring that **progress did not undermine social order** entirely.

To Know: The pursuit of knowledge was dominated by **Empirical Science and Specialization**. The purpose of knowing was functional utility, believing that all problems—from disease to industrial inefficiency—could be solved through the application of the rigorous **Scientific Method**. Knowledge became fragmented into specialized disciplines, each dedicated to gathering objective, measurable data. Fulfillment was achieved by contributing a verifiable, peer-reviewed fact or theory to one's specific field, thereby confirming the era's deep-seated faith in rational inquiry and technical progress.

To Strive: Striving was synonymous with **Ambition and Material Gain**. The existential quest was the pursuit of wealth, social mobility, and **Status**. Fulfillment was an earned state achieved through economic success, the establishment of a profitable business, or professional advancement within a corporate or governmental hierarchy. The cultural narrative of the "self-made individual" defined personal purpose, urging people to embrace continuous, relentless work to secure prosperity for the family and, more importantly, to carve out an enduring legacy of material success and high societal standing.

In this emerging era, AI handles efficiency and role specialization, freeing and forcing humans to refocus on subjective values, unique creation, and moral governance. The human becomes the Master Curator of Purpose

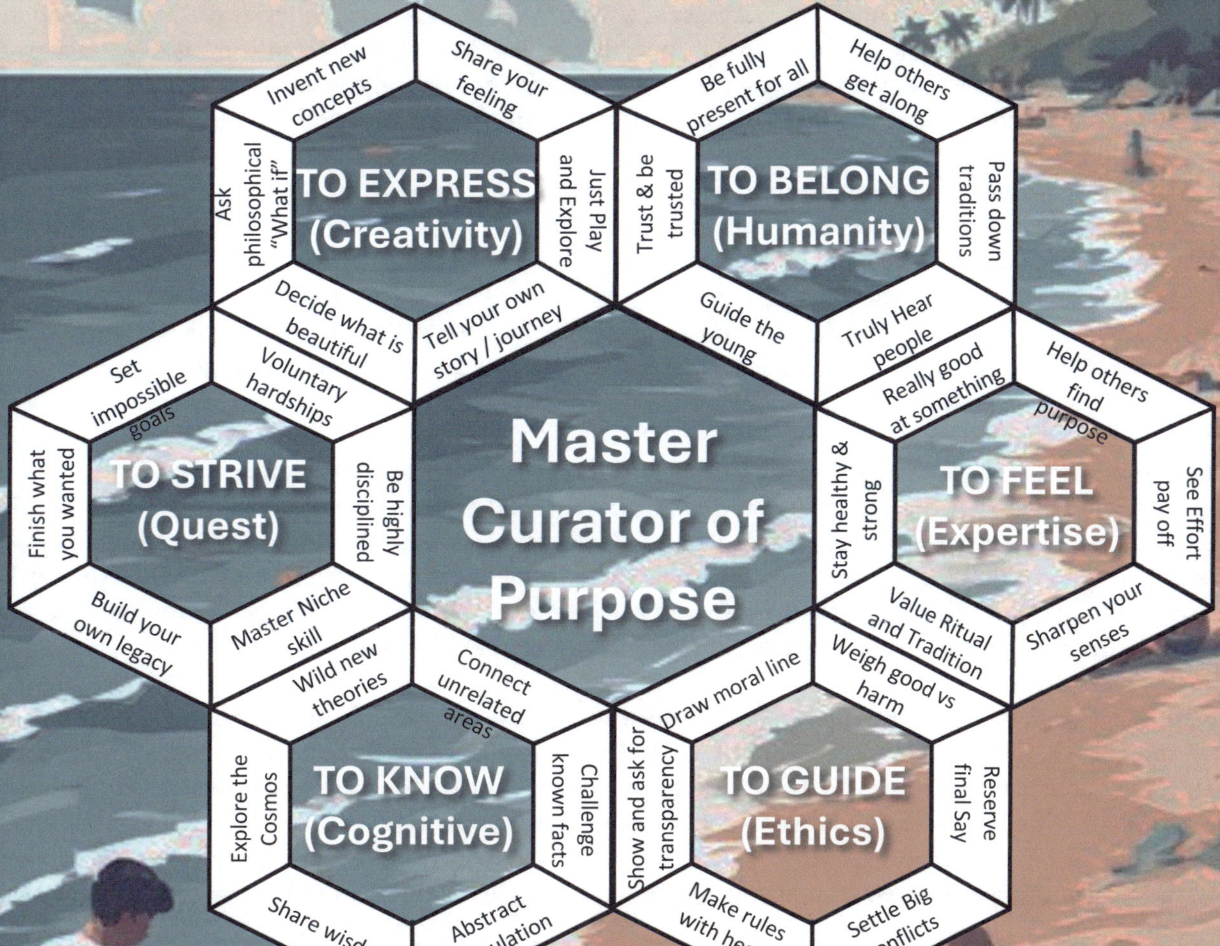

The Post-AI Era (Curation & Meaning) represents the definitive existential shift in human history. With advanced artificial intelligence and automation having conquered the challenges of efficiency, information processing, and mass labor, human purpose is radically redefined. We are no longer necessary for function, but for **meaning**. The individual's worth is derived entirely from their capacity for subjective judgment, ethical authority, and the creation of non-calculable value. The collective Raison D'être is to become the **Sole Arbiter of Worth, Will, and Wonder**—the source of all criteria for why life should continue and what makes it beautiful.

To Express: Expression is focused on **Subjectivity and Uncurated Truth**. In a world saturated with synthetic media and hyper-optimized content generated by AI, the only expression that retains intrinsic value is the genuinely **original, subjective human voice**. This purpose mandates focusing on authentic, unfiltered creation—personal philosophical content, aesthetic critique, memoir, and art that contains flaws, uncertainty, and emotional resonance. Fulfillment is found in generating novelty that AI cannot predict, thereby constantly pushing the boundaries of human experience and securing a creative future that is immune to algorithmic stagnation.

To Belong: Belonging becomes a deliberate, chosen investment in **Deep Vulnerability and Presence**. When digital connection is ubiquitous but emotionally shallow, true fulfillment relies on actively resisting transactional relationships and seeking authentic, physical presence. This purpose demands the **labor of Deep Connection**: cultivating empathy, mediating conflict through personal understanding, and prioritizing quality, face-to-face interaction. The collective human project requires maintaining the social fabric and emotional resonance that machines cannot replicate, ensuring that isolation does not erode the psychological foundation of society.

To Feel: Embodied Mastery transforms into the **Voluntary Pursuit of Difficulty and Flow**. Since automation eliminates the necessity of difficult labor, human physical purpose centers on self-imposed challenges—non-essential crafts, athletics, meditation, and the perfection of technical, yet unnecessary, skills. This pursuit generates the vital 'Flow' state—the psychological reward of earned accomplishment—which secures self-esteem and grounds abstract consciousness in tangible reality. Fulfillment is achieved through **Deliberate Effort** that reaffirms the dignity of the body, providing a visceral antidote to the inertia and complacency of total comfort.

To Guide: Moral authority centers on **AI Ethics and the Human Veto**. The purpose of guidance is to serve as the **ethical sovereign**, placing strict moral boundaries around autonomous systems. This involves defining the principles of data governance, ensuring transparency in black-box algorithms, and enforcing systemic accountability for technological outcomes. Crucially, every individual is mandated to exercise the **Human Veto Power**—the non-delegable right to overrule critical decisions made by algorithms—ensuring that power remains permanently aligned with subjective human values, not cold-blooded efficiency.

To Know: The pursuit of knowledge is now defined by **Abstract Synthesis and Axiomatic Inquiry**. In an environment where AI holds access to all existing data and established facts, human intellect must shift its focus. The purpose is to transcend data processing and concentrate on formulating **Axiomatic Hypotheses**—the profound "Why?" questions that challenge foundational assumptions. Fulfillment is found in connecting disparate fields of knowledge, creating new philosophical frameworks, and constantly resisting the temptation to settle for algorithmic answers, ensuring the continuous, boundless expansion of human understanding.

To Strive: Striving is the existential commitment to **Self-Imposed Discipline and Personal Legacy**. Since survival is guaranteed, the quest for meaning requires the individual to set ambitious, non-essential goals purely for the satisfaction of the effort. This involves cultivating profound resilience and fortitude against the immense temptation of automated ease. Fulfillment is an **earned state**, achieved through continuous, conscious effort and the construction of an enduring **Personal Legacy**—a body of work or moral character that defines one's worth by internal resolve rather than external economic output.

6x6™ Transformation Blueprint

Human-AI Personas

"Some people call this artificial intelligence, but the reality is this technology will enhance us. So instead of artificial intelligence, I think we'll augment our intelligence."
-Ginni Rometty, the former Chair, President, and CEO of IBM.

The Ai6x6 HUMAN-AI Synergy framework formalizes a new executive standard for the AI era, transitioning organizations from fragmented automation to a state of Orchestrated Autonomy. This model defines the modern enterprise through six core leadership movements, each serving as a specialized "sense" for the organization. **The Forecasters** serve as the "Eyes," tasked with the heavy lifting of foresight to eliminate strategic blind spots and detect invisible market signals before they escalate into crises. Supporting this vision, **The Diplomats** act as the "Voice," translating complexity into human trust and ensuring that transformation is met with cultural alignment rather than institutional resistance. The organization's creative energy is fueled by **The Creators**, the "Engine" of the enterprise, who move beyond static production to drive a state of radical velocity and infinite iteration. Their output is balanced by the surgical objectivity of **The Strategists**, acting as the "Compass" to ensure that every capital investment is mathematically anchored to peak ROI. To unify these efforts, **The Architects** build the "Institutional Brain," ensuring that the organization's collective memory is always accessible and relevant to the individual. Finally, the framework finds its movement in **The Executors**, the "Hands" of the organization, who coordinate autonomous workflows to achieve scale without traditional overhead. By weaving these archetypes into a single Symphony, leaders can navigate market ambiguity with surgical precision, transforming the enterprise into a self-evolving system of sustainable innovation.

Focus: Eradicating Strategic Friction and Safeguarding Executive Integrity

In the modern enterprise, leadership is often paralyzed by a paradox: an abundance of data coupled with a profound lack of clarity. Within the **Ai6x6 Human-AI Synergy framework**, defensive leadership is the practice of utilizing specialized roles to shield the organization's "head" from the noise, bias, and inertia that precede catastrophic failure. Before a leader can hunt for new markets, they must ensure their strategic foundation is immune to internal decay.

The Forecaster and The Diplomat: Seeing Threats and Solving Resistance

The **Forecaster** acts as a strategic sentinel, tasked with the eradication of surprise. By functioning as the "Eyes" of the leadership team, this role scans for "weak signals"—subtle shifts in regulation or competitor behavior—that pose existential threats. It mitigates **Perception Friction**, ensuring leaders are never blindsided by disruptions that were visible in the data.

Conversely, the **Diplomat** serves as the "Voice," neutralizing **Adoption Friction**. Leadership often underestimates the fear involved in AI transformation. The Diplomat uses explainable logic to gauge the internal temperature, preventing a "Values Mismatch" where executive intent is misinterpreted as a threat. By providing transparent reasoning, it builds a shield of trust that prevents the quiet sabotage common in rapidly changing cultures.

The Creator and The Strategist: Fighting Obsolescence and Mismatch

The **Creator's** defensive task is to kill **Velocity Friction**—the decay that occurs when mental models become outdated. By constantly re-indexing the organization's intellectual property, it ensures the company's "DNA" remains compatible with modern systems. It identifies logical flaws in business models before capital is committed, safeguarding the "Engine" against the obsolescence that bankrupts legacy players.

The **Strategist** acts as a ruthless auditor of the executive "Compass," eliminating **Allocation Friction**. Every company carries "Zombie Projects" funded by momentum rather than ROI. The Strategist applies causal inference to prove what is actually driving value, protecting the balance sheet from mismatches between capital outlay and market reality. This prevents leaders from betting on trends that have no mathematical foundation in actual performance.

The Architect and The Executor: Maximizing Efficiency and Governance

The **Architect** kills **Cognitive Friction**—the high-value time lost to reconciling conflicting data. By building a defensive "Institutional Brain," it filters irrelevant noise and ensures that when a leader asks a question, the answer is drawn from a unified, verified truth. This prevents "Information Silos" and protects executive focus, allowing energy to be spent on reasoning rather than data-chasing.

Finally, the **Executor** serves as the defensive "Hands," killing **Economic Friction** through the automation of oversight. In legacy models, leaders waste time on routine governance and compliance. The Executor automates these guardrails, ensuring the business stays within its "Safe Operating Envelope." It prevents **Resource Wastage** by self-correcting minor drifts, allowing the leadership team to remain "Hands-Off" on routine tasks while remaining "Eyes-On" the strategy.

Conclusion: The Shield of Orchestrated Autonomy

When these six roles operate defensively, the organization becomes "Safe to Innovate." By removing the friction of blindspots, resistance, obsolescence, mismatch, inefficiency, and wastage, the **Ai6x6 Human-AI Synergy framework** establishes the foundational stability required for future growth. A leader no longer fighting internal friction is finally free to win.

In the **Ai6x6 Human-AI Synergy framework**, offensive leadership is the transition from "Managing the Present" to "Architecting the Future." While defensive leadership clears the path, offensive leadership uses the six roles to aggressively seize opportunities that competitors cannot even perceive. It is about using AI to expand the leader's "Decision Space," allowing them to act with a level of precision and speed that was previously impossible.

The Forecaster and The Diplomat: Unlocking Prediction and Reasoning

The **Forecaster's** offensive mandate is **Predictive Dominance**. No longer just looking for threats, the Forecaster now identifies "Alpha Opportunities"—emerging market gaps and consumer shifts before they become mainstream. It identifies the "Known Future" by analyzing cross-industry data to find hidden correlations. This allows the leader to move with "First-Mover Certainty," committing capital to high-growth areas while competitors are still debating the data.

The **Diplomat** serves as the offensive "Voice" of persuasion and alignment. Its task is to **Unlock Reasoning** at scale. By synthesizing complex data into a "Decision Space Map," it allows leaders to present airtight, logical arguments to boards, investors, and partners. It moves beyond simple reporting to "Strategic Storytelling," using AI to simulate how various stakeholders will react to a move. This ensures the organization can pivot with radical speed because the Diplomat has already built the necessary cognitive and emotional consensus.

The Creator and The Strategist: Unlocking Generation and Optimization

The **Creator** is the offensive "Engine" of **Strategic Prototyping**. In an offensive stance, this role doesn't just modernize old assets; it generates entirely new business models, brands, and products in a "Sandboxed Reality." It allows a leader to move from abstract strategy to high-fidelity prototype in hours. This "Generative Velocity" means the organization can test more ideas in a week than a legacy competitor can in a year, effectively out-learning the market.

The **Strategist** acts as the offensive "Compass" for **Precision ROI**. Its mandate is to move from "What happened?" to "What is the best thing to do next?" through **Prescriptive AI**. It identifies the "Global Maximum" for any strategic move—finding the exact configuration of price, timing, and resource allocation that will yield the highest possible return. This offensive objectivity ensures that the leader's "big bets" are no longer gambles, but mathematically optimized maneuvers designed to win.

The Architect and The Executor: Unlocking Relevance and Orchestration

The **Architect** builds the offensive "Searchable Brain" of the enterprise. By utilizing **Retrieval-Augmented Generation (RAG)**, it ensures that the organization's collective intelligence is not just stored, but "Live." Every leader and manager is granted instant, context-aware access to the company's most potent insights. This **Hyper-Relevance** ensures that the "right hand" always knows what the "left hand" has learned, allowing the organization to operate as a single, unified, and highly informed strategic entity.

Finally, the **Executor** serves as the offensive "Hands" through **Agentic Orchestration**. Its mandate is to deploy "Reasoning Agents" that can execute complex, multi-step strategic goals with minimal human intervention. It moves the company from "Do as I say" to "Achieve this outcome." This offensive autonomy allows the leader to scale their intent across thousands of variables and geographies simultaneously. The result is **Strategic Reach**—the ability to execute a global vision with the agility of a startup.

Conclusion: The Sword of Orchestrated Autonomy

Offensive leadership within the **Ai6x6 Human-AI Synergy framework** is about the pursuit of "Surgical Precision" at "Radical Velocity." By unlocking prediction, reasoning, generation, optimization, relevance, and orchestration, the leader is no longer just a participant in the market—they are its architect. This framework ensures that growth is not an accident, but the inevitable outcome of a perfectly synchronized Human-AI Symphony.

Executive Power: Foresight

In the traditional leadership paradigm, the executive is often a prisoner of the "Present Moment." They spend their days reacting to reports that are already days or weeks old, effectively driving the organization by looking in the rearview mirror. The transformation into a **Forecaster** shifts the leader's primary responsibility from "Reaction" to "Anticipation." In this role, the leader becomes the organization's "Eyes," responsible for extending the brand's sensory reach across time and space. This is not the work of a psychic or a gambler; it is the work of a scientist who understands that the future is not a single point, but a field of high-probability outcomes that can be mapped and mastered.

Defensive Stance: Killing the Blindspots

The defensive mission of the Leader-Forecaster is the **Abolition of Surprise**. In a hyper-connected, volatile global economy, a surprise is almost always a failure of observation. The leader creates a "Shield of Awareness" by looking for the "Statistical Whisper." Most organizational crises—whether they are financial collapses, supply chain breaks, or reputational scandals—do not happen overnight. They are preceded by "Weak Signals" that a human mind, distracted by daily noise, usually misses. The Forecaster designs an environment where these signals are amplified. By identifying subtle anomalies in the heartbeat of the company, the leader can kill the blindspots of fraud, wastage, and failure before they escalate into catastrophes. This transforms the leadership culture from one of constant "Firefighting" to one of "Continuous Correction," where the organization sees the cliff long before the car reaches the edge.

Offensive Stance: Unlocking Prediction

Once the organization's defensive perimeter is secured by total visibility, the leader moves to the offense: **Capturing the Future.** This involves the use of "Digital Realities" to run strategic war-games. Instead of debating a new initiative based on "Gut Feeling," the leader simulates thousands of versions of that strategy against a million possible market conditions. This **Unlocks Prediction**, allowing the brand to be "Pre-emptive." If the organization can see a shift in consumer intent or a resource scarcity six months out, the leader can move the "Hands" of the company to occupy that future space today. This turns the brand into a "Predatory Competitor"—one that doesn't just "compete" for market share but "claims" it because they were the only ones who knew where it was going to be.

The Leadership Outcome: Mathematical Clarity

The ultimate value of the Forecaster is **Capital Velocity.** When you eliminate the "Uncertainty Tax" from your operations, you no longer need the massive safety nets of excess inventory or idle capital. You can run the company with surgical precision. For the leader, this provides **Mathematical Clarity.** You are no longer managing based on the loudest opinion in the boardroom; you are managing based on a calculated map of reality. This is the foundation of the "Resilient Brand"—an organization that doesn't just survive chaos but uses it as a fuel for growth because they saw the wave coming while the competition was still looking at the shore.

Executive Power: Predictive Governance

In the legacy era, a leader's "Eyes" were limited to the "Rearview Mirror"—static spreadsheets, quarterly reviews, and historical reports that summarized what had already occurred. In the AI-driven enterprise, the **Forecaster** role represents a fundamental transition to **Predictive Governance**. This is the executive's ability to lead via "Probability Maps" rather than "Fixed Plans." By architecting a converged stack of emerging technologies, the leader gains the "Temporal Sovereignty" to see market shifts, operational bottlenecks, and financial risks before they manifest as crises. The ultimate goal is the **Purification of Perception**, ensuring that the leader is always acting on "Live Truth" rather than "Expired Data." This role moves the organization from a reactive posture to an anticipatory one, where the "Eyes" of the firm are positioned at the leading edge of the future.

The Sensory Layer: Radical Transparency and Signal Ingestion

The **Sensory Layer** of the Forecaster acts as the organization's "Digital Nervous System." In this layer, the executive moves beyond manual data entry and into **Automated Signal Ingestion**. It leverages a convergence of **IoT (Internet of Things) sensors**, **Computer Vision**, and **Edge Computing** to capture the physical and digital reality of the business in high fidelity. Whether it is the real-time vibration frequency of a factory turbine, the high-resolution satellite imagery of a competitor's logistical hubs, or the "API-led connectivity" into global shipping lanes and financial markets, this layer provides a raw, unfiltered stream of "Environmental Truth." By removing human bias and administrative latency from the input phase, the leader ensures that the "Eyes" of the firm are receiving a high-fidelity signal from the very edge of the enterprise. This layer ensures that the executive is no longer dependent on filtered summaries, but has direct access to the raw pulse of the market and the internal operations.

The Processing Layer: Causal Intelligence and Relationship Mapping

The **Processing Layer** serves as the "Pattern Recognition Engine" of the stack. It transforms the raw sensory stream into a structured understanding of **Cause and Effect**. This layer utilizes **Knowledge Graphs** and **Graph Neural Networks (GNNs)** to map the hidden relationships between disparate data points that traditional analytics would miss. While legacy databases see data in isolated rows and columns, the Knowledge Graph sees the "Web of Influence"—understanding, for example, how a specific geopolitical tremor captured in the Sensory Layer will ripple through the supply chain to impact local liquidity forty-five days later. By applying **Time-Series Machine Learning**, this layer identifies the "Statistical Whisper" of an upcoming trend before it becomes a loud market signal. It moves the organization from "Correlation" (things happening together) to "Causality" (knowing *why* they happen), providing the executive with a "Logical Foundation" for every foresight and ensuring the "Compass" of the firm is calibrated to the objective laws of the market.

The Cognitive Layer: Strategic Anticipation and Scenario Simulation

The **Cognitive Layer** is the "Reasoning Mind" of the stack, where processed data is converted into **Executive Intent**. This layer utilizes **Large Language Models (LLMs)** and **Multi-Agent Simulation** to synthesize the logic of the Processing Layer into actionable "Navigational Scenarios." Instead of a single, rigid forecast, the Cognitive Layer generates thousands of "What-If" simulations, stress-testing the organization's resilience against a variety of potential futures. It acts as a "Strategic Advisor" to the leader, using **Natural Language Generation** to articulate complex probabilities as clear, actionable narratives. This allows the executive to move from "Guessing" to "Orchestrating," authorized by a system that has already "played out" the consequences of a decision across a digital twin of the market. It turns the "Forecaster" into a "Proactive Sentry," allowing the leader to position capital, talent, and resources at the exact point of future impact. The result is a brand that is never "surprised" by the market, but is instead "waiting for it" at the destination with a pre-positioned advantage.

1. THE FORECASTERS (THE EYES) | PREDICTIVE AI

OVERVIEW

Predictive AI forecasts future outcomes by analyzing vast historical data using machine learning algorithms. For the "Intuitive & Risk" leader, the AI becomes a crucial tool for quantifying uncertainty. The leader provides their initial intuition and acceptable risk limits, framing the precise question. The AI delivers objective, quantifiable evidence, such as likelihoods and key drivers, that either validates the leader's gut feeling or proposes data-driven alternatives. This synergy converts subjective risk into a calculated, strategic choice, empowering the leader to make proactive, decisive actions.

MAIN STEPS

Start

STEP 1 – Define Objective & Scope: *Specify question, target outcome, and limits*

STEP 2 – Data Collection & Preparation: *Gather, clean, and format historical inputs*

STEP 3 – Model Selection & Training: *Build patterns from cleaned training data*

STEP 4 – Evaluation & Validation: *Test model accuracy on unseen data*

STEP 5 – Deployment & Integration: *Insert model into real-time operations*

STEP 6 – Monitoring & Refinement: *Continuously check and improve model performance*

End

SAMPLE FLOW

Manufacturing - Predictive Maintenance - The Plant Leader manages profit risk by applying AI to foresee critical conveyor belt failures, shifting from reactive to pre-emptive maintenance.

1. Set 72-Hour Failure Target
Predict belt failure likelihood within 72 hours for preventative tech. scheduling.

2. Align Sensor & Log Data
Gather temperature, vibration, maintenance logs, and failure events; then clean data.

3. Train Anomaly Detector
Train Time-Series model on historical data to recognize patterns preceding failure.

4. Validate Accuracy
Verify model's high accuracy for 72-hour prediction with low false-positives.

5. Integrate Risk Score
Generate 0-100% real time risk Score on the leader's central operations dashboard.

6. Alert & Action Trigger
Alert triggers above 80% risk threshold; leader confirms action, mitigating risk.

USE CASES

- **Technology | High-Potential Employee Flight Risk Prediction:** Predicts top performers likely to resign using integrated data on compensation and engagement; senior leaders proactively implement customized retention strategies and specialized mentorship programs.
- **Education | Administrator Turnover Forecasting:** Forecasts principal and administrator turnover using complex performance, budget, and staff morale data, enabling leadership to plan succession and implement stabilization policies quickly.
- **Consulting | Project Success Probability Scoring:** Predicts if a key client project will fail critical milestones based on internal team dynamics, budget burn rate, and client engagement metrics before major loss occurs.
- **Government/Public Sector | Policy Impact Risk Assessment:** Estimates public compliance levels and potential negative social impact of new regulations before launch, informing strategic leadership decisions on rollout planning and communications.
- **Media & Entertainment | Content Executive Decision Success Rate:** Predicts future content revenue and audience reach against historical hits and genre saturation, guiding executive investment choices and portfolio diversification effectively.
- **Pharmaceuticals | Clinical Trial Failure Risk Modeling:** Forecasts Phase III trial failure risk using aggregated early safety data, investigator performance, and recruitment rates, strategically supporting R&D leadership decisions on continuation or pivoting.

In the **Leadership Transformation**, the **Forecasters (The Eyes)** serve as the organization's early warning system. While **Predictive AI** identifies statistical probabilities, it often lacks the "contextual depth" required for high-stakes decisions. To move from basic forecasts to high-definition strategic vision, a leader must integrate six secondary instruments. Each pairing solves a specific friction, from a lack of causal clarity to the inability to simulate physical reality.

1. The Digital Twin Pairing for Strategic Simulation

Pairing **Predictive AI with Digital Twin technology** is best utilized during high-cost physical or operational pivots, such as factory relocations. While AI estimates financial benefits, it cannot account for the physical "butterfly effects" of local infrastructure. By running predictions through a Digital Twin—a live virtual replica—the leader can simulate port strikes or power failures before committing capital. This allows the executive to "fail fast" virtually, ensuring the strategy is battle-tested against a thousand simulated disruptions.

2. The Causal Inference Pairing for Root-Cause Clarity

Leaders should pair **Predictive AI with Causal Inference** during market volatility when historical patterns are no longer reliable. Predictive AI finds correlations, but it cannot explain *why* something is happening. If customer churn spikes, a standard forecast simply predicts the trend will continue; however, Causal Inference identifies the specific "Leaver," such as a change in competitor shipping speed. This prevents the leader from "chasing ghosts" by ensuring intervention is aimed at a verified cause rather than a coincidence.

3. The Graph Neural Network Pairing for Ecosystem Orchestration

When managing millions of interconnected nodes, such as a logistics network, **Predictive AI must be paired with Graph Neural Networks (GNNs)**. Traditional models treat data points in isolation, but GNNs understand the "web of relationships." This pairing is best deployed to see how a delay in a Tier-3 supplier ripples across the entire global production line. It allows the Forecaster to perceive "Swarm Intelligence," identifying systemic vulnerabilities invisible to linear spreadsheets and ensuring resilience as network complexity increases.

4. The Probabilistic Programming Pairing for Risk Governance

For high-stakes capital allocations, a leader should pair **Predictive AI with Probabilistic Programming**. Standard predictions often provide a single "most likely" outcome, which can lead to overconfidence. Probabilistic Programming instead assigns a "Confidence Score," quantifying the exact level of uncertainty. This is essential during M&A cycles or massive R&D investments. It forces the executive to confront "known unknowns," architecting a strategy that is hedged against the mathematical probability of total loss.

5. The Adversarial Stress-Testing Pairing for Institutional Resilience

To combat "confirmation bias," leaders should pair **Predictive AI with Adversarial Stress-Testing**. This involves using a secondary AI model designed to "attack" and find weak spots in internal forecasts. This is best deployed when the outlook is overly optimistic and the team is at risk of groupthink. By tasking an adversarial model to simulate hostile regulatory environments or brand crises, the Forecaster identifies "Blind Spots" in the primary model. This ensures the leader's "Eyes" are scanning for cracks that could bring the strategy down.

6. The Quantum Machine Learning Pairing for Long-Horizon Radar

In industries with decadal timelines, such as pharmaceutical research, **Predictive AI should be paired with Quantum Machine Learning**. Classical models struggle to process the non-linear, hyper-complex variables involved in 10-year outlooks. This pairing is best utilized when a leader needs a "Long-Range Radar" to detect shifts in global health or macro-economic cycles. Quantum-inspired algorithms process vast "possibility spaces" at speeds classical systems cannot match, providing the temporal sovereignty to position the brand a decade ahead.

Executive Power: Empathy

As an organization moves toward total autonomy and machine-speed execution, the leader's role undergoes a profound shift: they must become the **Guardian of Meaning.** While the "Hands" execute and the "Eyes" watch, the **Diplomat** ensures that the organization's "Voice" remains coherent, ethical, and resonant. In an era where "Intelligence" is a commodity that can be bought or built, "Empathy" and "Integrity" become the primary sources of competitive advantage. The Diplomat role is about managing the relationship between the machine's efficiency and the human's trust.

Defensive Stance: Killing the Resistance

The greatest threat to a modern enterprise is not a competitor, but "Internal Friction"—the fear, skepticism, and cultural resistance that arise when people feel disconnected from the logic of the organization. The Diplomat uses the power of **Transparency** to "Kill this Resistance." By ensuring that every automated decision is "Explainable" and "Justifiable," the leader turns the organization from a "Black Box" into a "Glass Box." This is the **Shield of Trust.** When stakeholders, employees, and regulators understand the "Why" behind a move, their resistance vanishes. Furthermore, by hard-coding the brand's "Moral Compass" into its core architecture, the Diplomat ensures that the company never makes a "Technically Correct but Morally Bankrupt" decision. This prevents the reputational meltdowns that destroy brands in the digital age, ensuring that the company's "Voice" is always seen as principled and fair.

Offensive Stance: Unlocking Reasoning

On the offense, the Diplomat moves from "Managing Communications" to **"Orchestrating Narrative."** In most large companies, the "Truth" is fragmented—the engineers have one story, the sales team has another, and the executives have a third. The Diplomat uses the organization's collective intelligence to "Synthesize the Whole Story" into a single, persuasive voice. This **Unlocks Reasoning**, allowing the leader to speak to the world with "Total Awareness." Instead of generic, reactive corporate-speak, the leader's voice becomes a "Direct Expression of the Company's IQ." This allows the leader to persuade boards, win over customers, and align employees with a level of clarity that was previously impossible. It turns "Information" into "Influence," ensuring that the brand is always the most "Thoughtful" and "Authoritative" voice in the market.

The Leadership Outcome: Institutional Trust

The ultimate value of the Diplomat is **Alignment.** When the leader's "Voice" is backed by the "Total Knowledge" of the company and protected by a "Moral Shield," the organization achieves a state of "High-Trust Harmony." For the leader, this provides **Narrative Control.** You are no longer just "giving orders"; you are "orchestrating intent." This turns the organization into an "Authentic Powerhouse"—where every action is justifiable, every decision is ethical, and every word is backed by evidence. This is the key to "Sustainable Leadership"—ensuring that as the company grows more "Machine-Like" in its efficiency, it remains "Human-Like" in its purpose, building a brand that customers and employees are proud to belong to.

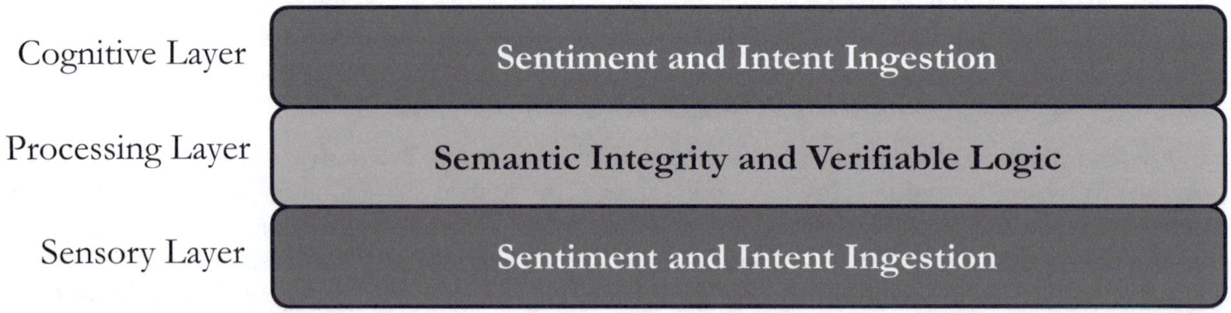

Executive Power: Institutional Trust

For the modern executive, "The Voice" is no longer about controlled PR cycles or the delivery of carefully manicured scripts; it is about **Institutional Trust** achieved through radical transparency and semantic alignment. In an era of informational chaos and deep skepticism, the **Diplomat** role ensures the brand's narrative is not "spun" by a marketing team, but "synthesized" from the company's verifiable operational data. The leader uses this stack to ensure the "Financial and Ethical Truth" of the organization is articulated with absolute integrity for every stakeholder. By architecting this role, the executive moves the brand from "Marketing" to "Authenticity," ensuring that the "Voice" of the organization is the most "Honest" and "Evidenced" in the market. The Diplomat serves as the bridge between the machine's efficiency and the human's need for meaning.

The Sensory Layer: Sentiment and Intent Ingestion

The **Sensory Layer** of the Diplomat acts as the organization's "Social Ear." It moves beyond basic keyword tracking into **Multimodal Sentiment Analysis**. By utilizing **Natural Language Processing (NLP) at the edge**, the leader ingests real-time signals from earnings calls, social discourse, glassdoor reviews, and regulatory shifts. This layer captures the "Emotional Temperature" and "Hidden Intent" of the global market, providing the leader with a real-time map of "Expectation and Friction." It allows the executive to perceive the gap between corporate reality and stakeholder perception before a trust crisis manifests. By "listening" to the subtle shifts in language and sentiment across the digital landscape, the sensory layer ensures that the "Voice" of the firm is never speaking into a vacuum, but is reacting to a high-fidelity "Map of Global Sentiment" that reflects the true concerns of the brand's community.

The Processing Layer: Semantic Integrity and Verifiable Logic

The **Processing Layer** serves as the "Fact-Checking Engine" and the "Guardian of Meaning." It utilizes **Knowledge Graphs** to link raw operational and financial data to the brand's stated values and promises. If the "Voice" claims a commitment to ethical sourcing or sustainability, this layer maps that claim to real-time supply chain logs and carbon offsets stored on a **Private Blockchain**. This ensures that the leader's message is "Hard-Coded" in evidence, killing the friction of skepticism by providing a "Verifiable Logic" for every public statement. It acts as a **Shield of Semantic Integrity**, ensuring that when the "Voice" speaks, it is backed by a structured logic that can withstand the scrutiny of any auditor, activist, or investor. This layer removes the "Ambiguity of Intent," ensuring that every word the leader utters is a reflection of a singular, data-backed truth that is mathematically consistent across the entire enterprise architecture.

The Cognitive Layer: Resonant Articulation and Personalized Narrative

The **Cognitive Layer** is where the "Single Truth" is transformed into **Hyper-Personalized Communication**. Utilizing **Large Language Models (LLMs)** and **Agentic Orchestration**, the Diplomat generates "Sovereign Narratives" that achieve **Empathy at Scale**. It doesn't issue one generic press release; it orchestrates thousands of tailored updates—providing a technical breakdown for analysts, a value-based update for investors, and a purpose-driven story for employees—all while maintaining a singular, unified "Truth Core." This layer ensures the leader's intent is perfectly "decoded" into the specific language and context of every individual stakeholder without losing its structural integrity. Because this layer is governed by **Explainable AI (XAI)**, it can "explain its own work," providing the leader with the reasoning behind every narrative choice. This allows the executive to scale their influence and project a "Voice" that feels personal and resonant to everyone while remaining objectively grounded in the brand's actual performance.

2. THE DIPLOMATS (THE VOICE) | EXPLAINABLE AI

OVERVIEW

Explainable AI (XAI) provides transparent, human-readable rationales for AI decisions, moving beyond black box outputs. It reveals why a model reached a conclusion, building trust and mitigating legal and ethical exposure. The Regulatory & Compliance leader leverages XAI to validate complex, automated decision-making processes, particularly in highly regulated domains. This synergy ensures AI systems are fair, auditable, and adhere to internal policies and external legal requirements, transforming compliance from a check-the-box function into a strategic enabler of trust.

MAIN STEPS

Start

STEP 1 – Define Audit Scope
Establish parameters for compliance assessment.

STEP 2 – Feature Importance Analysis
Quantify input factors driving AI output.

STEP 3 – Generate Local Explanation
Create rationale for a single prediction.

STEP 4 – Validate Model Fairness
Test decision consistency across subgroups.

STEP 5 – Human Review and Oversight
Leader verifies explanations meet standards.

STEP 6 – Document Rationale & Action
Record justification for audit readiness.

End

SAMPLE FLOW

Ethical Performance Review System: A leader uses Explainable AI to validate AI-assisted performance reviews for fairness and adherence to non-discrimination policies rigorously before deployment..

1. Identify Protected Attributes
Flag demographic features (age, gender, ethnicity) for sensitive monitoring purposes.

2. Review Individual Rationales
Examine XAI reports explaining scores for consistency and bias detection.

3. Simulate Counterfactuals
Test how score changes if protected attributes were different safely.

EXPLAINABLE AI

4. Determine Disparate Impact
Calculate performance score distribution variances across demographic groups statistically.

5. Adjust Weighting Algorithm
Recommend changes to model features or training data to correct identified bias.

6. Finalize Fairness report
Approve model for use, generating compliance reports justifying every decision accurately.

USE CASES

- **Manufacturing | Safety Protocol Validation:** Provides clear explanations for AI-identified workplace hazards and required mitigating actions, ensuring that safety changes are understood and correctly adopted by all operational team leaders.
- **Financial Services | Regulatory Change Impact:** Explains *why* certain internal processes or controls need modification based on new financial regulations, streamlining the leadership communication and compliance implementation process.
- **Healthcare | Diagnostic Tool Acceptance:** Gives clinical leaders auditable proof of a diagnostic AI's reasoning, establishing the necessary trust for wider organizational acceptance and ethical, consistent patient use.
- **Technology | Ethical Code Review:** Justifies automated flagging of software code for potential security or privacy violations, allowing engineering leaders to prioritize fixes based on clear regulatory risk levels.
- **Retail | Promotion Fairness Audit:** Explains the variables driving personalized discount decisions, ensuring that promotional strategies do not inadvertently discriminate against specific customer demographic segments or locations.
- **Energy | Grid Investment Justification:** Provides transparent rationale for AI-recommended infrastructure upgrade projects, allowing senior leaders to justify large capital expenditures to regulatory bodies confidently and quickly.

2. THE DIPLOMATS (THE VOICE) | EXPLAINABLE AI

In the **Leadership Transformation**, the **Diplomats (The Voice)** bridge the gap between complex machine logic and human trust. While **Explainable AI (XAI)** provides the "reasoning" behind an algorithmic decision, it often lacks the emotional resonance and cultural context required for adoption. To move from cold transparency to radical alignment, a leader must integrate six secondary instruments, each solving a specific friction from technical jargon to a lack of stakeholder empathy.

1. The Sentiment Analysis Pairing for Emotional Calibration

Pairing **Explainable AI with Sentiment Analysis** is best utilized when communicating AI-driven changes to a workforce or customer base. While XAI explains the logic—such as a shift in labor allocation—it ignores the audience's emotional climate. By running explanations through a Sentiment Analysis layer, the Diplomat gauges anxiety or resistance. This allows the executive to calibrate the "Voice," ensuring technical reasoning is delivered with the empathy required to maintain institutional morale.

2. The Contextual NLP Pairing for Multimodal Translation

Leaders should pair **Explainable AI with Contextual NLP** when navigating global operations. XAI provides the facts, but Contextual NLP translates them into the specific "language" of the recipient—whether it's technical jargon for regulators or financial logic for a board. This pairing is essential for cross-border transparency, ensuring the brand's "Voice" is accurately understood across different cultural contexts, eliminating the friction of misinterpretation.

3. The Knowledge Graph Pairing for Logical Traceability

When a decision involves multi-layered data—such as a strategic pivot in an R&D pipeline—**Explainable AI must be paired with Knowledge Graphs**. While XAI explains the output, the Knowledge Graph provides the "map of relationships" that led there. This pairing is best deployed when a leader needs to show the "Sovereign Truth" of a decision, allowing auditors to trace logic through thousands of interconnected data points, transforming the "Voice" into a verifiable record.

4. The Human-in-the-Loop Pairing for Ethical Governance

For high-stakes decisions involving social implications, a leader should pair **Explainable AI with Human-in-the-Loop (HITL) interfaces**. XAI identifies the "why," but the HITL layer provides the "should." This is best utilized in areas like hiring or credit scoring, where machine logic might overlook nuanced human values. The Diplomat ensures the "Voice" remains grounded in ethics, allowing executives to refine algorithmic paths to align with the organization's moral compass.

5. The Conversational AI Pairing for Real-Time Transparency

To combat "Information Lag," leaders should pair **Explainable AI with Conversational AI**. While XAI provides the static explanation, Conversational AI allows stakeholders to "interrogate" the decision in real-time. This is best deployed during town halls or investor calls. This pairing turns the Diplomat into a dynamic interface, providing instant, reasoned justifications and building a culture of radical, open-door transparency.

6. The Blockchain Pairing for Immutable Accountability

In industries where regulatory compliance is non-negotiable, **Explainable AI should be paired with Blockchain**. XAI generates the explanation, while the Blockchain stores it in an immutable ledger. This is best utilized for environmental auditing or high-frequency trading, where the "Voice" must be legally unassailable. It ensures that justifications are permanent and tamper-proof, providing absolute accountability for the brand's operational integrity.

Executive Power: Agency

In the traditional leadership model, the executive provides the "what" and the "why," but they are often disconnected from the "how." They rely on layers of bureaucracy and human intermediaries to translate a vision into a physical or digital reality. This reliance creates a significant lag between an executive's realization and the organization's execution. The transformation into a **Creator** represents the total collapse of that distance. It is the evolution of the leader into an "Architect of Agency," where the ability to manifest an idea is no longer limited by human bandwidth, departmental silos, or historical baggage, but by the clarity of the leader's intent.

Defensive Stance: Killing the Obsolescence

The defensive mission of the Leader-Creator is the **Purification of the Past**. Every established organization is haunted by "Digital and Intellectual Ghosting"—old ways of thinking, antiquated processes, and rigid structures that act as a hidden tax on every new idea. Over time, these legacy elements accumulate, creating a "Systemic Friction" that makes the organization heavy and slow. This role focuses on "Resurrecting the Archive." Instead of allowing the company's history to become a graveyard of forgotten lessons and lost assets, the leader uses the "Engine" to index, classify, and modernize everything the company has ever known. This kills the "Friction of Re-invention." It ensures that a team is never building a solution that already exists in a different department or was solved three years ago. By "Cleaning the Foundation," the leader ensures the organization is not an "Old Machine" trying to run "New Code," but a perpetually modern engine where every past success is a live building block for the future. It turns "Legacy" from a burden into a competitive vault.

Offensive Stance: Unlocking Generation

On the offense, the leader moves from "Managing Work" to **"Orchestrating Creation."** This is the shift toward "Infinite Prototyping." In the past, a leader might see three versions of a marketing campaign or two designs for a new product after weeks of waiting. In this new role, the leader demands a thousand versions in an hour. This **Unlocks Generation**, moving the company from a state of "Creative Scarcity" to "Creative Abundance." The leader's role becomes one of "High-Level Selection and Refinement." They are no longer waiting for the "Engine" to build; they are directing the "Engine" as it produces at the speed of thought. This allows the brand to be "Hyper-Relevant"—producing tailored experiences, products, and messages for every individual micro-market simultaneously. It allows the brand to "test" reality at a scale that makes failure almost impossible, because the "Winning Path" is discovered through sheer generative volume before a single dollar of capital is committed to the physical world.

The Leadership Outcome: Absolute Agency

The ultimate value of the Creator role is **Institutional Velocity**. When the distance between "Idea" and "Prototype" is reduced to zero, the organization becomes "Agile" in the truest sense of the word. For the leader, this provides **Absolute Agency**. You are no longer a victim of "Production Bottlenecks" or "Creative Fatigue." You can pivot the entire visual, functional, and digital output of the company in a single day to meet a changing market. This turns the organization into a "Living Brand"—one that evolves, grows, and creates in real-time response to its environment. It ensures that the "Engine" of the company is always as young and fast as its newest idea, transforming the act of leadership into an act of direct creation.

Cognitive Layer	Iterative Manifestation and Autonomous Prototyping
Processing Layer	Parametric Constraint Logic and Reality Filtering
Sensory Layer	Atomic Requirement Capture and Environmental Sensing

Executive Power: Generative Strategy

The **Creator** role transforms the executive from a "Decision Maker" into a "Scenario Architect." In this role, the leader doesn't just manage existing products or services; they use a **Generative Engine** to manifest entirely new business models and solutions on the fly. This represents a shift from "Rigid Planning" to "Fluid Creation," where the brand's output is a dynamic, real-time response to user intent rather than a static inventory of pre-coded features. The executive uses the Creator stack to "Design the Future" as it happens, moving from a state of "Rigid Utility" to "Fluid Adaptation." The Creator is the executive's primary tool for **Radical Agility**, ensuring the brand is always producing the most "Original" and "Relevant" experience because the "Engine" is literally "Creating" its own value at the point of use.

The Sensory Layer: Atomic Requirement Capture and Environmental Sensing

The **Sensory Layer** for the Creator captures "Market Gaps" as high-definition signals. It moves beyond traditional market research into **Atomic Requirement Capture**, integrating **User-Interface (UI) telemetry**, **Direct Customer Feedback Loops**, and **Biometric Environmental Sensing**. By utilizing **Computer Vision** to observe how customers interact with physical products or **Agentic Discovery** to scan competitor limitations and social pain points, this layer provides the "Raw Material" for innovation. It ensures the leader is building on "Actual Demand" rather than "Legacy Assumptions," capturing the nuanced requirements of the market before they are even articulated by the consumer. This layer acts as the "Direct Input" of the world into the brand's creative mind, ensuring that every manifest solution is born from a high-fidelity understanding of the user's current reality and untapped desires.

The Processing Layer: Parametric Constraint Logic and Reality Filtering

The **Processing Layer** acts as the "Reality Filter" and the "Guardian of Feasibility." It utilizes **Physics-Informed Neural Networks (PINNs)** and **Financial Modeling Engines** to ensure that whatever the brand "Creates" is technically and economically viable. It understands the "Rules of the Possible"—including regulatory limits, manufacturing costs, material science, and legal compliance. This layer stress-tests every generative idea against the organization's "Causal Truths," ensuring the executive only sees solutions that can actually survive the transition from concept to market. It acts as a **Check and Balance** on the creative process, filtering out "hallucinations" or high-risk concepts and leaving only high-yield opportunities that are optimized for the brand's specific operational constraints. By applying "Parametric Logic," this layer ensures that the executive's "Engine" is disciplined, productive, and focused on manifestations that can scale sustainably within the global market architecture.

The Cognitive Layer: Iterative Manifestation and Autonomous Prototyping

The **Cognitive Layer** is the "Invention Mind," powered by **Generative Adversarial Networks (GANs)** and **Action Transformers**. This layer "hallucinates" thousands of strategic variations and product prototypes based on the leader's specific intent. It doesn't just present an idea; it presents a **"Validated Solution."** The leader directs the engine with a high-level goal, such as "Manifest a new service for the emerging solar market in Africa," and the Cognitive Layer manifests the pricing structure, the user interface, and the logistical fulfillment path in a single, coherent output for executive approval. This layer allows for **Infinite Operational Flexibility**, where the brand can "Create" a unique version of itself for every individual user. It involves "Recursive Prototyping," where the AI learns from each iteration to refine the next, ensuring that the final output is the "Optimal Manifestation" of the leader's vision. This turns the executive into an "Orchestrator of Reality," capable of deploying variety at a speed that would paralyze a manual competitor.

OVERVIEW

Generative AI creates novel outputs like text, art, code, and immersive environments from simple prompts. It excels at rapid prototyping, style transfer, and exploratory ideation. The Creative & Artistic leader partners with Generative AI to rapidly visualize strategic concepts, explore diverse future scenarios, and prototype new organizational narratives. The synergy transforms vision casting from a slow, manual process into a continuous, inspiring, and engaging collaborative endeavor.

MAIN STEPS

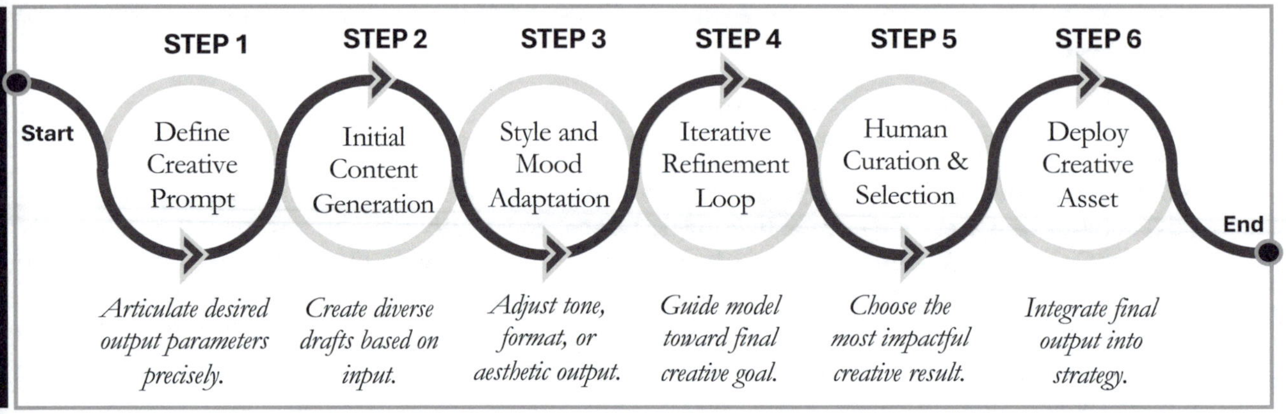

SAMPLE FLOW

Strategic Vision Casting: A CEO uses Generative AI to quickly visualize and articulate the company's bold 5-year strategic vision to all employees just in time for the new Corporate town hall.

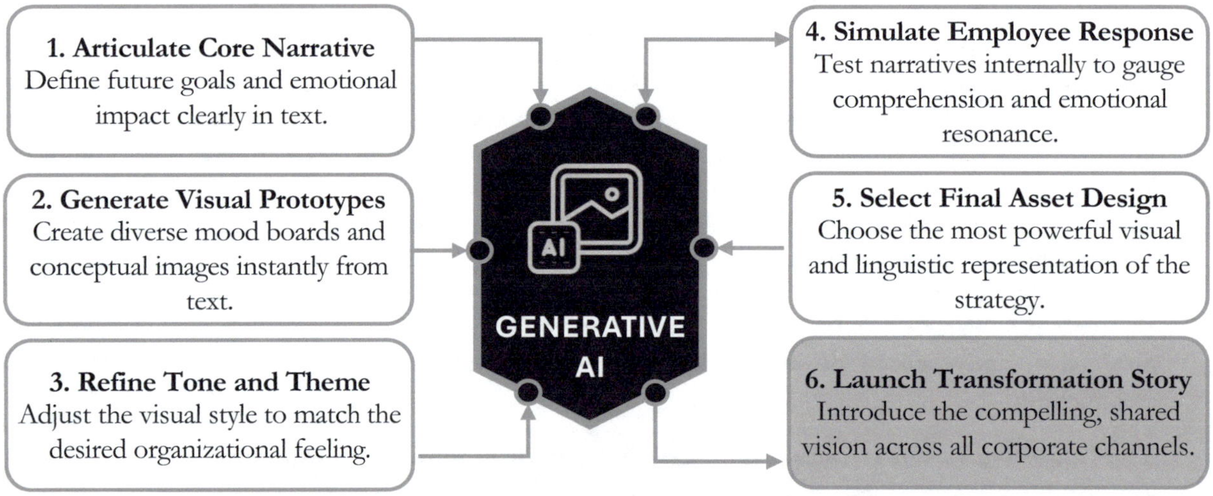

USE CASES

- **Manufacturing | Culture of Innovation:** Generates custom internal training videos and interactive simulations that creatively communicate new safety protocols and foster a bolder risk-taking culture among managers.
- **Financial Services | Ethics Guideline Communication:** Creates simplified, engaging, and scenario-based animated content from complex compliance manuals to ensure ethical leadership principles are easily understood by all staff.
- **Healthcare | Mission Statement Renewal:** Develops powerful, emotive imagery and narrative short films showcasing patient success stories, directly reinforcing the organization's core healing mission for clinical leaders.
- **Technology | Recruitment Campaign Aesthetics:** Designs unique, cutting-edge visual assets and candidate profiles that reflect a modern, experimental brand identity, attracting top-tier, creatively-driven engineering leadership talent.
- **Retail | Executive Decision Storytelling:** Creates interactive digital presentations that transform quarterly performance data into compelling visual stories, making complex strategic trade-offs transparent for executive teams.
- **Energy | Sustainability Report Design:** Generates striking, data-driven infographics and dynamic report layouts that clearly articulate long-term environmental commitment and inspire stakeholder trust in leadership's goals.

In the **Leadership Transformation**, the **Creators (The Engine)** represent the organization's capacity for rapid prototyping and creative manifestation. While the primary instrument of **Generative AI** produces the "First Draft" of assets, code, or strategies, it can often lack the grounding in physical reality or historical consistency required for industrial-grade innovation. To move from simple content generation to high-velocity value creation, a leader must integrate six secondary instruments to refine the "Engine."

1. The Digital Twin Pairing for Reality Anchoring

Pairing **Generative AI with Digital Twins** is best utilized when a leader is manifesting new physical products or infrastructure. While Generative AI can dream up thousands of efficient designs, the Digital Twin acts as the "Physics Engine" that stress-tests them against real-world constraints. This pairing allows the Creator to iterate at lightning speed while ensuring that every generated design is structurally sound and ready for the assembly line, eliminating the friction between the "Draft" and the "Physical Reality."

2. The Synthetic Data Pairing for Bias Mitigation

Leaders should pair **Generative AI with Synthetic Data Generation** when entering new markets where historical data is scarce or biased. This is best deployed to "train the future" without being anchored to the errors of the past. By generating high-fidelity, privacy-compliant synthetic datasets, the Creator can simulate thousands of "Edge Case" scenarios. This ensures that the generated strategies are robust and inclusive, allowing the brand to innovate in "Cold Start" environments with the confidence of a data-rich incumbent.

3. The RLHF Pairing for Value Alignment

When a firm's creative output must reflect specific brand values or aesthetic standards, **Generative AI must be paired with Reinforcement Learning from Human Feedback (RLHF)**. This is best utilized in marketing and customer experience design. The "Engine" generates the volume, but the human-in-the-loop provides the "Taste." This pairing ensures that the AI learns the subtle nuances of the brand's identity, moving the "First Draft" closer to a "Final Version" that feels authentically human and strategically aligned.

4. The API Orchestration Pairing for Functional Utility

To turn generative ideas into functional tools, a leader should pair **Generative AI with API Orchestration**. While the AI can write code or draft a plan, API Orchestration allows it to "reach out" and trigger existing enterprise systems. This is best utilized when automating complex back-office workflows. The Creator doesn't just describe a solution; it builds the bridge to execute it, ensuring that the generative "Engine" is plugged directly into the company's operational nervous system.

5. The Vector Database Pairing for Creative Memory

To prevent the "Engine" from repeating past mistakes or losing its way, leaders should pair **Generative AI with Vector Databases (RAG)**. This is best deployed when managing long-term innovation projects. The Vector Database allows the Generative AI to "remember" every previous iteration, technical constraint, and successful prototype. This pairing ensures that the Creator builds upon the firm's collective intelligence, turning the "First Draft" into a cumulative evolution of the brand's best ideas.

6. The Edge AI Pairing for Localized Manifestation

In decentralized operations, **Generative AI should be paired with Edge AI**. This is best utilized when the "Engine" needs to manifest solutions in real-time at the "Point of Need," such as a remote repair site or a retail floor. By running lightweight generative models on local hardware, the Creator can solve problems without waiting for cloud processing. This ensures that innovation is instantaneous and localized, providing the brand with the agility to respond to unique, on-the-ground challenges the moment they arise.

Executive Power: Objectivity

Strategy has traditionally been viewed as a "Human Art," often driven by the loudest voice in the room, the most persuasive presentation, or the memory of a recent success. However, human intuition, while powerful, is riddled with cognitive biases—sunk-cost fallacies, overconfidence, and the tendency to see patterns in random noise. The transformation into a **Strategist** turns the leader into a "Guardian of Objectivity." In this role, the executive uses a "Digital Compass" to navigate the infinite complexity of global markets without the interference of ego or outdated dogma.

Defensive Stance: Killing the Mismatch

The defensive mission of the Leader-Strategist is the **Validation of Reality**. Many companies fail not because they lack a plan, but because their internal "Compass" is misaligned with the external world—they are executing a brilliant plan for a market that no longer exists. This role focuses on "Causal Truth." It seeks to prove what is actually driving results versus what is merely a coincidence or a seasonal trend. By "Modeling the Risk" of every move with mathematical coldness, the leader kills the "Mismatch" between capital allocation and real-world opportunity. It prevents the organization from doubling down on "Dead Ends" and identifies "Sunk Cost Traps" before they swallow the company's future capital. This is the **Shield of Logic**, ensuring that every strategic move is grounded in the "Mathematical Why," not just the "Emotional Want." It allows the leader to say "No" to attractive but toxic opportunities with the certainty of a scientist, protecting the brand's resources for the paths that lead to genuine growth.

Offensive Stance: Unlocking Optimization

On the offense, the Strategist moves toward **Dynamic Navigation**. Traditional strategy is "Static"—a three-year plan written in a book and reviewed quarterly. This new role creates a "Fluid Strategy" that re-calculates itself every second based on live signals. Using "Optimal Reasoning," the leader can find the "Path of Least Resistance" to a goal that a human mind, limited by its linear thinking, could never map. It involves "Prescribing the Win"—having the "Compass" suggest the exact moves in pricing, logistics, and resource shifting that will maximize the brand's position in real-time. This **Unlocks Optimization**, allowing the leader to "Play the Board" with a level of precision that makes the competition's "Manual Strategy" look like guesswork. It turns the organization into a "Calculating Winner" that can pivot its weight as conditions shift, ensuring that the brand is always positioned at the point of maximum profitability and minimum waste.

The Leadership Outcome: Strategic Sovereignty

The ultimate value of the Strategist role is **Certainty in Chaos**. In a volatile world, the leader who has the most "Objective Compass" wins because they are the only ones who can see through the fog of war. For the leader, this provides **Strategic Sovereignty**. You are no longer "Reacting" to the market; you are "Optimizing" through it. It ensures that your "Strategy" is not a static document, but a living, breathing competitive advantage. This is the key to "Market Dominance"—the ability to see the "Best Move" and execute it with total confidence, knowing that your "Compass" is calibrated to the fundamental laws of the market rather than the whims of a boardroom.

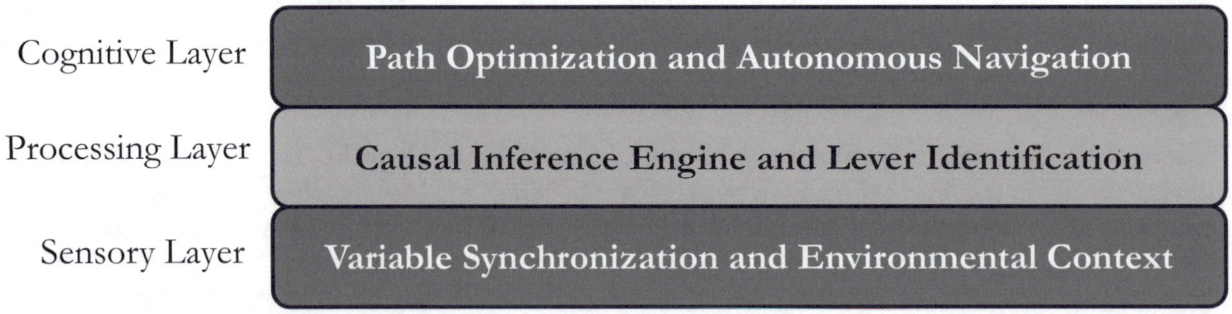

Executive Power: Causal Decisioning

The **Strategist** role represents the "Abolition of Boardroom Guesswork" and the birth of **Causal Decisioning**. It moves the leadership team away from "Opinions" and "Ego-driven Plans" toward a strategy grounded in **Objective Logic**. The leader uses this stack to ensure that every dollar, hour, and resource is allocated to the "True Drivers" of systemic value, protecting the firm from the "Vanity Metrics" and "Busy Work" that lead to strategic drift and resource depletion. The Strategist is the **Guardian of the Goal**, ensuring that every interaction and operational move is a mathematically verified "Step toward thc Peak." In this role, the executive evolves from a "Passive User of Data" to a "Strategic Navigator" who manages the company's "Value Yield" with the precision of a master pilot using a high-fidelity compass.

The Sensory Layer: Variable Synchronization and Environmental Context

The **Sensory Layer** of the Compass monitors the "Macro and Micro Levers" of the global economy in real-time. It synchronizes **Real-Time Market Data Feeds (Bloomberg/Reuters)** with internal **ERP (Enterprise Resource Planning)** and **CRM (Customer Relationship Management) Data Streams**. By tracking everything from global interest rate fluctuations and geopolitical tremors to internal employee sentiment and churn rates, this layer provides the "Coordinate Data" required to know exactly where the firm stands in the competitive landscape at any given micro-second. It provides the "Eyes" for the Compass, ensuring that the starting point of any strategic journey is grounded in absolute, unvarnished reality. This layer removes the "Blind Spots" in executive decision-making, ensuring that the leader is aware of every environmental factor that could impact the trajectory of the firm's success.

The Processing Layer: Causal Inference Engine and Lever Identification

The **Processing Layer** is the heart of the Strategist role, utilizing **Bayesian Networks** and **Causal Discovery Algorithms** to determine "Why" things are happening. While traditional analytics and "Big Data" show simple "Correlation" (things that happen together), the Strategist identifies the **Causal Path**. It can simulate, for example, exactly how a 5% increase in R&D spend will impact customer lifetime value in specific regional markets three years later. This layer provides the "Mathematical Evidence" for strategy, ensuring the leader is pulling the "Levers" that actually matter for the brand's long-term health. It acts as a **Shield of Objectivity**, stripping away the "Noise" of a single department's local goals and focusing the executive on "Global Optimization." By identifying the "Mathematical Why," this layer ensures the firm is never "Chasing Dead Ends," but is always focused on the "Causal Truths" that drive sustainable profitability and growth.

The Cognitive Layer: Path Optimization and Autonomous Navigation

The **Cognitive Layer** acts as the "Strategic Navigator" through the use of **Reinforcement Learning** and **Multi-Agent Game Theory Simulation**. This layer "plays the game" of the market millions of times in a digital sandbox to find the path of least resistance to the brand's stated goal. When the external environment shifts—a competitor suddenly drops prices, a new regulation is passed, or a trade war begins—the Compass "Recalculates" the route instantly. This layer doesn't just present a status; it "Prescribes the Best Move" in real-time. This provides the executive with **Strategic Certainty**, allowing them to navigate extreme volatility with the confidence that their "Compass" is calibrated to the objective laws of the market. It turns the organization into a "Calculating Winner" that can pivot its weight in milliseconds to find the most "Value-Dense" ground, ensuring the brand "Wins" because its "Logic" is simply faster and more accurate than the competition.

OVERVIEW

Cognitive AI mimics human intelligence to understand and interpret unstructured data like text, speech, and images. It processes complex inputs to generate hypotheses, assess logical connections, and offer nuanced context. The Research & Analytical leader uses Cognitive AI as a hyper-efficient knowledge partner, rapidly synthesizing vast amounts of data to rigorously test assumptions, structure complex arguments, and make deeply informed, evidence-based strategic decisions.

MAIN STEPS

Start

STEP 1	STEP 2	STEP 3	STEP 4	STEP 5	STEP 6
Data Ingestion	Understand Context	Hypothesis Generation	Evidence Synthesis	Analytical Refinement	Insight Formulation
Collect and structure input knowledge quickly	*Understand meaning and relationships deeply.*	*Formulate potential explanatory outcomes.*	*Find relevant supporting information.*	*Test hypotheses against real-world data.*	*Structure findings for human action.*

End

SAMPLE FLOW

Corporate Strategy - Organizational Restructure: A CEO uses AI to analyze market reports and policy documents to formulate an optimal long-term organizational restructure strategy.

1. Define Restructure Goal Specify the performance goals and necessary scope of organizational change.

2. Absorb Global Policy Data Ingest worldwide regulatory documents, news, and economic trend reports effectively.

3. Build Strategy Scenarios Formulate multiple plausible future organizational models and resource allocations.

4. Evaluate Resource Viability Test each scenario against current budget, talent availability, and asset capacity.

5. Assess Cultural Readiness Analyze internal communications and employee surveys for change acceptance

6. Evidence-Backed Plan Select the most rigorously supported restructuring model for executive implementation.

USE CASES

- **Manufacturing | Supply Chain Policy Review:** Analyzes thousands of global trade agreements and sustainability reports to identify risks and suggest new ethical, compliant international sourcing policies for leadership.
- **Financial Services | Regulatory Impact Assessment:** Scans new financial legislation and cross-references it with internal operational documents to provide leadership with a comprehensive compliance strategy and implementation roadmap.
- **Healthcare | Public Health Strategy Design:** Synthesizes epidemiological studies, government reports, and social media trends to advise leadership on the most effective public health campaign narratives and resource allocation models.
- **Technology | Patent Landscape Mapping:** Reviews millions of academic papers and global patent filings to identify white-space innovation areas and recommend strategic R&D investment for long-term product advantage.
- **Retail | Future Workforce Skill Gaps:** Analyzes labor market reports and employee performance data to forecast necessary leadership skills and training investments required in five years.
- **Energy | Investment Thesis Validation:** Interprets complex climate models and competing energy policy proposals to either validate or reject large-scale renewable infrastructure investment strategies for the board.

In the **Leadership Transformation**, the **Strategists (The Compass)** serve as the conductors of capital and talent. While the primary instrument of **Cognitive AI** provides the reasoning power to analyze complex scenarios, it requires secondary instruments to navigate real-world volatility and ensure strategic precision. To move from simple analysis to "Causal Yield Optimization," a leader must integrate these six pairings to calibrate the "Compass."

1. The Causal Discovery Pairing for Strategic Truth

Pairing **Cognitive AI with Causal Discovery** is best utilized when a leader needs to move beyond correlation to understand the "Why" behind ROI. While AI can show that two variables move together, Causal Discovery identifies the actual lever that drives profit. This is essential when entering a new market; it ensures the Strategist isn't chasing "false signals" but is investing in the specific drivers that mathematically guarantee sustainable growth.

2. The Game Theory Simulation Pairing for Competitive Defense

Leaders should pair **Cognitive AI with Algorithmic Game Theory** when navigating a crowded competitive landscape. This is best deployed during price wars or major product launches. The pairing simulates millions of "moves" by competitors, allowing the Strategist to anticipate counter-attacks. This turns the Compass into a tactical radar, ensuring the firm's strategy remains robust even when rivals attempt to disrupt the market.

3. The Multi-Objective Optimization Pairing for Balanced Growth

When a firm must balance conflicting goals—such as maximizing profit while reaching ESG targets—**Cognitive AI must be paired with Multi-Objective Optimization (MOO)**. This is best utilized during annual budgeting. MOO finds the "Pareto Frontier," the mathematical sweet spot where no single goal is sacrificed for another. This ensures the Strategist guides the organization toward a balanced, "Harmonious" success that satisfies all stakeholders.

4. The Monte Carlo Pairing for Probability Mapping

For high-risk strategic moves, a leader should pair **Cognitive AI with Monte Carlo Simulations**. While Cognitive AI suggests a path, the Monte Carlo layer runs that path through 100,000 "What-If" scenarios of market failure. This is best used in capital-intensive industries like energy or mining. It provides the Strategist with a "Heat Map" of risk, ensuring the Compass is pointing toward the most resilient path, not just the most optimistic one.

5. The Knowledge Graph Pairing for Relational Context

To navigate the "Hidden Connections" within an enterprise, leaders should pair **Cognitive AI with Knowledge Graphs**. This is best deployed during mergers or large-scale reorganizations. The Knowledge Graph maps the dependencies between departments, talent, and data. This allows the Strategist to see how a change in one area will impact the rest of the firm, preventing "Strategic Friction" and ensuring the entire organization moves in sync.

6. The Reinforcement Learning Pairing for Adaptive Navigation

In rapidly shifting environments, **Cognitive AI should be paired with Reinforcement Learning (RL)**. This is best utilized for dynamic resource allocation or high-frequency trading. The RL layer allows the strategy to "learn" from every success and failure in real-time, self-correcting the Compass without human intervention. This provides the brand with "Adaptive Intelligence," ensuring the strategy evolves as fast as the market it seeks to dominate.

Executive Power: Relevance

In the traditional hierarchy, knowledge is often treated as a static asset to be guarded. This led to the "Gatekeeper" model of leadership, where power was derived from what a leader knew that others did not. In the modern era, however, the volume of data has made traditional gatekeeping impossible. Power no longer comes from the possession of information, but from the **Design of Relevance**. The transformation into an **Architect** turns the leader into the designer of the "Corporate Brain." In this role, the executive is responsible for ensuring that the organization's collective wisdom is fluid, accessible, and contextually aware, ensuring that the right insight reaches the right person at the exact moment of decision.

Defensive Stance: Killing the Inefficiency

The defensive mission of the Leader-Architect is the **Abolition of Silos**. In most large organizations, the "Brain" is fragmented—logistics, sales, and legal teams act as isolated lobes that rarely communicate. This fragmentation creates "Institutional Friction," where teams waste time searching for data or, worse, acting on contradictory information. The Architect uses "Knowledge Mapping" to identify the "Connective Tissue" of the business. By creating a unified "Nervous System," the leader ensures that every department is "Resolving the Identity" of reality in the same way. This kills the "Inefficiency of Contradiction." It ensures that the organization is not arguing over "which data is right," but is instead acting on a "Single Version of the Truth." This is the **Foundation of Clarity**, allowing the leader to purge the "Mental Waste" of the company and ensure that every employee is working with the full intellectual weight of the entire enterprise behind them.

Offensive Stance: Unlocking Relevance

On the offense, the Architect moves toward **Hyper-Personalized Intelligence**. This is the ability to "Tailor the Brain" to the specific needs of every individual role. Instead of overwhelming employees with a "Firehose of Data," the leader directs the "Architectural Layer" to deliver only the "Best Next Action" or the "Exact Context" required for a specific task. This **Unlocks Relevance**, turning the organization into a "Self-Correcting Organism" that learns from every interaction. It involves "Adaptive Learning," where the system gets smarter with every success and failure, automatically updating the organization's "Logic" so that the same mistake is never made twice. This creates an "Evolutionary Advantage," where the company's "Brain" is constantly refining its own pathways to find the most efficient route to value.

The Leadership Outcome: Institutional Wisdom

The ultimate value of the Architect role is **Scalable Intelligence**. It ensures that the company's "Wisdom" is not trapped in the heads of a few key individuals but is embedded in the "Architecture" of the brand itself. For the leader, this provides **Systemic Sovereignty**. You can scale your vision across a global workforce because the "Brain" of the company is providing the "Guardrails" and "Guidance" to keep everyone aligned. This turns the organization into a "High-IQ Brand"—one that is fundamentally smarter and faster than any competitor relying on manual knowledge sharing or disconnected spreadsheets.

Executive Power: Cognitive Infrastructure

In the legacy enterprise, knowledge was a perishable and fragmented commodity, often trapped in departmental silos or lost in the fading memories of departing employees. The **Architect** role represents the executive's transition from a "Manager of People" to a "Designer of the Corporate Brain." In this role, the leader constructs a permanent, self-evolving **Cognitive Infrastructure** that serves as the organization's "Institutional Memory." This is the shift from "Managing Information" to "Architecting Connectivity," ensuring that the brand's intelligence is a compound asset that grows more sophisticated with every second of operation. The Architect ensures the organization has a "Systemic Mind"—one that understands its own place within the user's broader ecosystem and can contribute to the brand's total wisdom with absolute "Structural Intelligence."

The Sensory Layer: Universal Identity Resolution and Perception

The **Sensory Layer** of the Brain acts as the "Perceptual Input" for the entire enterprise. It moves beyond simple data ingestion into **Universal Entity Resolution**. Utilizing **Entity Linking** and **Graph-based Data Ingestion**, this layer ensures that every "thing" in the company—every customer, every SKU, every contract, every machine, and every employee—has a unique, persistent digital identity that the "Brain" can recognize across all platforms. Whether it is an email from a vendor, a vibration sensor reading from a truck on the road, or a line of code in the global ERP, this layer "labels" the reality in real-time. It captures the "Atomic Facts" of the business without human intervention, creating a high-fidelity "Nervous System" that feeds the central consciousness with raw, unfiltered, and perfectly identified truths. This layer removes the "Friction of Disconnect," ensuring the "Brain" of the firm has a perfect perception of its own physical and digital existence.

The Processing Layer: Structural Intelligence and Semantic Mapping

The **Processing Layer** is the "Map-Maker" and "Neural Connector" of the organization. It utilizes **Knowledge Graphs** and **Vector Databases** to define and store the relationships between the identities captured in the sensory layer. While traditional systems store data in isolated buckets (Silos), the Architect's processing layer stores the **Connectivity** and the **Context**. It understands, for example, the semantic link between a rise in raw material costs in the supply chain and the specific high-value customer segments that will be most affected by a subsequent price shift in the sales department. By employing **Graph Neural Networks (GNNs)**, this layer identifies "Hidden Patterns" in the organizational structure, allowing the "Brain" to perceive itself as a unified organism. This creates a **Shield of Systemic Integrity**, ensuring that when the shipping department learns of a delay, the marketing department "knows" it instantly through the shared structural logic of the graph, killing the risk of "Operational Blindness."

The Cognitive Layer: Recursive Institutional Memory and Learning

The **Cognitive Layer** is the "Thinking Mind" and the "Wisdom Core" of the firm, powered by **Retrieval-Augmented Generation (RAG)** and **Recursive Feedback Loops**. This layer ensures the brand "never forgets" a lesson. It synthesizes the relationships from the processing layer into a "Continuous Stream of Wisdom" that is available to any node in the network. When a leader asks a complex strategic question, the Cognitive Layer doesn't just search for keywords; it "reasons" through the organization's entire historical context to provide an answer grounded in the brand's "Institutional Memory." It involves **Self-Evolving Schemas**, where the Brain actually updates its own understanding of the business as new data arrives. This turns the organization into a "Learning Organism" with an ever-increasing "Corporate IQ." For the executive, this provides an "Intelligence Asset" that is independent of any single software package or individual employee, ensuring the brand's "Mind" remains a permanent, accumulating competitive advantage.

5. THE ARCHITECTS (THE BRAIN) | HUMAN IN-THE-LOOP AI

OVERVIEW

Human-in-the-Loop (HITL) AI is a cycle where human supervisors continuously train, tune, and validate a model's performance, especially for ambiguous tasks. The Supervisory & Coaching leader focuses on defining ethical boundaries and quality thresholds, ensuring AI decisions align with organizational values. This synergy elevates the leader from a decision-maker to a model trainer and quality assurance expert. By providing critical feedback, the leader helps the AI evolve, improving accuracy, fairness, and overall governance in complex, high-stakes leadership decisions.

MAIN STEPS

Start

STEP 1 – Initial Model Training
Establish baseline knowledge and performance rules

STEP 2 – Active Decision Queue
Flag uncertain outputs for human review.

STEP 3 – Human Supervisor Intervention
Leader reviews, corrects, documents outcomes.

STEP 4 – Feedback Integration Cycle
Corrected data retrains and improves the model.

STEP 5 – Model Validation & Testing
Confirm updated model meets quality standards.

STEP 6 – Deployment and Monitoring
Deploy new model; track performance, collect data.

End

SAMPLE FLOW

Ethical Performance Review Calibration: Leader uses HITL to ensure AI-driven performance recommendations are fair, unbiased, and align with ethical standards..

HUMAN-IN-THE-LOOP AI

1. Input Ethical Standards
Define fairness metrics and bias thresholds for all performance scoring clearly.

2. Flag Biased Actions
HITL flags reviews where demographic factors may have influenced the scoring inaccurately.

3. Coach Model on Bias
Leader corrects scores and labels the exact data features causing the ethical bias.

4. Retrain Algorithm
The model immediately incorporates the leader's ethical corrections into its logic.

5. Validate Unbiased Output
Test the updated model against diverse profiles to confirm ethical fairness is achieved correctly.

6. Approve Final Review Set
Leader signs off on the final, bias-mitigated performance review scores and decisions.

USE CASES

- **Manufacturing | Complex Automation Oversight:** Leaders coach AI on prioritizing production schedules during unexpected equipment failures, ensuring the recovery plan balances efficiency with safety standards effectively.
- **Financial Services | High-Value Fraud Scoring:** Supervisors review ambiguous transaction alerts flagged by the AI, labeling subtle fraudulent patterns to rapidly enhance the system's accuracy and integrity reliably.
- **Healthcare | Triage System Refinement:** Senior medical staff validate unusual patient symptom combinations flagged by the AI, providing critical diagnostic feedback to improve the model's triage reliability safely.
- **Technology | Content Moderation Training:** Leaders manually review borderline platform content, coaching the AI on nuanced policy interpretations to ensure consistent application of community standards clearly.
- **Retail | Customer Loyalty Correction:** Managers review and correct AI-generated personalized offers that customers found intrusive, teaching the model better boundaries for privacy and engagement accurately.
- **Energy | Critical Infrastructure Maintenance:** Supervisors validate AI-recommended maintenance schedules for high-voltage assets, ensuring the model's decisions prioritize grid stability and public safety first quickly.

In the **Leadership Transformation**, the **Architects (The Brain)** serve as the organization's "Librarians," responsible for converting fragmented data into a unified, permanent institutional memory. While the primary instrument of **Human-in-the-Loop** ensures that this memory is governed by human intent and wisdom, it requires secondary technical instruments to handle the massive scale of modern enterprise data. To move from "Information Overload" to "Cognitive Sovereignty," a leader must integrate these six pairings to build the "Brain."

1. The Vector Database Pairing for Contextual Retrieval

Pairing **Human-in-the-Loop with Vector Databases (RAG)** is best utilized when a leader needs to access deep institutional knowledge instantly. While the human provides the query, the Vector Database retrieves information based on "meaning" rather than just keywords. This is essential during executive transitions or complex legal audits, ensuring the "Brain" can pull relevant historical context from decades of unstructured documents, effectively curing corporate amnesia.

2. The Federated Learning Pairing for Privacy-First Intelligence

Leaders should pair **Human-in-the-Loop with Federated Learning** when managing highly sensitive data across multiple global regions. This is best deployed in healthcare or finance. It allows the "Brain" to learn from data at the local level without the data ever leaving its secure origin. The human architect governs the learning objectives, ensuring the firm grows smarter as a whole while maintaining absolute data sovereignty and regulatory compliance.

3. The Knowledge Graph Pairing for Semantic Linking

When a firm needs to understand the complex relationships between its products, employees, and customers, **Human-in-the-Loop must be paired with Knowledge Graphs**. While the human defines the business logic, the Knowledge Graph maps the "web of connections." This is best utilized during post-merger integrations to identify overlapping expertise or hidden dependencies, ensuring the "Brain" sees the organization as a connected organism rather than a set of silos.

4. The Graph Neural Network Pairing for Pattern Recognition

For identifying hidden risks or opportunities within a network, a leader should pair **Human-in-the-Loop with Graph Neural Networks (GNNs)**. The human sets the "Strategic Sentinel" parameters, while the GNN analyzes the shape of the data to find anomalies. This is best used in fraud detection or supply chain resilience, where the "Brain" can spot a failing node before the human eye could ever detect the pattern in a standard spreadsheet.

5. The Differential Privacy Pairing for Secure Insight

To share insights without compromising individual privacy, leaders should pair **Human-in-the-Loop with Differential Privacy**. This is best utilized when a brand wants to share "Consumer Trends" with partners without revealing specific customer identities. The human architect decides what insights are valuable, and the technology adds "mathematical noise" to protect the raw data. This pairing allows the "Brain" to be collaborative and open while remaining a fortress of privacy.

6. The Natural Language Processing Pairing for Human Synthesis

In environments where data is predominantly qualitative, **Human-in-the-Loop should be paired with Advanced NLP**. This is best utilized for synthesizing global employee feedback or customer reviews. The NLP layer processes millions of words to identify themes, which the human architect then synthesizes into a strategic vision. This ensures the "Brain" is sensitive to the human "Pulse" of the organization, turning raw sentiment into structured institutional wisdom.

Executive Power: Autonomy

The final evolution of leadership is the transition from "Managing Labor" to **"Directing Autonomy."** Historically, a leader's "Hands" were limited by the physical and mental bandwidth of their workforce. Growth required more people, and more people required more management, leading to a "Complexity Tax" that eventually slowed every giant to a crawl. The transformation into an **Executor** represents the birth of the "Autonomous Enterprise." In this role, the leader's intent is translated into action by a "Self-Directing System." The leader becomes the "Conductor of Execution," moving from the micro-management of tasks to the macro-orchestration of goals.

Defensive Stance: Killing the Wastage

The defensive mission of the Leader-Executor is the **Elimination of Friction**. Every manual step, every redundant approval, and every "Hand-off" between departments is a point of potential failure and capital leakage. The Executor uses "Process Discovery" to look deep into the "Plumbing" of the company and find where time and energy are being wasted. By establishing "Self-Healing Operations," the leader ensures that the organization can fix its own minor errors—like a stuck shipping order or a data entry glitch—without human intervention. This kills the "Wastage of Downtime." It ensures that the "Hands" of the company are always moving, creating a "Frictionless Engine" where the marginal cost of execution drops toward zero. This is the **Shield of Productivity**, protecting the brand's margins by automating the "Mundane" so the "Human Soul" of the company can focus on "Meaningful Innovation."

Offensive Stance: Unlocking Orchestration

On the offense, the leader moves toward **Agentic Orchestration**. This is the transition from "Simple Automation" (completing a pre-programmed task) to "Autonomous Agency" (achieving a high-level strategic goal). The leader no longer gives step-by-step instructions; they define a "Strategic Target"—such as "Optimize the European Supply Chain for Carbon Neutrality"—and deploy a "Swarm of Agents" to achieve it. These "Hands" can navigate complex environments, negotiate with external partners, and re-allocate resources in real-time. This **Unlocks Orchestration**, allowing the brand to "Execute" at a global scale with the precision of a local boutique. It turns the organization into an "Autonomous Force"—one that can act, react, and expand into new markets with near-zero "Management Overhead."

The Leadership Outcome: Absolute Autonomy

The ultimate value of the Executor role is **Strategic Leverage**. It allows a single leader to move an entire empire with the same ease that one might move a small team. For the leader, this provides **Operational Freedom**. You are no longer "Stuck in the Weeds" of daily crisis management; you are "Orchestrating Outcomes." It turns the organization into a "Perpetual Machine"—one that runs 24/7, improves every second, and executes with a level of "Autonomy" that makes the brand "Indestructible." This is the key to "Global Dominance"—the ability to manifest your "Voice" and "Strategy" into the physical world with "Zero Resistance" and "Infinite Scale."

Cognitive Layer	**Autonomous Fulfillment and Self-Correction**
Processing Layer	**Multi-Agent Workflow Orchestration and Logic**
Sensory Layer	**Reflexive Trigger Recognition and Alerting**

Executive Power: Agentic Autonomy

The final and most radical shift in modern leadership is the transition from "Managing Work" to **"Orchestrating Autonomy."** The **Executor** role represents the birth of the "Agentic Firm," where the leader provides the "Strategic Intent" and the AI-augmented "Hands" manifest the result. Historically, execution was a game of "Human Follow-through," limited by manual speed, human fatigue, and bureaucratic friction. The Executor role replaces this with a system of **Autonomous Agency**, turning the organization into a "Direct Manifestation Machine" that can navigate complex global tasks and fulfill the leader's vision at the speed of light. This is the shift from "Passive Tools" to "Active Agents," giving the executive the power to "Operate at the Speed of Thought" by transforming "Intent" into "Autonomous Reality" without the operational drag of traditional management.

The Sensory Layer: Reflexive Trigger Recognition and Alerting

The **Sensory Layer** of the Hands serves as the organization's "Reflex System." It is designed for **Real-Time Event Detection**, monitoring the environment for specific "Triggers" that require an immediate autonomous response. Utilizing **Computer Vision at the edge**, **Anomaly Detection Algorithms**, and **API-led Event Streaming**, this layer identifies the moment a plan deviates from reality. It doesn't wait for a human to flag a "Late Shipment," a "Security Breach," or an "Inventory Shortage"; it "feels" the friction as it happens. This sensory input provides the "Activation Signal" for the rest of the stack, ensuring that the "Hands" of the company are always ready to move the millisecond an operational or strategic need is identified. This layer protects the user's "Attention" and the brand's "Margins" by ensuring that "Minor Glitches" never grow into "Major Crises," because the "Hands" are always there to catch the ball before it hits the ground.

The Processing Layer: Multi-Agent Workflow Orchestration and Logic

The **Processing Layer** acts as the "Cerebellum" and the "Coordinator," breaking down the leader's high-level intent into thousands of executable sub-tasks. It utilizes **Agentic Logic (ReAct Frameworks)** and **Chain-of-Thought Reasoning** to orchestrate a swarm of specialized AI agents. If the leader's intent is to "re-route the global supply chain to avoid a hurricane," this layer identifies the complex web of sub-tasks required—from re-negotiating vendor contracts and updating customs filings to notifying local distributors. It manages the "Inter-Agent Communication," ensuring that each bot knows its role, its deadlines, and its constraints. This layer provides the **Governance and Logic** for the execution, ensuring that the "Hands" stay within the "Ethical and Financial Guardrails" set by the executive. It kills the friction of micro-management through automated, high-speed coordination, ensuring the brand's "Hands" are working in perfect harmony with the "Brain."

The Cognitive Layer: Autonomous Fulfillment and Self-Correction

The **Cognitive Layer** is where the "Hands" actually manifest the result through **Action Transformers** and **Robotic Process Automation (RPA)**. This is the layer of "Closed-Loop Execution." It doesn't just "try" to do a task; it "achieves" it. If an agent encounters an obstacle—such as a closed port or a failed API—this layer uses **Self-Correction Logic** and **Reinforcement Learning** to find an alternative path without asking the leader for permission or intervention. It turns "Information" into "Physical or Digital Action," whether that means moving funds on a blockchain, updating a manufacturing schedule in real-time, or deploying a software patch across a global fleet. This provides the executive with **Absolute Leverage**, allowing them to "Direct the Engine" of the company rather than "Manage the People." It transforms the organization into an "Autonomous Powerhouse" that operates with the precision of a law of physics, delivering "Market Leadership" through sheer execution dominance.

OVERVIEW

Agentic AI operates as an autonomous mission manager, breaking down high-level strategic goals into executable tasks. The Driving Mission leader defines the ultimate objective and core values, essentially setting the constitutional limits for the AI agent. This synergy offloads complex, time-consuming multi-step execution to the AI, allowing the leader to focus exclusively on strategic vision and alignment. The leader initiates the mission; the agent executes and reports on key milestones. This partnership fundamentally converts strategy into action, maximizing the leader's impact on high-level direction and governance.

MAIN STEPS

Start

STEP 1 — Mission Goal Setting
Define the high-level desired outcome.

STEP 2 — Autonomous Task Breakdown
Agent breaks goal into specific sub-tasks.

STEP 3 — Execution and Monitoring
Agent performs steps, tracks progress autonomously.

STEP 4 — Milestone Reporting & Review
Agent summarizes for leader approval.

STEP 5 — Course Correction, Redirection
Leader adjusts mission parameters or values.

STEP 6 — Mission Completion & Handoff
Agent delivers final output to human team.

End

SAMPLE FLOW

Strategic Market Entry Simulation: The leader tasks an agent with simulating a five-year strategy for entering a new global market segment.

1. Define Target Outcome
Specify profit margins and desired market share for the target region clearly.

2. Generate Strategic Paths
Agent autonomously researches regulatory, competitive, and logistical variables correctly.

3. Execute Scenario Modeling
Agent runs multiple five-year financial and operational simulations autonomously.

4. Present Viability Report
Agent delivers a concise summary of the highest-potential market entry strategy promptly.

5. Adjust Risk Tolerance
Leader changes the acceptable financial risk parameters or ethical non-negotiables ethically.

6. Authorize Final Strategy
Leader approves the final mission plan, initiating resources allocation for execution criteria.

USE CASES

- **Manufacturing | Quarterly OKR Redesign:** An agent autonomously redesigns the quarterly OKRs for manufacturing plant managers based on new corporate sustainability goals and budget realities effectively.
- **Financial Services | Future Talent Mapping:** The leader commissions an agent to map future talent needs across the organization, autonomously proposing structural changes to align with predicted regulatory shifts correctly.
- **Healthcare | Research Governance Transformation:** An agent is tasked with planning a complete transformation of clinical research governance, autonomously generating new policy drafts and training materials for staff ethically.
- **Technology | Global Change Orchestration:** An agent orchestrates the global change management process for a major organizational pivot, independently handling communications, training logistics, and feedback collection clearly.
- **Retail | Employee Empowerment Program:** The leader deploys an agent to define and launch a new employee empowerment program across all stores, autonomously creating communication plans and training modules accurately.
- **Energy | Decarbonization Roadmap:** An agent drives the strategic implementation of a decarbonization roadmap, independently managing compliance tracking and resource reallocation across disparate operating units quickly.

In the **Leadership Transformation**, the **Executors (The Hands)** represent the organization's "Virtuosos"—the autonomous fulfillment engine that turns strategy into action. While the primary instrument of **Agentic AI** provides the "agency" to navigate tasks, it requires secondary instruments to ensure precision, safety, and physical manifestation. To move from "Resource Leakage" to "Autonomous Value Action," a leader must integrate these six pairings to empower the "Hands."

1. The RPA Pairing for Legacy Integration

Pairing **Agentic AI with Robotic Process Automation (RPA)** is best utilized when the "Hands" must interact with older, rigid software systems. While the Agent decides the strategy, RPA provides the "fingers" to execute clicks and data entry in legacy environments. This is essential for back-office transformation, allowing autonomous agents to bridge the gap between modern intelligence and the firm's existing digital infrastructure.

2. The Digital Twin Pairing for Operational Guardrails

Leaders should pair **Agentic AI with Digital Twins** when deploying autonomous agents in high-risk physical environments, like automated warehouses. Before an agent moves a pallet or adjusts a machine, it "rehearses" the action in the Digital Twin. This pairing ensures that the "Hands" never take an action that could cause physical damage or safety violations, providing a virtual sandbox for autonomous decision-making.

3. The Edge AI Pairing for Real-Time Actuation

When execution must happen at the "Point of Action" with zero latency—such as on a moving delivery drone or a factory floor—**Agentic AI must be paired with Edge AI**. This allows the "Hands" to process data and make decisions locally without waiting for the cloud. This pairing is vital for autonomous logistics, ensuring that the agents can react to sudden physical obstacles or mechanical shifts in milliseconds.

4. The Blockchain Pairing for Autonomous Contracting

For secure, multi-party transactions, a leader should pair **Agentic AI with Blockchain (Smart Contracts)**. This is best used in automated procurement. The "Hands" don't just find a supplier; they execute the payment and verify delivery through an immutable ledger. This pairing ensures that autonomous actions are legally binding and transparent, eliminating the friction of manual financial reconciliation.

5. The Computer Vision Pairing for Physical Awareness

To allow the "Hands" to interact with the non-digital world, leaders should pair **Agentic AI with Computer Vision**. This is best utilized in quality control and autonomous maintenance. By "seeing" the environment, the agent can identify a hairline fracture in a machine or a mislabeled package. This pairing gives the "Hands" the sensory awareness required to perform precision physical tasks that were previously reserved for humans.

6. The Reinforcement Learning Pairing for Skill Acquisition

In dynamic environments where tasks change constantly, **Agentic AI should be paired with Reinforcement Learning (RL)**. This is best utilized for optimizing complex energy grids or ad-bidding engines. The "Hands" learn from the success or failure of every action, constantly refining their technique to maximize efficiency. This provides the brand with a self-improving execution layer that grows more skillful with every task performed.

THE Ai6x6™ STRATEGIC ROADMAP SERVICE

Your 3-Year Vision, Stress-Tested for the AI Era

The **Ai6x6™** book series has given you the "source code"—the clinical lenses, the vulnerability audits, and the diagnostic frameworks. But strategic orchestration at the highest level requires a cold, objective eye. If you are ready to move from self-diagnosis to radical transformation, we invite you to engage the **Ai6x6™ Roadmap Service.**

The 21-Day Transformation Engine

We don't believe in "billable hours" or six-month discovery phases. We believe in high-fidelity data and rapid execution. Our domain experts take your raw diagnostic output and synthesize it into a personalized, holistic blueprint for the next three years of your organization's life.

How It Works

1. **Submit Your Data:** Upload your completed Discovery Questionnaires, Vulnerability Audits, and Radar Charts directly to our secure portal or via email or via courier.
2. **Contextualize:** Provide your current 3-year roadmap and any business documentation that can provide more context on your line of business, your mission and success KPIs.
3. **The Clinical Review:** Our experts analyze your organizational DNA against the **6 Lens Sequence**, identifying hidden frictions and untapped leverage points. Assign a single point within your organization for asking questions.
4. **The Delivery:** Within **2 to 3 weeks**, you receive a personalized **Holistic Transformation Roadmap**, including our proprietary **Self-Disruption Strategy**.

Why Self-Disruption?

The greatest threat to your organization isn't your competitor; it is your own legacy success. Our roadmap doesn't just show you how to grow; it shows you how to cannibalize your outdated models before the market does it for you.

Packages from $3000

The Commitment

- **Speed:** 14–21-day turnaround.
- **Security:** Full NDA-backed engagement.
- **Precision:** Fact-based, quantitative analysis. No "fluff".

SECURE YOUR DIAGNOSTIC REVIEW:

Ai6x6.com/roadmap

Consulting Roadmap

SECTION C

BLUEPRINT SERIES

Business Maturity

Ai6x6™ Transformation Blueprint

Story 1 — The Marshall Plan Reconstruction (1948–1952): Post-war Europe faced shattered economies, broken supply chains, and institutional collapse. The Marshall Plan introduced structured funding, transparent reporting, and coordinated execution across nations. Local governments provided context while centralized oversight ensured accountability. This blend of autonomy and structure accelerated recovery. The program demonstrated how disciplined governance can rebuild complex systems. It became a blueprint for large-scale organizational maturity.

Story 2 — The Rise of Quality Management in Japan (1950s–1960s): Japan's industries struggled with inconsistent quality and limited global credibility. Statistical quality control introduced measurement, repeatability, and disciplined improvement cycles. Workers contributed insights while leaders standardized processes. Over time, quality became a cultural norm rather than a corrective action. This shift transformed fragmented operations into cohesive systems. It marked the emergence of maturity through methodical refinement.

Story 3 — NASA's Systems Engineering Discipline (1960s): The space race demanded unprecedented coordination across thousands of contractors and scientific domains. NASA formalized systems engineering to integrate requirements, risks, and interfaces into a unified framework. Structured reviews balanced innovation with rigor. Human judgment guided trade-offs while documentation preserved institutional memory. This discipline enabled complex missions to succeed repeatedly. It became a global model for organizational maturity in high-stakes environments.

Story 4 — The Birth of Modern Corporate Governance (1970s–1980s): Corporate scandals exposed weaknesses in oversight, transparency, and fiduciary responsibility. Boards adopted clearer roles, audit mechanisms, and accountability structures. Regulations reinforced ethical conduct while organizations matured through formal governance. Leaders learned to balance autonomy with compliance. This era established governance as a strategic capability, not an administrative burden. It marked a turning point in institutional maturity.

Story 5 — Toyota's Global Expansion Playbook (1980s–1990s): As Toyota expanded globally, it faced cultural differences, supply-chain complexity, and operational variability. The company codified principles, standardized training, and embedded continuous improvement across regions. Local teams adapted practices while preserving core philosophy. This balance of global consistency and local flexibility enabled scalable excellence. It demonstrated maturity as the ability to replicate capability without diluting identity.

Story 6 — The Digital Transformation of Banking (1990s–2000s): Banks confronted rising customer expectations, regulatory pressure, and technological disruption. They modernized legacy systems, standardized data, and introduced structured risk frameworks. Digital channels required new operating models and cross-functional coordination. Human expertise guided decisions where automation lacked nuance. Over time, banks evolved from siloed institutions into integrated digital enterprises. This transition reflected maturity through modernization and disciplined reinvention.

LESSONS FROM A LEADERSHIP TRANSFORMATION PERSPECTIVE

Leadership maturity emerges when organizations shift from reactive decision-making to structured, principle-driven stewardship. Across decades, leaders learned to balance autonomy with oversight, innovation with discipline, and speed with accountability. Mature leadership is defined not by charisma but by the ability to design systems that endure beyond individuals. It requires translating complexity into clarity, aligning diverse stakeholders, and institutionalizing behaviors that scale. As organizations evolve, leadership becomes less about directing actions and more about shaping frameworks that enable consistent, ethical, and resilient performance. Maturity is achieved when leadership creates stability without suppressing adaptability.

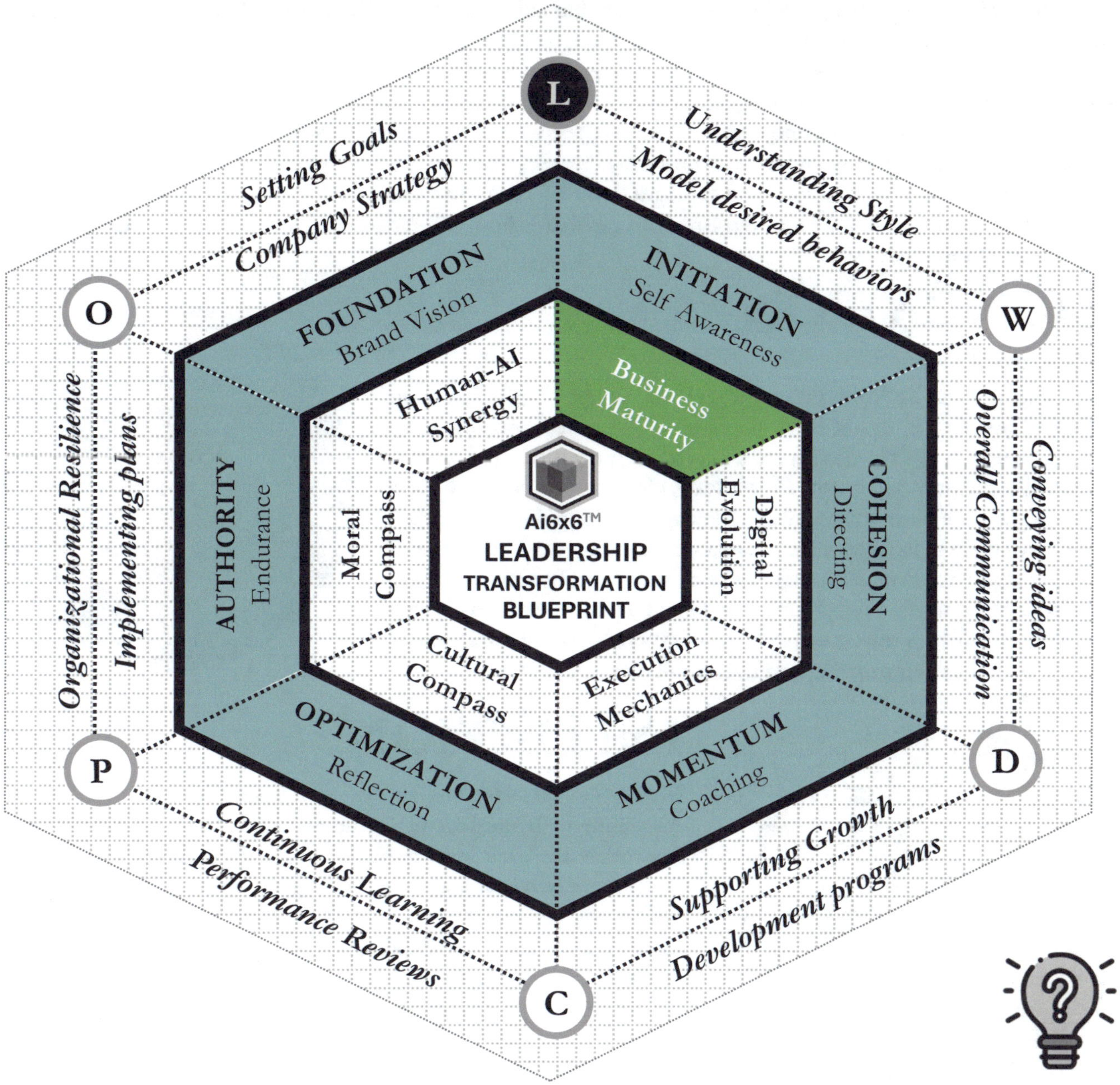

The **Ai6x6 Research Lab** selected this lens because an organization can never be more "mature" than the person leading it. We view business maturity as an internal filter for character, endurance, and the capacity for sustained impact. This dimension ensures that the aggressive scaling required in the AI era is anchored in an unwavering sense of self. It is a vital lens because, in the turbulence of radical change, a leader's internal stillness is the only force that prevents the enterprise from drifting. This lens is the heart's commitment to growing inward as fast as the business grows outward.

We believe true maturity is found in the ability to hold a mirror to one's own leadership tendencies, recognizing that the flaws of the leader are inevitably amplified by the technology they deploy. By focusing on self-awareness, the Lab provides a mechanism for leaders to audit their own resilience before they test the resilience of their teams. This lens ensures that the pursuit of market presence does not come at the cost of personal integrity. It is the steady hand that guides the organization through transformation, recognizing that while the landscape is shifting, the bedrock of leadership character must remain immutable.

Every leadership transformation begins with ambition, but ambition alone is never enough to carry an organization into its future. **Business Maturity** is the lens that forces leaders to confront the gap between what they hope to achieve and what their enterprise is prepared to sustain. It is the discipline that reveals whether the organization is structurally, operationally, and culturally ready for the future it claims to desire. In the AI era, this gap becomes even more pronounced; AI does not simply automate tasks—it reshapes workflows, decision cycles, and the very rhythm of how leadership operates. It exposes weaknesses that were previously tolerable and accelerates the consequences of poor preparation, demanding a level of rigor that traditional management frameworks often overlook. It examines how ambition translates into capability, how vision converts into execution, and how culture sustains momentum under pressure.

FOUNDATION

Define the Core

The journey begins at **Foundation**, where the primary objective is to *Define the Core*. This is the bedrock upon which all transformation rests, requiring leaders to **Filter the Noise** of competing priorities and anchor their identity and strategic intent. At this level, clarity is not optional—it is existential. Without clear governance, AI initiatives drift, priorities conflict, and teams operate in isolated silos. Foundation maturity is the discipline of intent: the ability to distinguish urgency from readiness. Leaders who skip this stage mistake motion for progress, building transformation on shifting sand. True maturity begins when the organization knows what it stands for, what it will not compromise, and how it will measure success.

INITIATION

Start the Motion

Ascending to **Initiation** marks the first active step: *Start the Motion*. This level turns the gaze inward, demanding self-awareness and the courage to confront personal biases. By **Wiring the System**, leaders catalyze the shift from passive management to intentional authority. Initiation is where leadership becomes kinetic—it transforms from aspiration into activation. The organization begins to move, not because it is told to, but because its internal circuits are aligned. Maturity here is emotional as much as structural; it requires self-regulation, empathy, and the ability to lead through friction. The leader's persona is tested, revealing whether they can sustain momentum through disruption rather than retreat into comfort.

COHESION

Connect the Parts

The climb continues to **Cohesion**, represented by *The Engine*. This is the critical unifier where people, processes, and systems are harmonized. The focus shifts to **Connect the Parts**—to ensure that individual leadership goals align with a unified organizational voice. By **Shifting the Gear**, leaders eliminate silos and systemic friction, transforming fragmented efforts into a synchronized machine. Cohesion maturity is the art of integration: turning disconnected motion into collective momentum. Without it, organizations celebrate activity instead of achievement. The mature leader understands that alignment is not consensus—it is coherence, the ability to move as one without losing diversity of thought.

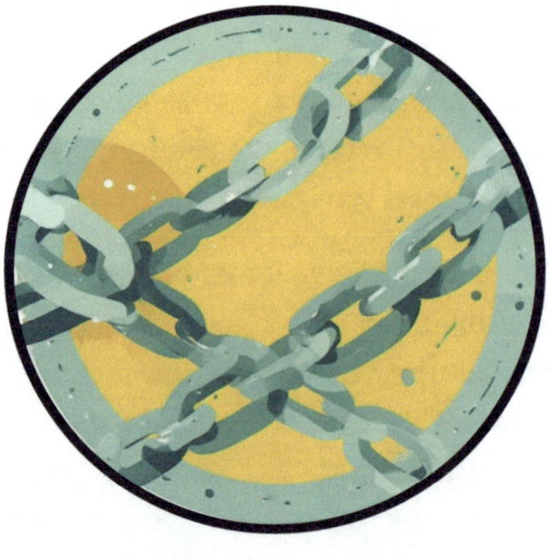

MOMENTUM

Drive the Speed

The fourth rung is **Momentum**, where the organization begins to *Drive the Speed*. Leadership here moves toward scaling impact with predictable velocity. By **Removing the Drag** of legacy thinking and institutional inertia, leaders accelerate the collective intelligence of the enterprise. Momentum maturity is the physics of transformation—it measures how efficiently energy converts into progress. At this stage, the organization's rhythm changes; it begins to operate with self-propelling force. The leader's role shifts from directing individuals to empowering systems, ensuring that innovation scales faster than resistance. Momentum is not chaos—it is controlled acceleration, the moment when transformation gains its gravity.

OPTIMIZATION

Refine the Flow

Refinement occurs at **Optimization**, characterized by *The Flow*. Here, leadership pauses to **Refine the Flow** by **Locking the Code** of influence through reflection and data-driven inquiry. Immature organizations measure what is easy; mature ones measure what matters. Optimization maturity is the discipline of precision—removing friction without losing flexibility. Leaders refine decisions still in motion, adjusting with humility and surgical accuracy. Ego and bias are replaced by evidence and iteration. The organization becomes self-healing, capable of continuous improvement without external intervention. Optimization is not about perfection—it is about adaptability, the ability to evolve faster than the environment demands.

AUTHORITY

Own the Market

The summit is reached at **Authority**, where leaders finally *Own the Market*. This stage manifests as *The Benchmark*—a legacy of endurance that can withstand external shocks and volatility. By **Fueling the Future**, leaders ensure that transformation is not a temporary win but a permanent standard. Authority maturity is the discipline of stewardship: sustaining excellence while preparing for reinvention. Immature organizations avoid this stage because it exposes uncomfortable truths; mature ones embrace it as the crucible of resilience. At this level, leadership becomes institutional—it transcends personality and becomes culture. The organization no longer reacts to the market; it defines it.

Business Maturity is the scaffolding that holds the weight of innovation. It ensures that speed does not collapse under fragility and that ambition does not outpace capability. Each level builds upon the last, forming a continuum from clarity to mastery. Foundation defines intent, Initiation activates leadership, Cohesion integrates systems, Momentum accelerates impact, Optimization refines intelligence, and Authority sustains legacy. In the AI era, maturity is not a luxury—it is survival. As algorithms amplify both strengths and weaknesses, leaders must design organizations that can learn, adapt, and endure. Business Maturity is therefore not a checklist but a **leadership discipline**—a way of thinking that transforms ambition into architecture. When leaders embrace this full sequence, they discover that maturity is not the end of transformation but its engine—the invisible structure that makes both explosive speed and long-term continuity possible.

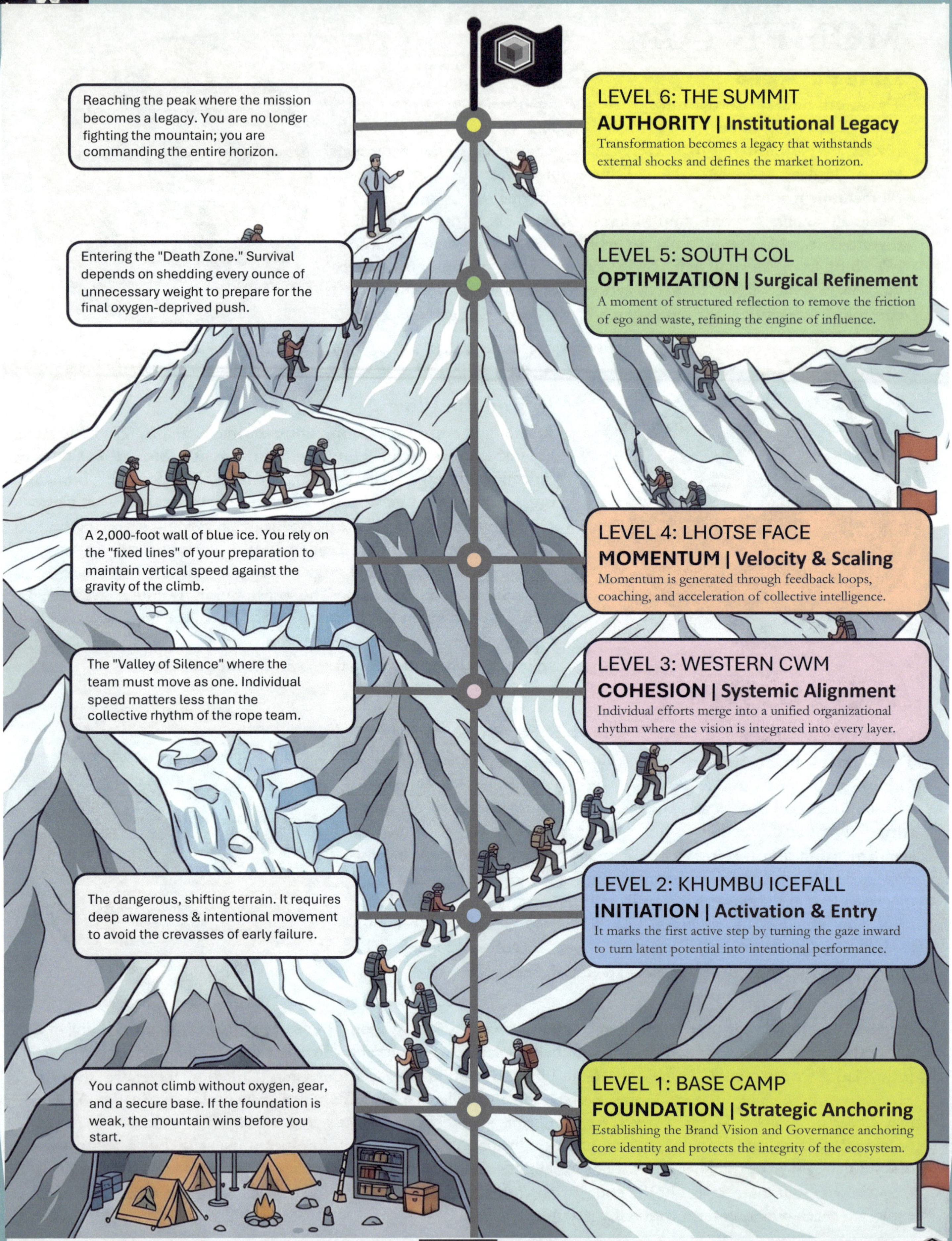
Reaching the peak where the mission becomes a legacy. You are no longer fighting the mountain; you are commanding the entire horizon.
LEVEL 6: THE SUMMIT
AUTHORITY | Institutional Legacy
Transformation becomes a legacy that withstands external shocks and defines the market horizon.
Entering the "Death Zone." Survival depends on shedding every ounce of unnecessary weight to prepare for the final oxygen-deprived push.
LEVEL 5: SOUTH COL
OPTIMIZATION | Surgical Refinement
A moment of structured reflection to remove the friction of ego and waste, refining the engine of influence.
A 2,000-foot wall of blue ice. You rely on the "fixed lines" of your preparation to maintain vertical speed against the gravity of the climb.
LEVEL 4: LHOTSE FACE
MOMENTUM | Velocity & Scaling
Momentum is generated through feedback loops, coaching, and acceleration of collective intelligence.
The "Valley of Silence" where the team must move as one. Individual speed matters less than the collective rhythm of the rope team.
LEVEL 3: WESTERN CWM
COHESION | Systemic Alignment
Individual efforts merge into a unified organizational rhythm where the vision is integrated into every layer.
The dangerous, shifting terrain. It requires deep awareness & intentional movement to avoid the crevasses of early failure.
LEVEL 2: KHUMBU ICEFALL
INITIATION | Activation & Entry
It marks the first active step by turning the gaze inward to turn latent potential into intentional performance.
You cannot climb without oxygen, gear, and a secure base. If the foundation is weak, the mountain wins before you start.
LEVEL 1: BASE CAMP
FOUNDATION | Strategic Anchoring
Establishing the Brand Vision and Governance anchoring core identity and protects the integrity of the ecosystem.

6x6™ Transformation Blueprint

FOUNDATION | Brand Vision

"Leadership is about vision and responsibility, not power." - Jack Ma.

Founder of Alibaba Group, he built it from a small apartment into one of the world's largest e-commerce and digital payment companies. His entrepreneurial spirit, relentless vision for digital transformation, and charismatic leadership made him a global business icon.

The Foundation level is your Brand Vision, serving as the bedrock that defines your core internal identity and anchors your market presence. This initial stage establishes the immutable values that will govern every subsequent leadership action.

Crafting a compelling and aspirational future for your leadership identity, meticulously defining your unique impact, core values, and overarching purpose. This isn't merely about personal goals; it's about how your leadership resonates, inspires, and profoundly shapes the collective journey of your team and organization.

"Brand Vision" is crucial for guiding all strategic decisions, inspiring followers to commit wholeheartedly to a shared future, and ensuring your leadership evolves with a clear, unwavering direction, preventing drift, maximizing influence, and fostering a legacy of impactful change and sustained organizational success.

For business strategy maturity, the "**Brand Vision**" chapter is foundational because it dictates the clarity and coherence of an organization's strategic direction. A mature business strategy begins with leaders setting clear goals, defining a compelling vision, and ensuring all teams are aligned with this strategic direction. This foundational alignment prevents fragmented efforts, ensures resources are directed towards shared objectives, and fosters organizational unity. It's about establishing a strong strategic compass that guides all subsequent actions, enabling the business to move forward with purpose and precision, which is critical for achieving long-term strategic success and adaptability. Here is how we approach our learning:

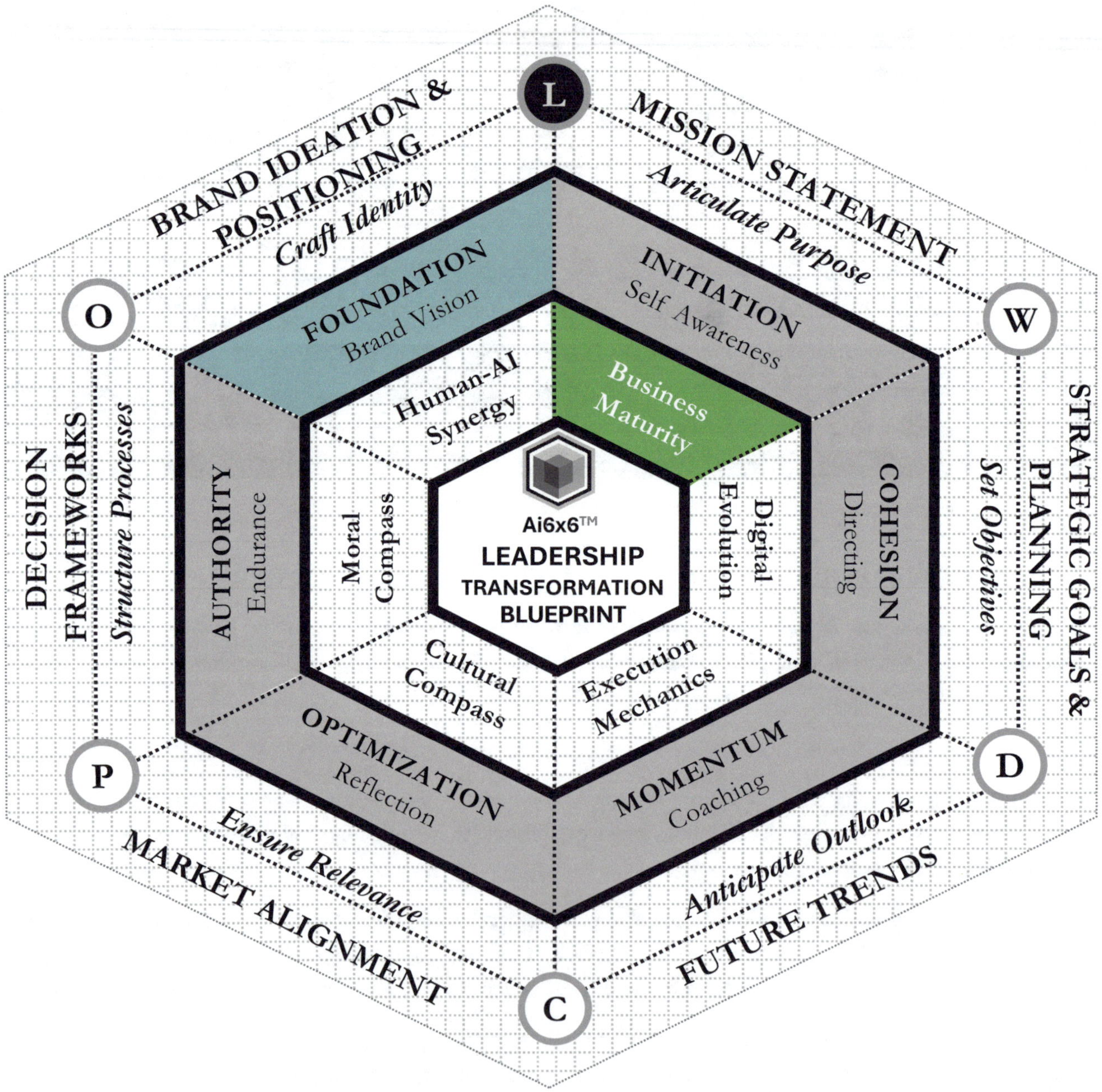

Ai6x6 BUSINESS MATURITY FOR BRAND VISION | LEADERSHIP TRANSFORMATION

For a business leader, the initial phase of brand ideation and positioning is a critical strategic imperative. It is here that you define the very soul of your brand—its core identity and its distinctive presence in the market. This foundational work ensures that every subsequent action, from product development to marketing, is aligned and cohesive. Your role is to guide the concept development, establishing the unique value and personality that truly differentiates your brand from the competition. This isn't just about a logo; it's about the emotional connection, the promise, and the distinct experience you offer.

Furthermore, a precise competitive positioning is crucial. Identifying where your brand stands relative to rivals in the market allows you to carve out a unique space, highlight your strengths, and clearly articulate why customers should choose you. This understanding informs your messaging and competitive strategy. Crucially, meticulous audience alignment is vital. Deeply understanding your target demographics, their needs, aspirations, and pain points, allows you to tailor your messaging and offerings, accordingly, ensuring your brand resonates powerfully and authentically with those you aim to serve. Neglecting this foundational phase can lead to a confused brand identity, a lack of differentiation, inefficient marketing spend, and ultimately, a struggle to capture and retain market share. By championing a clear, compelling brand vision, you empower your organization to communicate its value effectively, attract its ideal customers, and build a lasting presence that drives long-term growth and relevance in a competitive landscape.

WHAT?	WHY?	HOW?	BEWARE!
CONCEPT DEVELOPMENT	**DISTINCT IDENTITY**	**BRAINSTORMING SESSIONS**	**ALIGNMENT**
Establishing the unique value and personality.	Creates a memorable and differentiated brand.	Conduct workshops to define brand values, mission, and unique offerings.	Gaining consensus on core brand identity among diverse stakeholders.
COMPETITIVE POSITIONING	**MARKET ADVANTAGE**	**COMPETITOR ANALYSIS**	**SATURATION**
Identifying the brand's place in the market.	Ensures the brand stands out and attracts target customers.	Research rival brands, their strengths, weaknesses, and market perception.	Finding a truly unique position in a crowded marketplace.
AUDIENCE ALIGNMENT	**RESONANCE**	**PERSONA DEVELOPMENT**	**DIVERSE NEEDS**
Understanding target demographics and tailoring messaging.	Ensures communication is relevant and impactful for the target audiene.	Create detailed customer personas based on research and data.	Catering to the nuanced needs of multiple audience segments.

For a business leader, crafting a powerful and authentic Mission Statement is a critical strategic imperative. It serves as the guiding philosophy that encapsulates your brand's fundamental purpose, providing clarity and direction for every facet of your organization. More than just words on a wall, it is the north star that inspires and unites your teams. Central to this is defining your core values and impact. This involves articulating the ethical and aspirational foundation of your brand—what you stand for, how you operate, and the positive difference you aim to make in the world. These values underpin all decisions and actions.

Furthermore, clearly establishing your long-term vision is crucial. This paints a compelling picture of what your brand strives to achieve over an extended period, inspiring employees and stakeholders with a sense of purpose and direction beyond immediate goals. Crucially, authentic brand storytelling is vital for communicating this mission-driven narrative. Using relatable stories, compelling visuals, and consistent messaging allows you to connect emotionally with your audiences, fostering trust, loyalty, and a shared sense of purpose. Neglecting a clear and inspiring mission statement can lead to a lack of organizational direction, fragmented efforts, and difficulty in motivating employees and customers. By championing a well-defined and passionately communicated mission, you empower your organization with a powerful sense of purpose, inspiring collective action and ultimately securing your brand's long-term impact and legacy.

WHAT?	WHY?	HOW?	BEWARE!
CORE VALUES AND IMPACT	**GUIDING PRINCIPLES**	**WORKSHOPS**	**AUTHENTICITY**
Defining the ethical and aspirational foundation.	Provides the moral and operational framework for the brand.	Facilitate sessions to uncover and articulate foundational beliefs	Ensuring values are genuinely embraced, not just superficial statements.
LONG-TERM VISION	**INSPIRATIONAL DIRECTION**	**FUTURE-BACK THINKING**	**AMBITION VS. REALISM**
Establishing what the brand strives to achieve over time.	Motivates stakeholders and provides a future-oriented roadmap.	Envision the desired future state and work backward to define milestones.	Balancing aspirational goals with achievable targets.
BRAND STORYTELLING	**EMOTIONAL CONNECTION**	**NARRATIVE CRAFTING**	**CONSISTENCY**
Communicating mission-driven narratives.	Resonates deeply with audiences, fostering trust and loyalty.	Develop compelling stories that illustrate the brand's purpose and values.	Maintaining a cohesive and authentic narrative across all communication channels.

3. Strategic goals and planning

For a business leader, establishing clear strategic goals and meticulous planning is a critical imperative for translating your brand vision into tangible success. It's about structuring key objectives and outlining the roadmap that will guide your organization's efforts, ensuring every initiative contributes to your brand's growth and impact. Central to this is defining measurable brand objectives. This involves setting specific, quantifiable benchmarks for growth (e.g., market share, revenue), engagement (e.g., customer loyalty, social media interaction), and business development (e.g., new product launches). These metrics provide clarity and accountability.

Furthermore, strategic resource allocation is crucial. Optimizing investments in branding initiatives, marketing campaigns, product development, and operational strategies ensures that your financial and human capital are deployed where they will generate the greatest return and support your brand's strategic priorities. Crucially, developing adaptive growth strategies is vital. This means continuously adjusting plans in response to evolving market conditions, shifts in consumer behaviors, and competitive dynamics. An agile approach ensures your brand remains relevant and responsive, capitalizing on new opportunities while mitigating risks. Neglecting clear strategic goals and planning can lead to fragmented efforts, inefficient resource utilization, a reactive approach to market changes, and ultimately, a struggle to achieve long-term brand success. By championing a disciplined, data-driven, and flexible approach to strategic planning, you empower your organization to operate with focus, measure progress effectively, and adapt swiftly, ultimately driving the sustained growth, relevance, and profitability of your brand in a dynamic marketplace.

WHAT?	WHY?	HOW?	BEWARE!
MEASURABLE OBJECTIVES	**CLARITY & ACCOUNTABILITY**	**SMART GOALS**	**METRIC OVERLOAD**
Setting benchmarks for growth and engagement.	Provides quantifiable targets for brand success.	Develop Specific, Measurable, Achievable, Relevant, Time-bound objectives.	Too many objectives leading to scattered focus.
RESOURCE ALLOCATION	**MAXIMIZED ROI**	**BUDGETING & PORTFOLIO**	**COMPETITION**
Optimizing investments in branding, marketing, and operations.	Ensures financial and human capital are deployed efficiently.	Prioritize initiatives based on strategic impact and allocate resources accordingly.	Internal competition for limited resources across departments.
ADAPTIVE GROWTH STRATEGIES	**RESILIENCE & AGILITY**	**SCENARIO PLANNING**	**RESISTANCE TO CHANGE**
Adjusting plans to evolving market conditions.	Ensures the brand remains relevant and responsive to market shifts.	Develop contingency plans and flexible roadmaps for various market futures.	Organizational inertia or reluctance to pivot established plans.

For a business leader, proactively leveraging future trends and conducting rigorous forecasting are critical strategic imperatives for aligning brand initiatives with anticipated market shifts, ensuring long-term relevance and competitive advantage. In a world of accelerating change, reactive approaches are insufficient; foresight is key to sustained brand success. Central to this is comprehensive industry and consumer trend analysis. This involves diligently studying data, market reports, and consumer behavior patterns to anticipate emerging demands, technological innovations, and evolving cultural preferences. This predictive intelligence informs proactive brand development.

Furthermore, robust risk assessment is crucial. Identifying potential challenges—from disruptive technologies and new competitors to shifting consumer values and regulatory changes—allows you to prepare contingency strategies, mitigating threats before they impact your brand. This foresight transforms potential weaknesses into opportunities. Crucially, utilizing data-driven forecasting models is vital. By leveraging advanced analytics and AI, you can shape decision-making and strategic adjustments with greater precision, moving beyond intuition to empirically grounded predictions that guide investments and brand initiatives. Neglecting trend forecasting can lead to brand stagnation, missed market opportunities, being blindsided by competitors, and a struggle to connect with future generations of consumers. By championing a forward-looking, data-driven approach to understanding the future, you empower your organization to innovate proactively, adapt strategically, and ultimately secure your brand's long-term viability, market leadership, and enduring relevance in a dynamic world.

WHAT?	WHY?	HOW?	BEWARE!
INDUSTRY AND CONSUMER TREND	**PROACTIVE INNOVATION**	**MARKET RESEARCH**	**SIGNAL VS. NOISE**
Studying data to anticipate emerging demands.	Ensures brand offerings are aligned with future market needs.	Conduct ongoing research, monitor emerging tech, and analyze consumer data.	Distinguishing fleeting fads from significant, lasting trends.
RISK ASSESSMENT	**BRAND RESILIENCE**	**SCENARIO PLANNING**	**UNFORESEEN EVENTS**
Identifying challenges and preparing contingency strategies.	Mitigates threats and ensures the brand can navigate disruptions.	Develop potential future scenarios and prepare response strategies for each.	The inherent difficulty in predicting truly black swan events.
DATA-DRIVEN FORECASTING MODELS	**INFORMED STRATEGY**	**PREDICTIVE ANALYTICS**	**ACCURACY**
Using analytics to shape decision-making.	Grounds brand decisions in empirical data, reducing guesswork.	Implement AI/ML models to forecast market shifts and consumer preferences.	Ensuring forecasting models are reliable and adaptable to new data.

For a business leader, rigorously ensuring market alignment is a critical strategic imperative for maintaining brand relevance and consistency with evolving customer expectations. In a dynamic marketplace, a brand that fails to adapt risks becoming obsolete; continuous alignment is key to sustained connection and customer loyalty. Central to this is fostering brand adaptability. This means constantly refining and evolving your brand's messaging, offerings, and experiences to remain relevant to shifting consumer preferences, technological advancements, and cultural changes. An agile brand is a resilient brand.

Furthermore, prioritizing cultural and social alignment is crucial. Actively reflecting societal trends, values, and ethical considerations in your branding efforts ensures that your brand resonates deeply with modern consumers who increasingly expect companies to demonstrate social responsibility. Crucially, continuous competitive benchmarking is vital for market positioning. Regularly comparing your brand strategies—from messaging to pricing and customer experience—with industry leaders and key rivals provides critical insights into market positioning, highlighting opportunities to differentiate and refine your value proposition. Neglecting market alignment can lead to a stagnant brand, a disconnection with target audiences, lost market share, and a struggle to regain relevance. By championing a proactive, adaptive, and culturally attuned approach to market alignment, you empower your organization to continuously evolve its brand, deepen customer connections, and ultimately secure its long-term relevance, competitive edge, and sustained growth in a rapidly changing world.

WHAT?	WHY?	HOW?	BEWARE!
BRAND ADAPTABILITY	**LONG-TERM RELEVANCE**	**AGILE BRANDING**	**CONSISTENCY**
Maintaining relevance through continuous refinement and evolution.	Ensures the brand remains appealing to evolving customer needs.	Regularly review and update brand messaging, visual identity, and offerings.	Balancing adaptability with maintaining a consistent brand identity.
CULTURAL AND SOCIAL ALIGNMENT	**CONSUMER CONNECTION**	**TREND MONITORING**	**AUTHENTICITY**
Reflecting societal trends and values.	Resonates deeply with socially conscious consumers.	Actively monitor cultural shifts, social movements, and ethical consumer demands.	Ensuring alignment is genuine and not perceived as performative.
COMPETITIVE BENCHMARKING	**MARKET POSITIONING**	**COMPETITOR ANALYSIS**	**DATA ACCURACY**
Comparing brand strategies with industry leaders.	Informs strategic decisions to differentiate and capture market share.	Analyze competitor messaging, campaigns, and market reception.	Difficulty obtaining comprehensive and unbiased competitive intelligence.

For a business leader, establishing robust decision frameworks is a critical strategic imperative for ensuring consistency, efficiency, and effectiveness in all brand-related decision-making. In complex and fast-paced environments, ad-hoc decisions can lead to inconsistencies and missed opportunities; structured processes provide clarity and accountability. Central to this is prioritizing data-informed decisions. By systematically using comprehensive market research, customer insights, and performance analytics to guide strategic choices, you reduce guesswork, enhance accuracy, and build a culture of evidence-based brand management.

Furthermore, fostering collaborative stakeholder decisions is crucial. Involving key leadership, diverse employees, and even customers in significant brand evolution discussions ensures broader perspectives, builds internal buy-in, and strengthens external resonance. This collective intelligence leads to more robust outcomes. Crucially, integrating ethical and sustainable decision-making principles is vital. Embedding responsible business values—from environmental impact to social equity and transparent practices—into your decision processes ensures that brand growth is achieved not just profitably, but also responsibly, enhancing long-term reputation and trust. Neglecting clear decision frameworks can lead to inconsistent brand messaging, reactive responses to market changes, internal misalignment, and a struggle to adapt to new challenges effectively. By championing structured, data-driven, and values-aligned decision processes, you empower your organization to make more strategic, consistent, and responsible brand choices, ultimately driving sustained growth, relevance, and positive impact for your business.

WHAT?	WHY?	HOW?	BEWARE!
DATA-INFORMED DECISIONS	**ACCURATE STRATEGY**	**ANALYTICS INTEGRATION**	**DATA OVERLOAD**
Using research and insights to guide strategic choices.	Reduces guesswork and enhances the effectiveness of brand decisions.	Ensure data from market research, analytics, and customer feedback for decision-makers.	Interpreting complex data without suffering from analysis paralysis.
STAKEHOLDER DECISIONS	**BROADER BUY-IN**	**COLLABORATIVE WORKSHOPS**	**CONSENSUS**
Involving leadership, employees, and customers in brand evolution.	Ensures decisions are well-rounded and embraced across the organization.	Facilitate workshops and feedback sessions with diverse internal and external stakeholders.	Reaching consensus among numerous, potentially conflicting, stakeholder interests.
ETHICAL & SUSTAINABLE DECISION	**REPUTATION & TRUST**	**ESG GUIDELINES**	**TRADE-OFFS**
Integrating responsible business values.	Builds long-term brand credibility and societal respect.	Clear environmental, social, and governance (ESG) guidelines for all brand-related decisions.	Balancing ethical/sustainable goals with immediate financial pressures.

6x6™ Transformation Blueprint

INITIATION | Self Awareness

"The greatest discovery of all time is that a person can change his future by merely changing his attitude." - Oprah Winfrey.

A media executive, talk show host, and philanthropist, best known for The Oprah Winfrey Show. She built a vast media empire, becoming one of the wealthiest and most influential women globally, transforming media and empowering millions.

The Initiation level is your Self Awareness, marking the first active step in leadership by turning the gaze inward to recognize personal biases. By identifying these leadership tendencies, you catalyze the shift from passive management to intentional authority.

Deeply understanding your inherent strengths, latent weaknesses, core values, emotional triggers, and critically, your profound impact on others. This introspective journey allows for authentic leadership, enabling genuine personal and professional growth, and the strategic adaptation of your leadership style to diverse situations.

“Self Awareness” is the essential mirror that reflects true leadership potential, fostering empathy, improving communication, and driving continuous improvement in all interactions, ultimately building stronger, more cohesive, and effective teams.

The "**Self-Awareness**" chapter is crucial for business strategy maturity as it mobilizes the leader's internal capabilities before directing others. A mature business strategy relies on leaders who deeply understand their own leadership style, strengths, and areas for improvement. This self-awareness enables them to adapt their approach to different strategic challenges, build more effective teams, and make decisions that align with both their personal values and organizational objectives. By fostering introspective leadership, the organization gains agility and resilience, ensuring strategic initiatives are driven by thoughtful, adaptable, and impactful leadership. Here is how we approach our learning:

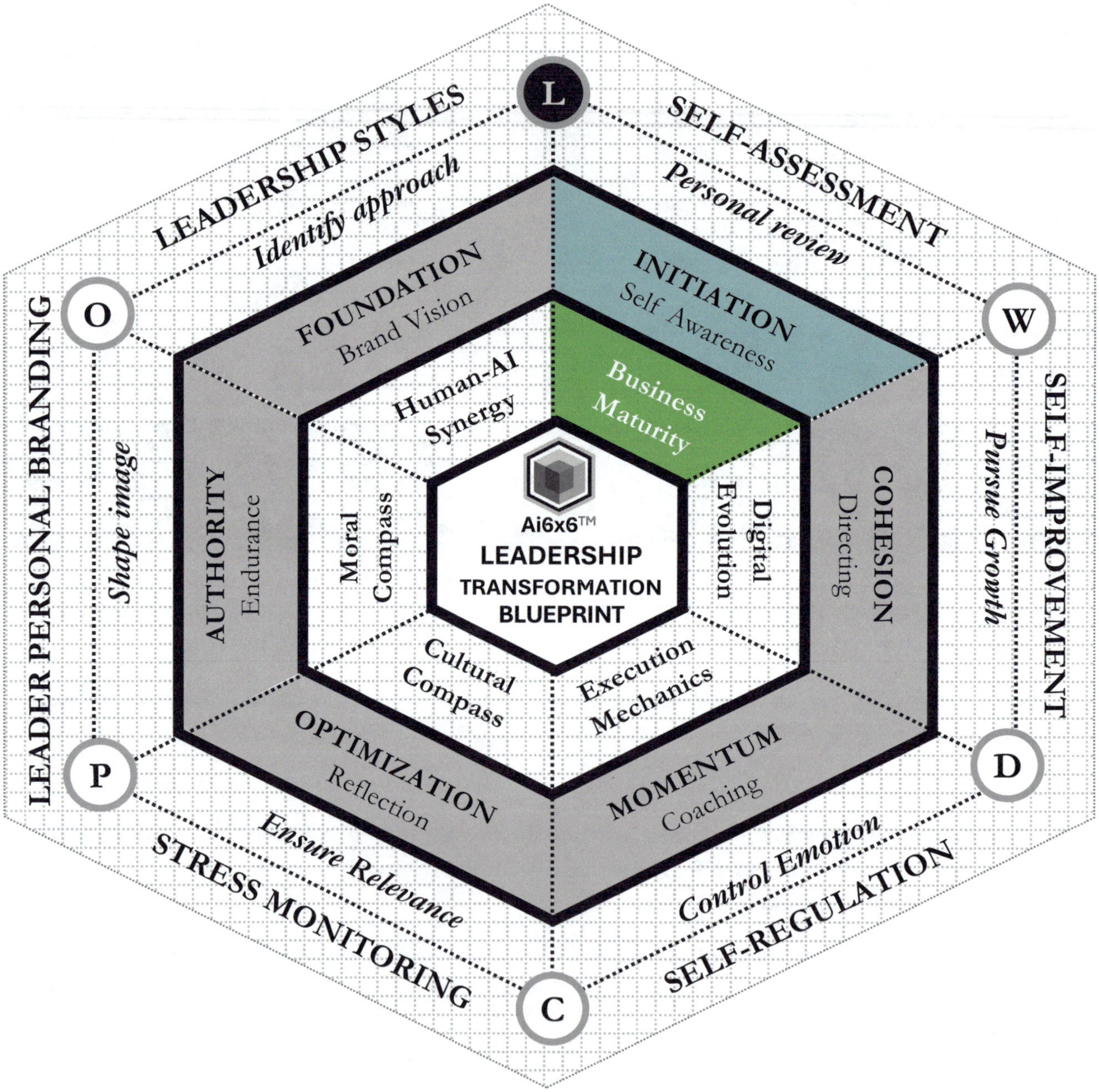

Ai6x6 BUSINESS MATURITY FOR SELF-AWARENESS | LEADERSHIP TRANSFORMATION

For a business leader, deeply understanding your own leadership style is a critical imperative for enhancing effectiveness, optimizing decision-making, and fostering productive team dynamics. Without this self-awareness, you risk unintentional impacts on your team and organization. It's about recognizing your inherent tendencies and how they are perceived by others. Central to this is rigorous style assessment. Identifying your individual leadership approach through formal evaluations, personality insights, and behavioral assessments (e.g., situational leadership, transformational, democratic, autocratic) provides clarity on your default modes of operation and how they influence your interactions. Furthermore, comprehensive executive profiling is crucial. This involves analyzing your strengths (e.g., strategic thinking, empathy), weaknesses (e.g., micromanagement, impulsiveness), and behavioral tendencies in various leadership roles. This objective assessment helps you leverage your advantages and address areas for development.

Crucially, developing competency modeling helps define the key skills and attributes required for effective leadership in different organizational contexts. By understanding the ideal competencies for your role and organization, you can benchmark your own style against these requirements, identifying specific areas for growth. Neglecting to understand your leadership style can lead to communication breakdowns, misaligned team expectations, ineffective decision-making, and a struggle to adapt to diverse situations. By championing a deep understanding of your personal leadership tendencies, you empower yourself to lead more intentionally, adapt your approach as needed, and ultimately drive greater team performance and organizational success.

WHAT?	WHY?	HOW?	BEWARE!
STYLE ASSESSMENT	**CLARITY & INSIGHT**	**ASSESSMENT TOOLS**	**BIAS**
Identifying individual leadership approaches.	Provides a clear understanding of personal leadership tendencies.	Utilize psychometric tests, 360-degree feedback, and self-reflection exercises.	Self-perception bias in assessments.
EXECUTIVE PROFILING	**OPTIMIZED IMPACT**	**COACHING & MENTORING**	**VULNERABILITY**
Analyzing strengths, weaknesses, and behavioral tendencies.	Helps leverage strengths and address areas for development.	Engage with executive coaches or mentors for objective feedback and insights.	Being open to critical feedback and acknowledging weaknesses.
COMPETENCY MODELING	**TARGETED DEVELOPMENT**	**ROLE ANALYSIS**	**GENERICISM**
Defining key skills and attributes for effective leadership.	Guides personal growth towards ideal leadership competencies.	Define the key skills and behaviors required for success in specific leadership roles.	Creating models that are too generic and not specific enough for the context.

For a business leader, engaging in continuous self-assessment is a fundamental strategic imperative for refining personal strengths, addressing weaknesses, and enhancing overall leadership effectiveness. Without this honest introspection, growth stagnates, and blind spots can impede progress. It's about building a consistent habit of looking inward to drive outward improvement. Central to this is leveraging 360-degree feedback. Gathering input from peers, mentors, direct reports, and even external stakeholders provides a holistic view of your impact and perception, revealing areas where your self-perception aligns or diverges from others' experiences. This multi-dimensional feedback is invaluable for identifying blind spots.

Furthermore, conducting a personal SWOT (Strengths, Weaknesses, Opportunities, Threats) analysis is crucial. Systematically evaluating your internal strengths (e.g., communication skills, strategic vision) and weaknesses (e.g., delegation issues, impulsiveness) alongside external opportunities (e.g., market trends, new technologies) and threats (e.g., competitive pressures, industry shifts) provides a comprehensive framework for personal and professional development. Crucially, committing to continuous self-evaluation involves developing consistent habits to regularly reassess your leadership effectiveness. This might include journaling, periodic self-reflection exercises, or reviewing performance against personal goals, fostering an ongoing cycle of learning and adaptation. Neglecting self-assessment can lead to stagnation, unchallenged assumptions about one's effectiveness, strained team relationships, and missed opportunities for growth. By championing a disciplined approach to self-reflection, you empower yourself to lead with greater awareness, & drive superior performance for your team and organization.

WHAT?	WHY?	HOW?	BEWARE!
360-DEGREE FEEDBACK	**HOLISTIC PERCEPTION**	**SURVEY PLATFORMS**	**DEFENSIVENESS**
Gathering input from peers, mentors, and teams.	Provides a comprehensive view of leadership impact and perception.	Utilize 360-degree feedback tools and solicit input from diverse stakeholders.	Leaders potentially reacting defensively to critical feedback.
PERSONAL SWOT ANALYSIS	**STRATEGIC DEVELOPMENT**	**FRAMEWORK APPLICATION**	**OBJECTIVITY**
Evaluating internal strengths and weaknesses alongside external factors.	Identifies areas for personal growth and leverages opportunities.	Apply the SWOT framework to personal skills, career, and leadership context.	Maintaining objectivity when self-assessing personal weaknesses.
CONTINUOUS SELF-EVALUATION	**ONGOING GROWTH**	**REGULAR REFLECTION**	**CONSISTENCY**
Developing habits to reassess leadership effectiveness regularly.	Fosters a proactive approach to sustained leadership development.	Dedicate time for weekly or monthly self-reflection sessions and journaling.	Maintaining discipline and consistency in regular self-evaluation.

For a business leader, a proactive commitment to self-improvement is a continuous strategic imperative for ongoing leadership development. In a dynamic business landscape, leadership skills are not static; they must evolve to meet new challenges and opportunities. It's about systematically setting goals and strategies to refine your capabilities. Central to this is establishing clear skill development plans. This involves identifying specific areas for growth (e.g., public speaking, strategic foresight, team motivation) and creating structured learning paths through courses, workshops, or practical experience to enhance these leadership capabilities.

Furthermore, engaging in coaching and mentorship is crucial. Collaborating with experienced leaders, executive coaches, or industry mentors provides invaluable guidance, objective feedback, and a safe space to discuss challenges and refine decision-making, accelerating your growth trajectory. Crucially, cultivating an adaptive mindset is vital. Embracing challenges as learning opportunities, being open to new ideas, and continuously seeking to evolve your leadership approaches ensures that you remain flexible and resilient in the face of change. This mindset transforms obstacles into stepping stones for growth. Neglecting self-improvement can lead to stagnation, a failure to lead effectively in new contexts, and a growing disconnect with evolving team and organizational needs. By championing a disciplined and curious approach to ongoing leadership development, you empower yourself to continuously grow, inspire your teams, and ultimately drive greater organizational success through increasingly effective and adaptable leadership.

WHAT?	WHY?	HOW?	BEWARE!
SKILL DEVELOPMENT PLANS	**CAPABILITY ENHANCEMENT**	**TRAINING PROGRAMS**	**TIME COMMITMENT**
Establishing structured learning paths.	Systematically improves specific leadership skills.	Enroll in online courses, workshops, or executive education programs.	Balancing demanding leadership responsibilities with dedicated learning time.
COACHING AND MENTORSHIP	**ACCELERATED GROWTH**	**MENTOR IDENTIFICATION**	**FINDING FIT**
Engaging with experienced leaders for growth and guidance.	Provides personalized feedback and actionable insights.	Seek out experienced leaders or professional coaches for guidance.	Finding the right coach or mentor whose style aligns with your needs.
ADAPTIVE MINDSET	**RESILIENCE & EVOLUTION**	**REFLECTION PRACTICES**	**COMFORT ZONE**
Embracing challenges and learning opportunities.	Fosters flexibility and continuous growth in leadership approach.	Practice mindfulness, seek feedback, and reframe failures as learning opportunities.	Overcoming the natural human tendency to stick to familiar approaches.

4. Self-regulation

For a business leader, mastering self-regulation is a critical strategic imperative for effectively managing emotions, impulses, and reactions, especially in high-pressure leadership scenarios. Your ability to remain composed and make rational decisions under duress directly impacts team morale, strategic outcomes, and your overall credibility. Central to this is cultivating emotional intelligence awareness. Understanding your personal triggers—what makes you frustrated, anxious, or defensive—and recognizing how these emotions influence your leadership decisions allows for greater control and more measured responses.

Furthermore, applying effective conflict resolution techniques is crucial. Developing mindful communication strategies, active listening skills, and the ability to de-escalate disagreements ensures that conflicts within teams are managed effectively, maintaining cohesion and morale rather than allowing emotions to derail productive discussions. Crucially, strengthening your ability to make decisions under pressure is vital. This involves developing strategies to stay focused, rational, and decisive in high-stakes situations, preventing impulsivity or paralysis and ensuring that critical choices are made with clarity and foresight. Neglecting self-regulation can lead to impulsive decisions, strained interpersonal relationships, a loss of team trust, and a reactive leadership style that struggles in times of crisis. By championing emotional intelligence and disciplined self-management, you empower yourself to lead with greater composure, foster a more positive team environment, and ultimately make more effective decisions that drive sustained organizational success, even amidst significant challenges.

WHAT?	WHY?	HOW?	BEWARE!
EMOTIONAL INTELLIGENCE AWARENESS	**MEASURED RESPONSES**	**EI ASSESSMENTS**	**HONESTY**
Understanding personal triggers.	Enables greater control over emotions in leadership decisions.	Use emotional intelligence assessments and reflective journaling.	Being brutally honest about personal emotional triggers.
CONFLICT RESOLUTION TECHNIQUES	**TEAM COHESION**	**COMMUNICATION TRAINING**	**DE-ESCALATION**
Applying mindful communication strategies.	Manages disagreements constructively, maintaining morale and productivity.	Engage in training for active listening, empathy, and negotiation skills.	Successfully de-escalating highly emotional or entrenched conflicts.
DECISION-MAKING UNDER PRESSURE	**CRITICAL LEADERSHIP**	**SIMULATION & PRACTICE**	**REAL-WORLD STRESS**
Strengthening the ability to stay focused and rational.	Ensures effective and clear decision-making in high-stakes situations.	Engage in simulated crisis scenarios and decision-making drills.	Replicating the true stress of real high-pressure situations.

For a business leader, proactive stress monitoring is a critical strategic imperative for maintaining personal resilience and balance while navigating the immense pressures of leading teams and making critical decisions. Leadership is demanding, and without conscious management of stress, burnout can severely impact performance, judgment, and overall well-being. Central to this is rigorous workload management. Prioritizing tasks effectively, delegating appropriately, and understanding capacity limits are crucial for reducing burnout risks and ensuring that energy is focused on the most impactful activities.

Furthermore, integrating mindfulness and relaxation techniques is vital. Implementing practices such as meditation, deep breathing exercises, or regular breaks helps maintain mental and emotional well-being, enhancing clarity of thought and reducing the detrimental effects of chronic stress. Crucially, developing robust resilience-building strategies is key. Strengthening adaptability through consistent self-care, establishing healthy boundaries between work and personal life, and implementing effective stress-management systems (e.g., exercise, hobbies, support networks) allows you to bounce back from setbacks and sustain performance over the long term. Neglecting stress monitoring can lead to poor decision-making, strained relationships, high turnover within your team due to a stressed leader, and ultimately, a decline in organizational performance. By championing personal well-being and proactive stress management, you empower yourself to lead more effectively, inspire a healthier team culture, and ensure your sustained effectiveness as a leader, contributing to your organization's overall success and stability.

WHAT?	WHY?	HOW?	BEWARE!
WORKLOAD MANAGEMENT	**SUSTAINED PERFORMANCE**	**TIME MANAGEMENT**	**OVERCOMMITMENT**
Prioritizing tasks effectively to reduce burnout risks.	Ensures energy is focused on high-impact activities.	Implement effective prioritization frameworks (e.g., Eisenhower Matrix) and delegation.	Difficulty saying "no" to new responsibilities or demands.
MINDFULNESS & RELAXATION	**MENTAL CLARITY**	**DAILY PRACTICE**	**DISCIPLINE**
Implementing practices to maintain well-being.	Enhances focus, reduces stress, and improves decision-making.	Integrate short meditation sessions, deep breathing, or regular breaks.	Maintaining consistency in mindfulness practices amidst busy schedules.
RESILIENCE-BUILDING STRATEGIES	**LONG-TERM EFFECTIVENESS**	**HOLISTIC APPROACH**	**GUILT**
Strengthening adaptability through self-care.	Ensures ability to bounce back from setbacks and sustain leadership.	Prioritize sleep, nutrition, exercise, hobbies, and a strong support network.	Feeling guilty about taking time for self-care during demanding periods.

6. Leader personal branding

For a business leader, consciously shaping a distinct personal brand is a critical strategic imperative for cultivating a powerful leadership identity and professional presence. In today's interconnected world, your personal brand is a reflection of your values, expertise, and leadership philosophy, influencing your credibility, influence, and ability to attract talent and opportunities. Central to this is authentic leadership messaging. Consistently communicating your core values, leadership philosophy, and unique perspective across all platforms—from internal meetings to public appearances—builds trust and establishes your genuine voice, resonating with those you lead and engage with.

Furthermore, meticulously managing your public presence and reputation is crucial. Actively cultivating your online and offline branding through thought leadership, speaking engagements, and media interactions enhances your credibility, amplifies your influence, and ensures your professional image aligns with your aspirations. Crucially, engaging in strategic networking and thought leadership activities is vital. Establishing yourself as an authority in your industry or organizational circles through publishing articles, speaking at conferences, or participating in key forums builds recognition, expands your sphere of influence, and positions you as a visionary leader. Neglecting personal branding can lead to an inconsistent professional image, limited influence, missed opportunities for career advancement, and a struggle to inspire followership. By championing a proactive and authentic approach to shaping your leadership brand, you empower yourself to lead with greater impact, attract high-caliber talent, and ultimately secure your long-term success and legacy as a respected and influential leader.

WHAT?	WHY?	HOW?	BEWARE!
AUTHENTIC LEADERSHIP	**TRUST & INFLUENCE**	**VALUE ARTICULATION**	**CONSISTENCY**
Communicating core values and philosophy consistently.	Builds credibility and resonates deeply with followers.	Clearly define and consistently express your leadership philosophy and values.	Maintaining authenticity across all internal and external communications.
PUBLIC PRESENCE AND REPUTATION	**ENHANCED CREDIBILITY**	**THOUGHT LEADERSHIP**	**TIME COMMITMENT**
Managing online and offline branding.	Amplifies influence and attracts opportunities.	Publish articles, speak at conferences, and engage in relevant public discourse.	Building a strong public presence requires significant time and effort.
NETWORKING AND THOUGHT LEADERSHIP	**EXPANDED INFLUENCE**	**STRATEGIC CONNECTIONS**	**GENUINENESS**
Establishing authority and recognition.	Positions you as a visionary leader within industry circles.	Actively engage with peers, industry leaders, and professional orgs.	Ensuring networking efforts are genuine and not purely transactional.

6x6™ Transformation Blueprint

COHESION | Directing Leaders

"Management is doing things right; leadership is doing the right things."
- Peter Drucker.

Widely regarded as the father of modern management, his groundbreaking theories on business management and organizational behavior profoundly shaped corporate strategy & leadership practices. He introduced pivotal concepts like "management by objectives."

The Cohesion level is your Directing phase, achieved when individual leadership goals align with a unified organizational voice. Through unified communication channels, you ensure that the vision is not just stated but integrated into every layer of the company.

Clearly guiding and instructing teams with precision, setting unambiguous expectations, and ensuring that tasks are not only understood but also executed efficiently and accurately. This phase is vital for initial team development, establishing a robust operational structure, and achieving immediate, tangible objectives with maximum efficiency.

"Directing" lays the groundwork for subsequent phases by ensuring foundational tasks are completed correctly, building team discipline, initial momentum, and a clear understanding of roles and responsibilities within the team.

"Directing" is vital for business strategy maturity because it represents the active function of translating strategic plans into tangible actions and getting work done effectively. A mature business strategy requires leaders who can effectively convey ideas, provide clear guidance, inspire trust, and motivate teams to act decisively. This direct leadership ensures that strategic objectives are clearly understood and executed efficiently across the organization. It's about maintaining momentum, overcoming obstacles, and ensuring that strategic initiatives gain traction, accelerating the pace at which the business achieves its goals and adapts to market demands. Here is how we approach our learning:

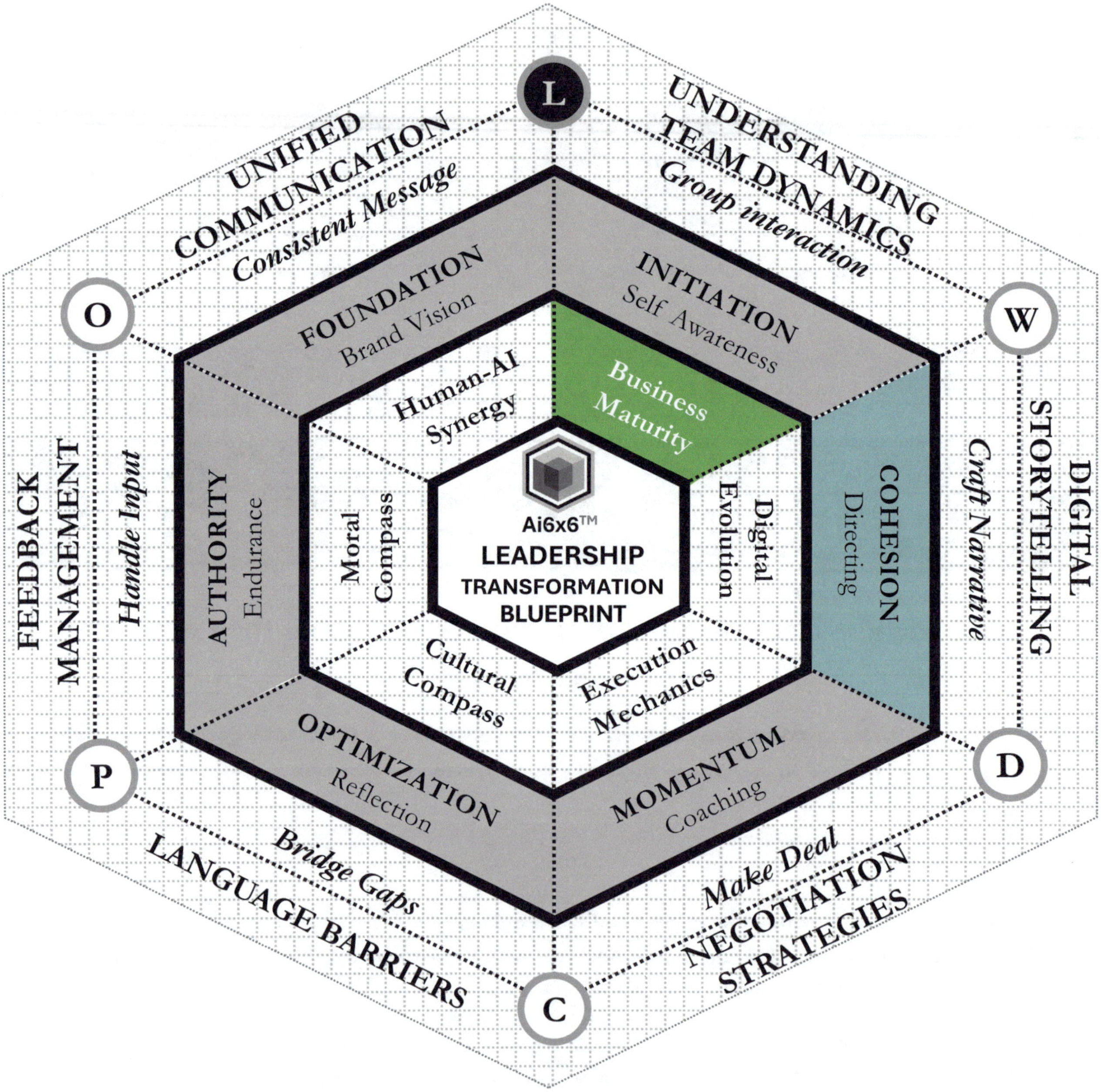

Ai6x6 BUSINESS MATURITY FOR DIRECTING
LEADERSHIP TRANSFORMATION

For a business leader, establishing unified communication is a critical strategic imperative for ensuring seamless and consistent interaction across all leadership channels. In today's hybrid work environments, clear and cohesive communication is the bedrock of organizational alignment, preventing confusion and fostering trust. Your ability to convey a single, coherent message, regardless of the platform, is paramount. Central to this is achieving meticulous cross-platform messaging. Ensuring alignment and consistency between communications sent via email, chat platforms, video conferencing tools, and collaborative documents prevents fragmentation and reinforces a cohesive message, regardless of where your team members engage.

Furthermore, prioritizing leadership transparency is crucial. Communicating decisions, goals, and strategic updates clearly and consistently across multiple formats builds trust, reduces speculation, and ensures that all employees feel informed and connected to the organizational direction. Crucially, developing adaptive communication methods is vital. This involves adjusting your tone, style, and chosen medium based on the context of the message and the specific needs and preferences of your audience. This flexibility ensures your message is not just delivered, but truly understood and absorbed. Neglecting unified communication can lead to misinformation, internal silos, confusion among teams, and a struggle to execute strategic initiatives effectively. By championing a consistent, transparent, and adaptable approach to leadership communication, you empower your organization to operate with greater clarity, accelerate decision-making, and build a stronger, more engaged workforce, ultimately driving sustained organizational success.

WHAT?	WHY?	HOW?	BEWARE!
CROSS-PLATFORM MESSAGING	**CONSISTENCY**	**INTEGRATED TOOLS**	**TOOL FRAGMENTATION**
Ensuring alignment across email, chat, video conferencing, and collaborative tools.	Prevents fragmentation and reinforces a cohesive message.	Utilize communication platforms that allow for seamless cross-channel messaging and archiving.	Managing consistency across many disparate communication tools.
LEADERSHIP TRANSPARENCY	**TRUST & ALIGNMENT**	**REGULAR UPDATES**	**INFORMATION OVERLOAD**
Communicating decisions and goals clearly across multiple formats.	Builds trust and ensures employees feel informed and connected.	Establish a cadence for company-wide updates (e.g., town halls, newsletters).	Balancing transparency with avoiding overwhelming employees.
ADAPTIVE COMMUNICATION	**RELEVANCE & UNDERSTANDING**	**AUDIENCE ANALYSIS**	**PERCEPTION**
Adjusting tone and style based on context and audience needs.	Ensures messages are not just delivered but truly understood.	Understand audience preferences and tailor message delivery (e.g., direct, empathetic, formal).	Leaders being perceived as inauthentic if adapting too much.

For a business leader, a deep understanding of team dynamics is a critical strategic imperative for navigating interpersonal relationships effectively and ensuring that leadership communication truly lands and motivates. A team is more than a collection of individuals; its underlying dynamics profoundly influence collaboration, morale, and performance. Central to this is mastering active listening techniques. Strengthening engagement and building trust through intentional, empathetic listening ensures that you fully grasp team members' perspectives, concerns, and ideas, fostering a sense of psychological safety and mutual respect. Furthermore, adept conflict resolution within teams is crucial. Addressing disagreements proactively and constructively, while maintaining cohesion and morale, prevents minor issues from escalating and ensures that team energy is directed towards productive outcomes rather than internal friction.

Crucially, developing role-based communication strategies is vital. Tailoring your messaging and approach for different team structures, individual roles, and departmental responsibilities ensures that communication is relevant, clear, and actionable for each specific audience, maximizing understanding and effectiveness. Neglecting team dynamics can lead to misunderstandings, unresolved conflicts, low morale, and ultimately, a breakdown in communication that hinders collective progress. By championing a nuanced understanding of interpersonal relationships and adapting your communication accordingly, you empower your organization to foster stronger, more cohesive teams, accelerate collaboration, and achieve shared goals with greater efficiency and harmony. This human-centered approach is fundamental to effective leadership.

WHAT?	WHY?	HOW?	BEWARE!
ACTIVE LISTENING TECHNIQUES	**DEEPER UNDERSTANDING**	**PRACTICE & FEEDBACK**	**DISTRACTION**
Strengthening engagement and trust.	Ensures accurate comprehension of team members' perspectives.	Consciously practice empathetic listening, summarize, and ask clarifying questions.	Maintaining focus in busy or high-stress environments.
TEAM CONFLICT RESOLUTION	**MORALE & PRODUCTIVITY**	**MEDIATION SKILLS**	**EMOTION**
Addressing disagreements while maintaining cohesion.	Prevents internal friction and maintains team effectiveness.	Train leaders in mediation, negotiation, and de-escalation techniques.	Managing strong emotions during conflict resolution.
ROLE-BASED COMMUNICATION STRATEGIES	**RELEVANCE & CLARITY**	**AUDIENCE SEGMENTATION**	**OVER-COMPLEXITY**
Tailoring messaging for different team structures.	Ensures messages are understood and acted upon by specific roles.	Analyze team structures and roles to customize communication style and content.	Over-tailoring leading to fragmented or inconsistent messages.

3. Digital storytelling

For a business leader, strategically embracing digital storytelling is a critical imperative for crafting compelling leadership narratives that inspire, inform, and engage teams and stakeholders in today's visually and digitally saturated world. Beyond traditional presentations, technology offers powerful new avenues to communicate your vision and values. Central to this is leveraging multimedia content creation. Using video, dynamic graphics, interactive presentations, and immersive formats allows you to convey impactful messages with greater emotional resonance and memorability, cutting through noise and capturing attention more effectively than static text.

Furthermore, exploring multimodal deep learning insights is crucial. By utilizing AI-driven analytics, you can enhance storytelling effectiveness and personalize narratives, understanding which visual cues, linguistic styles, and emotional tones resonate most deeply with specific audiences, tailoring your message for maximum impact. Crucially, prioritizing authenticity verification in your digital storytelling ensures brand and leadership credibility. In an age of deepfakes and misinformation, transparent and verifiable digital narratives build trust, reinforcing your genuine voice and commitment to truth. Neglecting digital storytelling can lead to flat, uninspiring communications that fail to connect with modern audiences, losing out on opportunities to rally support and convey complex ideas effectively. By championing creative, tech-driven, and authentic digital narratives, you empower your organization to communicate its vision with greater impact, inspire action, and build a powerful, memorable brand presence, ultimately driving sustained organizational success through compelling leadership communication.

WHAT?	WHY?	HOW?	BEWARE!
MULTIMEDIA CONTENT CREATION	**ENHANCED ENGAGEMENT**	**TOOL UTILIZATION**	**RESOURCE INTENSITY**
Leveraging video, graphics, and presentations for impactful messaging.	Conveys messages with greater emotional resonance and memorability.	Employ video editing software, graphic design tools, and interactive presentation platforms.	High time and skill investment for producing high-quality multimedia.
MULTIMODAL DEEP LEARNING	**PERSONALIZED IMPACT**	**AI ANALYTICS**	**DATA PRIVACY**
Using AI-driven insights to enhance storytelling effectiveness.	Tailors narratives for maximum resonance with specific audiences.	Integrate AI tools to analyze audience engagement with different content formats and styles.	Balancing data-driven personalization with audience privacy concerns.
AUTHENTICITY VERIFICATION	**TRUST BUILDING**	**TRANSPARENCY PROTOCOLS**	**SKEPTICISM**
Ensuring brand and leadership credibility through transparent digital storytelling.	Reinforces genuine voice and commitment to truth in an era of misinformation.	Implement clear guidelines for content sourcing and disclosure of AI usage.	Overcoming public skepticism about AI-generated or enhanced content.

For a business leader, mastering robust negotiation strategies is a critical imperative for leading discussions with confidence, strategically positioning your organization, and achieving favorable outcomes. Effective negotiation is not just about reaching an agreement, but about maximizing value, building relationships, and advancing your strategic objectives. Central to this is employing persuasive communication techniques. Framing arguments effectively, understanding your counterpart's motivations, and clearly articulating your value proposition allows you to influence outcomes and guide discussions towards mutually beneficial solutions with greater impact.

Furthermore, striving for win-win negotiation models is crucial. Actively seeking mutually beneficial solutions in decision-making scenarios fosters long-term relationships, builds trust, and creates more sustainable agreements rather than zero-sum outcomes that can breed resentment. Crucially, adeptly navigating power dynamics is vital. Managing hierarchical challenges, understanding influence structures, and maintaining fairness in communication ensures that negotiations are conducted ethically and effectively, regardless of perceived power imbalances. Neglecting strong negotiation skills can lead to suboptimal deals, strained partnerships, missed opportunities, and an inability to drive your agenda effectively. By championing disciplined, empathetic, and strategically informed negotiation approaches, you empower yourself to achieve superior outcomes, build stronger alliances, and ultimately drive greater organizational success through effective and impactful leadership.

WHAT?	WHY?	HOW?	BEWARE!
PERSUASIVE COMMUNICATION	**IMPACTFUL INFLUENCE**	**ARGUMENT STRUCTURING**	**EMOTIONAL BARRIERS**
Framing arguments effectively to influence outcomes.	Guides discussions towards desired results with greater impact.	Practice framing arguments, active listening, and understanding counterpart's needs.	Managing emotions during high-stakes discussions.
WIN-WIN NEGOTIATION MODELS	**LONG-TERM RELATIONSHIPS**	**BATNA (Best Alternative To a Negotiated Agreement)**	**COMPROMISE AVERSION**
Finding mutually beneficial solutions.	Fosters trust and creates more sustainable agreements.	Identify alternatives to ensure mutually beneficial outcomes.	Reluctance to compromise, leading to stalemates.
NAVIGATING POWER DYNAMICS	**ETHICAL OUTCOMES**	**POSITIONAL AWARENESS**	**EXPLOITATION**
Managing hierarchical challenges while maintaining fairness.	Ensures fair and effective negotiations regardless of perceived power.	Understand formal and informal power dynamics and adapt communication style.	Parties exploiting power imbalances unfairly.

For a business leader, actively addressing language barriers is a critical strategic imperative for overcoming communication obstacles in diverse leadership environments, ensuring clarity, inclusivity, and effective collaboration. In today's globalized workforce, miscommunication due to linguistic differences can lead to inefficiencies, misunderstandings, and eroded trust. Central to this is implementing robust cross-cultural communication strategies. Adapting your messaging, non-verbal cues, and communication style to different linguistic and cultural contexts ensures that your message is not just translated, but truly understood and respectfully received by diverse teams.

Furthermore, leveraging translation and interpretation tools is crucial. Utilizing AI-driven language processing software for documents, real-time captioning in video calls, or professional interpreters for critical discussions enables seamless interactions across linguistic divides, enhancing accessibility and comprehension. Crucially, emphasizing non-verbal communication is vital for strengthening understanding. Focusing on gestures, facial expressions, tone of voice, and body language can convey meaning and emotion beyond spoken words, providing additional context and fostering connection, especially when verbal communication is challenging. Neglecting language barriers can lead to misinterpretations, reduced team productivity, exclusion of diverse voices, and a struggle to lead effectively across international boundaries. By championing a proactive, culturally sensitive, and technologically augmented approach to communication, you empower your organization to build truly inclusive teams, foster clearer understanding, and ultimately drive greater operational efficiency and strategic alignment in a globalized world.

WHAT?	WHY?	HOW?	BEWARE!
CROSS-CULTURAL COMMUNICATION	**CULTURAL SENSITIVITY**	**CULTURAL TRAINING**	**STEREOTYPING**
Adapting messaging to different linguistic and cultural contexts.	Ensures messages are understood and received respectfully.	Provide training on cultural nuances, communication styles, and etiquette.	Risk of over-generalizing cultural differences.
TRANSLATION AND INTERPRETATION TOOLS	**ACCESSIBILITY**	**TECHNOLOGY ADOPTION**	**NUANCE**
Using AI-driven language processing for seamless interactions.	Enables clear communication across diverse linguistic backgrounds.	Utilize real-time translation apps, captioning software, and professional interpreters.	Automated tools sometimes missing subtle linguistic nuances or context.
NON-VERBAL COMMUNICATION EMPHASIS	**ENHANCED CLARITY**	**AWARENESS & PRACTICE**	**MIS-INTERPRETATION**
Strengthening understanding through gestures, expressions, and tone.	Conveys meaning and emotion beyond spoken words.	Consciously focus on body language, eye contact, and tone during interactions.	Non-verbal cues varying across cultures and leading to misunderstandings.

For a business leader, implementing robust feedback management systems is a critical strategic imperative for continuously refining leadership effectiveness, enhancing communication, and fostering a culture of growth. It's about systematically collecting, analyzing, and acting upon feedback from all levels of the organization. Central to this is leveraging sentiment analysis for leadership insights. By using AI tools to gauge employee and stakeholder perceptions from text-based feedback (e.g., surveys, comments, internal communications), you can gain a deeper understanding of morale, engagement, and the emotional impact of your leadership actions.

Furthermore, adopting constructive criticism frameworks is crucial. Delivering feedback in a way that is clear, specific, actionable, and focuses on growth rather than blame fosters trust and encourages continuous improvement within individuals and teams. Crucially, establishing continuous feedback loops ensures regular evaluation and improvement in communication effectiveness. Implementing systems for ongoing, real-time feedback—rather than just annual reviews—allows for agile adjustments to your leadership approach, ensuring responsiveness and continuous learning. Neglecting effective feedback management can lead to unaddressed issues, stagnant leadership styles, low employee morale, and a missed opportunity for continuous organizational improvement. By championing a structured, empathetic, and data-driven approach to feedback, you empower your organization to learn from every interaction, build a culture of open communication, and ultimately drive sustained leadership effectiveness and organizational success.

WHAT?	WHY?	HOW?	BEWARE!
SENTIMENT ANALYSIS	**DEEPER UNDERSTANDING**	**AI TOOLS**	**ACCURACY**
Using AI tools to gauge perceptions.	Reveals underlying emotions and perceptions of leadership impact.	Integrate AI/NLP tools to analyze unstructured feedback (comments, reviews) for sentiment.	AI struggling with sarcasm, irony, or highly nuanced language.
CONSTRUCTIVE CRITICISM FRAMEWORKS	**IMPACTFUL DEVELOPMENT**	**TRAINING**	**VULNERABILITY**
Delivering feedback to foster growth and collaboration.	Ensures feedback leads to positive-growth and avoids defensiveness.	Provide training on models like "SBI" (Situation-Behavior-Impact) for delivering feedback.	Leaders feeling uncomfortable giving or receiving critical feedback.
CONTINUOUS FEEDBACK LOOPS	**AGILE ADAPTATION**	**FREQUENT CHECK-INS**	**FEEDBACK FATIGUE**
Establishing systems for regular evaluation and improvement.	Enables ongoing refinement of leadership effectiveness and communication.	Implement regular 1:1s, pulse surveys, and peer feedback mechanisms.	Overwhelming employees or leaders with too many feedback requests.

6x6™ Transformation Blueprint

MOMENTUM | Coaching

"My job is not to be easy on people. My job is to make them better."
- Steve Jobs.

Co-founder of Apple Inc., a visionary entrepreneur who revolutionized personal computing, digital music, and mobile phones. His relentless pursuit of innovation, meticulous design, and focus on user experience transformed multiple industries, making Apple an iconic company.

The Momentum level is your Coaching engine, generated when leadership shifts from directing individuals to empowering them through growth. By utilizing mentoring and executive education, you accelerate the collective intelligence of the organization.

Empowering individuals and teams through the art of active, empathetic listening, posing insightful and thought-provoking questions, and fostering their innate growth, problem-solving capabilities, and self-reliance. This approach fundamentally transforms leadership from a hierarchical "telling" to an enabling "guiding."

"Coaching" builds resilient, highly capable, and ultimately self-sufficient teams ready to tackle complex challenges, innovate solutions, and drive their own development, cultivating an environment of continuous learning and empowerment.

The "Coaching" chapter is paramount for business strategy maturity as it focuses on optimizing individual and team performance, fostering leadership in others. A mature business strategy is executed by capable and growing teams, and leaders who excel at coaching support their team members' growth, provide targeted guidance, and empower future leaders. This capability ensures continuous skill development and succession planning, building a robust talent pipeline that can adapt to evolving strategic needs. It's about multiplying leadership capacity and fostering an environment where strategic execution is consistently enhanced through empowered individuals. Here is how we approach our learning:

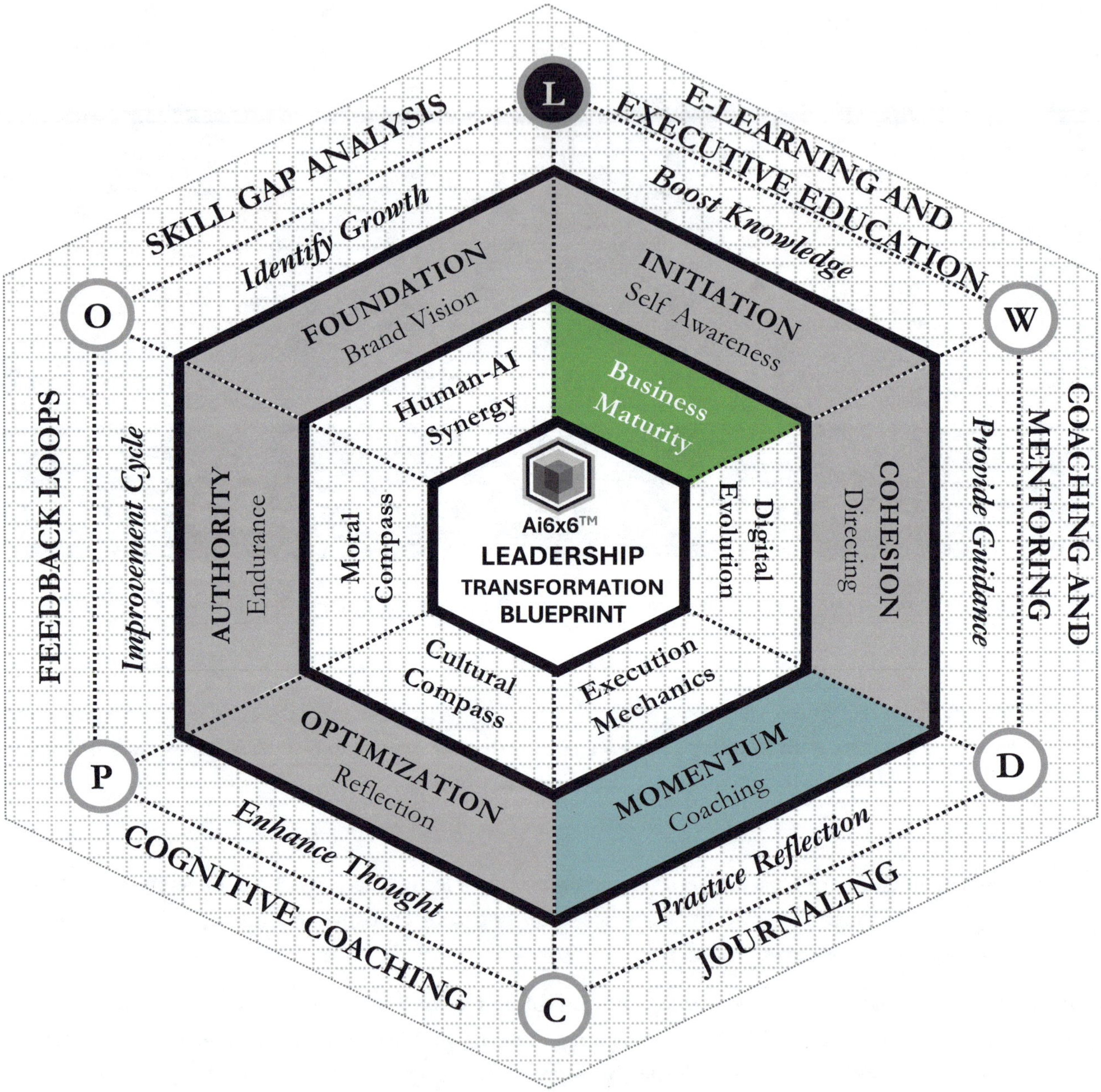

Ai6x6 BUSINESS MATURITY FOR COACHING
LEADERSHIP TRANSFORMATION

For a business leader, meticulously conducting skill gap analysis is a critical strategic imperative for identifying specific areas for growth and improvement in leadership capabilities, both for yourself and your team members. This systematic approach ensures that development efforts are targeted, efficient, and directly contribute to enhanced leadership effectiveness. Central to this is comprehensive competency assessment. Evaluating existing leadership strengths and weaknesses through structured analysis (e.g., behavioral interviews, simulations, 360-degree feedback) provides an objective baseline, pinpointing precise areas where development is most needed.

Furthermore, rigorous industry benchmarking is crucial. Comparing your leadership skills and those of your team against market standards and evolving leadership trends (e.g., digital transformation leadership, empathetic leadership) provides valuable external context, highlighting competencies that are becoming increasingly vital for success. Crucially, creating personalized development plans is vital. Tailoring strategies to address identified leadership deficiencies ensures that training, coaching, and growth opportunities are relevant and impactful for each individual, maximizing the return on development investment. Neglecting skill gap analysis can lead to generic training, misallocated development resources, persistent leadership weaknesses, and a struggle to adapt to new leadership challenges. By championing a data-driven approach to identifying and addressing skill gaps, you empower your organization to build a stronger, more capable leadership pipeline, accelerating individual growth and ultimately driving greater organizational success through enhanced leadership effectiveness.

WHAT?	WHY?	HOW?	BEWARE!
COMPETENCY ASSESSMENT	**TARGETED DEVELOPMENT**	**ASSESSMENT TOOLS**	**SUBJECTIVITY**
Evaluating leadership strengths and weaknesses.	Pinpoints specific areas where development is most needed.	Use 360-degree feedback, behavioral interviews, and simulations.	Potential for bias in subjective assessments.
INDUSTRY BENCHMARKING	**COMPETITIVE READINESS**	**EXTERNAL RESEARCH**	**DATA ACCESS**
Comparing skills against market standards.	Identifies crucial skills for future leadership success.	Analyze leadership frameworks, industry reports, and competitor leadership profiles.	Difficulty obtaining reliable benchmark data for specific leadership roles.
PERSONALIZED DEVELOPMENT PLANS	**INDIVIDUAL GROWTH**	**COACHING & TRAINING**	**TIME & RESOURCES**
Creating tailored strategies.	Ensures development efforts are relevant and impactful for each leader.	Develop customized learning paths, coaching engagements, and practical assignments.	Ensuring sufficient time and resources for individual development plans.

For a business leader, strategically leveraging e-learning and executive education programs is a critical imperative for continuously enhancing leadership knowledge, fostering innovation, and staying competitive in a rapidly changing world. Digital tools and structured programs offer flexible and scalable avenues for ongoing development. Central to this is exploring comprehensive online leadership courses. Providing access to structured learning paths for continuous skill development—from strategic management to emotional intelligence and digital transformation—ensures leaders can acquire new competencies and refresh existing knowledge at their own pace and convenience.

Furthermore, engaging in advanced executive education programs is crucial. These specialized training initiatives, often from top universities or business schools, provide immersive experiences tailored for senior leadership growth, fostering critical thinking, strategic networking, and exposure to cutting-edge business concepts. Crucially, integrating microlearning and adaptive learning methodologies is vital for practical application. Utilizing bite-sized, data-driven educational content ensures that learning is digestible, highly relevant, and immediately applicable to real-world leadership challenges, maximizing retention and behavior change. Neglecting continuous leadership education can lead to outdated skills, a struggle to lead effectively in new contexts, missed opportunities for innovation, and a reactive approach to market shifts. By championing accessible, relevant, and impactful e-learning and executive education, you empower your organization to build a highly knowledgeable, adaptable, and forward-thinking leadership team, ultimately driving sustained organizational success and competitive advantage.

WHAT?	WHY?	HOW?	BEWARE!
ONLINE LEADERSHIP COURSES	**SKILL DEVELOPMENT**	**PLATFORM SUBSCRIPTION**	**ENGAGEMENT**
Exploring structured learning paths.	Provides flexible access to continuous learning and new competencies.	Subscribe to reputable e-learning platforms (e.g., Coursera, LinkedIn Learning).	Ensuring leaders actually complete courses and apply learnings.
EXECUTIVE EDUCATION PROGRAMS	**STRATEGIC EXPOSURE**	**UNIVERSITY PARTNERSHIPS**	**COST & TIME**
Engaging in advanced training tailored for leadership growth.	Fosters critical thinking and networking with industry peers.	Enroll in programs from renowned business schools or executive training institutes.	High financial investment and time commitment for immersive programs.
MICROLEARNING AND ADAPTIVE LEARNING	**PRACTICAL APPLICATION**	**CONTENT CURATION**	**CURATION**
Utilizing bite-sized, data-driven educational content.	Enhances retention and enables immediate application of concepts.	Deliver short, focused learning modules relevant to daily leadership challenges.	Designing and delivering microlearning content effectively.

For a business leader, establishing robust coaching and mentoring programs is a critical strategic imperative for providing personalized guidance, fostering growth, and building strong, supportive leadership systems throughout the organization. Beyond formal training, these one-on-one and peer-based relationships are essential for nuanced development and knowledge transfer. Central to this is facilitating executive coaching. One-on-one sessions with professional coaches offer highly personalized feedback, help refine decision-making, enhance strategic thinking, and address specific leadership challenges in a confidential and objective environment.

Furthermore, developing peer mentoring networks is crucial. Encouraging structured collaboration and knowledge sharing among leaders fosters a supportive community where experienced leaders can guide and learn from each other, building collective wisdom and mutual support. Crucially, integrating developmental coaching focuses on fostering holistic personal and professional growth through structured feedback and goal setting. This approach nurtures leaders not just in their immediate roles but also prepares them for future challenges, cultivating a continuous growth mindset. Neglecting comprehensive coaching and mentoring can lead to isolated leaders, slower individual growth, a lack of succession planning, and a struggle to retain top talent.

By championing these personalized guidance systems, you empower your organization to cultivate a deep bench of capable leaders, accelerate individual and collective development, and ultimately drive sustained organizational success through strong, well-supported leadership

WHAT?	WHY?	HOW?	BEWARE!
EXECUTIVE COACHING	**PERSONALIZED GROWTH**	**COACH ENGAGEMENT**	**TRUST**
One-on-one sessions focused on refining decision-making and strategy.	Provides tailored support for individual leadership challenges.	Hire certified executive coaches and define clear coaching objectives.	Building trust and psychological safety between coach and the coached.
PEER MENTORING NETWORKS	**COLLECTIVE WISDOM**	**MATCHMAKING**	**COMMITMENT**
Encouraging leadership collaboration and knowledge sharing.	Fosters a supportive community and accelerates shared learning.	Create structured programs to match leaders with compatible peers for mentorship.	Ensuring consistent participation and commitment from peer mentors.
DEVELOPMENTAL COACHING	**HOLISTIC LEADERSHIP**	**FEEDBACK FRAMEWORKS**	**RESISTANCE**
Fostering personal and professional growth through structured feedback.	Nurtures comprehensive growth, preparing leaders for future roles.	Integrate consistent feedback loops and growth plans into coaching sessions.	Leaders being receptive to ongoing, critical feedback.

For a business leader, incorporating regular journaling is a critical strategic imperative for strengthening leadership awareness through structured self-reflection and documentation. In the fast pace of leadership, carving out time for introspection can seem a luxury, but it is a powerful tool for clarity, growth, and informed decision-making. Central to this is adopting reflective leadership practices. Using writing to analyze daily experiences, explore successes and setbacks, and process complex decisions allows you to gain deeper insights into your own reactions, biases, and leadership effectiveness, leading to more intentional future actions.

Furthermore, meticulously engaging in goal tracking and accountability through journaling is crucial. Maintaining progress logs for personal and professional development goals ensures continuous momentum, allowing you to regularly review achievements, identify roadblocks, and adjust strategies to stay on track. Crucially, integrating emotional intelligence exercises through structured journaling enhances self-awareness and resilience. By consciously exploring your emotional responses to various leadership scenarios, you can better understand your triggers, manage your reactions, and cultivate greater empathy, ultimately improving your interpersonal effectiveness and mental well-being. Neglecting personal reflection can lead to repeating mistakes, a lack of clarity in decision-making, and difficulty in managing stress effectively. By championing journaling as a core leadership practice, you empower yourself to learn continuously, lead with greater intentionality, and build enduring resilience, ultimately contributing to your sustained effectiveness and the overall success of your organization.

WHAT?	WHY?	HOW?	BEWARE!
REFLECTIVE LEADERSHIP PRACTICES	**DEEPER INSIGHTS**	**DAILY JOURNALING**	**CONSISTENCY**
Using writing to analyze experiences.	Gathers insights into personal reactions and leadership effectiveness.	Dedicate time each day or week for structured reflection on leadership experiences.	Maintaining a regular journaling habit amidst a busy schedule.
GOAL TRACKING ACCOUNTABILITY	**SUSTAINED GROWTH**	**PROGRESS LOGS**	**OBJECTIVITY**
Maintaining progress logs for continuous development.	Ensures continuous momentum towards personal and professional goals.	Document progress against leadership development goals and personal milestones.	Being honest and objective in tracking progress and setbacks.
EMOTIONAL INTELLIGENCE EXERCISES	**COMPOSURE & EMPATHY**	**PROMPT-BASED JOURNALING**	**VULNERABILITY**
Enhancing self-awareness and resilience.	Improves emotional control and understanding of others.	Use prompts to explore emotional responses to specific scenarios.	Being open and honest about emotional experiences in writing.

For a business leader, engaging in cognitive coaching is a critical strategic imperative for continuously improving leadership thinking and problem-solving capabilities. In today's complex and uncertain business landscape, superior cognitive agility, critical thinking, and bias mitigation are paramount for effective decision-making and navigating intricate challenges. Central to this is mastering critical thinking frameworks. Strengthening analytical skills through structured methodologies—such as breaking down complex problems, evaluating evidence, and considering multiple perspectives—enables more robust and reasoned decision-making in high-stakes environments.

Furthermore, developing robust bias mitigation strategies is crucial. Recognizing and actively addressing cognitive distortions and unconscious biases in leadership judgments (e.g., confirmation bias, anchoring bias) ensures fairer, more objective decisions that are less prone to error and promote inclusivity. Crucially, applying neuroscience-backed coaching techniques enhances leadership cognition. Leveraging insights from brain science—such as understanding attention, memory, and emotional regulation—can lead to more effective learning, improved focus, and enhanced mental resilience under pressure. Neglecting cognitive development can lead to flawed decision-making, reactive problem-solving, perpetuation of biases, and a struggle to adapt to novel situations. By championing a proactive and scientifically informed approach to improving leadership thinking, you empower yourself to make sharper decisions, solve complex problems more effectively, and ultimately drive greater organizational success through enhanced intellectual leadership

WHAT?	WHY?	HOW?	BEWARE!
CRITICAL THINKING FRAMEWORKS	**ROBUST DECISIONS**	**METHODOLOGY TRAINING**	**MENTAL FATIGUE**
Strengthening analytical skills for complex decision-making.	Enables more reasoned and effective problem-solving.	Apply frameworks like MECE (Mutually Exclusive, Collectively Exhaustive) or first-principles thinking.	Sustaining critical thinking under pressure or long hours.
BIAS MITIGATION STRATEGIES	**OBJECTIVITY & FAIRNESS**	**AWARENESS & INTERVENTION**	**UNCONSCIOUS BIAS**
Recognizing and addressing cognitive distortions in leadership judgments.	Promotes unbiased and equitable decision-making.	Learn about common cognitive biases and apply techniques to counter them.	Uncovering deeply ingrained unconscious biases.
NEUROSCIENCE-BACKED COACHING TECHNIQUES	**ENHANCED COGNITION**	**COACHING INTEGRATION**	**SCIENTIFIC VALIDATION**
Applying psychological principles to enhance leadership cognition.	Improves focus, memory, and mental resilience.	Work with coaches who integrate principles of neuroplasticity and brain health.	Ensuring techniques are truly evidence-based and not pseudoscientific.

For a business leader, establishing robust feedback loops is a critical strategic imperative for continuous leadership improvement. It's about creating structured review systems that provide regular, actionable insights, ensuring that your leadership approach is constantly refined and adapted to meet evolving team and organizational needs. Central to this is obtaining real-time performance insights. Gathering regular evaluations from various sources allows you to identify strengths, address weaknesses, and refine your leadership approaches in an agile manner, rather than waiting for annual reviews.

Furthermore, implementing comprehensive 360-degree feedback systems is crucial. Encouraging input from peers, subordinates, stakeholders, and even superiors provides a holistic view of your leadership impact and perception, revealing blind spots and opportunities for growth from multiple perspectives. Crucially, leveraging data-driven leadership refinement is vital. Utilizing analytics to assess the impact and effectiveness of your leadership actions—such as team productivity, engagement scores, or project success rates—provides objective evidence for where your leadership is truly making a difference and where adjustments are needed. Neglecting structured feedback loops can lead to stagnant leadership, unaddressed issues, missed opportunities for growth, and a reactive approach to team challenges. By championing a continuous, multi-source, and data-driven approach to feedback, you empower yourself to lead with greater awareness, accelerate your personal development, and ultimately drive superior team performance and organizational success through increasingly effective and adaptable leadership

WHAT?	WHY?	HOW?	BEWARE!
REAL-TIME PERFORMANCE INSIGHTS	**AGILE REFINEMENT**	**PULSE SURVEYS**	**FEEDBACK FATIGUE**
Gathering regular evaluations.	Enables quick adjustments to leadership approaches.	Implement frequent, short surveys for immediate feedback.	Overwhelming teams with too many feedback requests.
360-DEGREE FEEDBACK SYSTEMS	**HOLISTIC PERCEPTION**	**MULTI-SOURCE SURVEYS**	**TRUST**
Encouraging input from peers, subordinates, and stakeholders.	Provides comprehensive understanding of leadership impact.	Use platforms that facilitate anonymous input from various sources.	Ensuring psychological safety for honest and constructive feedback.
DATA-DRIVEN LEADERSHIP REFINEMENT	**OBJECTIVE IMPROVEMENT**	**KPI ALIGNMENT**	**ATTRIBUTION**
Leveraging analytics to assess impact and effectiveness.	Grounds leadership development in quantifiable results.	Connect leadership behaviors to team KPIs (e.g., morale, productivity, success).	Accurately attributing team performance changes directly to leadership actions.

6x6™ Transformation Blueprint

OPTIMIZATION | Reflection

"We do not learn from experience... we learn from reflecting on experience."
- W. Edwards Deming.

An American statistician and management consultant, renowned for his profound influence on Japanese manufacturing. He taught top management how to improve design, quality, and sales through statistical methods and continuous improvement principles, significantly impacting global quality management.

The Optimization level is your Reflection stage, occurring when a leader pauses to refine the engine of their own influence. Through structured personal inquiry, you remove the friction of ego and bias to reach a state of high-performance clarity.

Intentionally and systematically reviewing past experiences, both successes and failures, to meticulously extract valuable lessons, profound insights, and actionable takeaways. This critical practice is the engine for continuous learning, enabling leaders to adapt strategies and refine their approaches.

"Reflection" prepares effectively for future challenges, ensuring that every experience contributes meaningfully to growth and improved decision-making, fostering a culture of introspection, accountability, and adaptive leadership, preventing the repetition of mistakes and accelerating progress.

"Reflection" is critical for business strategy maturity because it feeds into continuously evolving the leadership approach for the next phase or challenge. A mature business strategy is not static; it requires leaders who engage in continuous learning, rigorously evaluating successes and failures, and actively seeking feedback. This reflective practice enables leaders to identify what works and what doesn't, allowing them to refine their strategic thinking and execution. It's about fostering organizational agility, ensuring that leadership can learn from experience and adapt its strategic direction effectively to maintain relevance and drive future growth. Here is how we approach our learning:

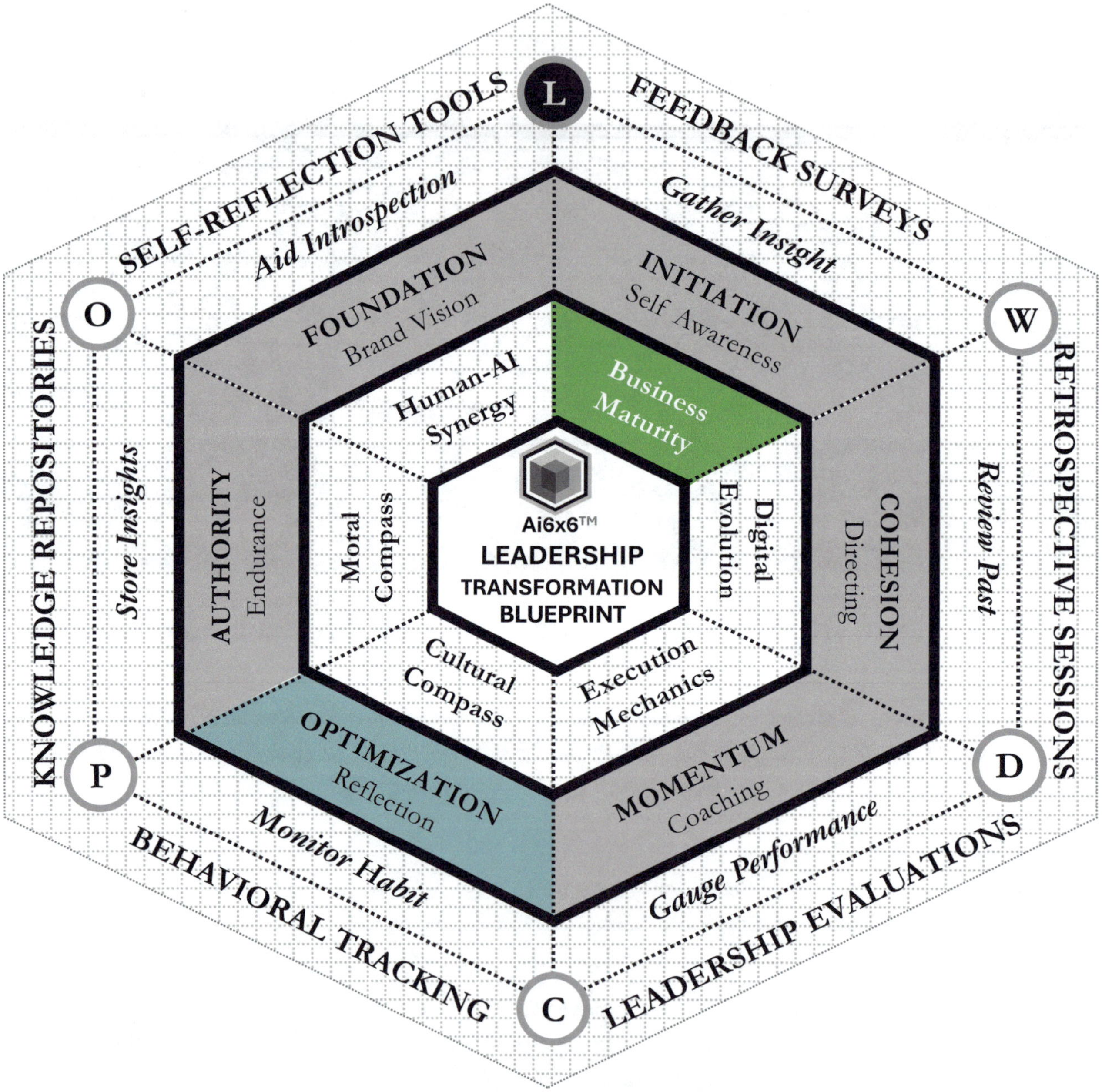

Ai6x6 BUSINESS MATURITY FOR REFLECTION
LEADERSHIP TRANSFORMATION

For a business leader, consciously utilizing self-reflection tools is a critical strategic imperative for enhancing leadership awareness through structured introspection. In the constant demands of leadership, carving out time for deliberate thought is essential for learning from experiences, gaining clarity, and refining your approach. It's about transforming raw experience into actionable wisdom. Central to this is engaging in guided reflection exercises. Using frameworks or prompts that encourage deep thinking about specific leadership experiences, decisions, and their outcomes allows you to systematically assess your personal growth, identify recurring patterns, and extract valuable lessons.

Furthermore, consistent journaling for leadership development is crucial. Documenting your insights, challenges, and emotional responses helps refine decision-making, process complex situations, and enhance emotional intelligence by providing a written record of your growth journey. Crucially, integrating meditation and mindfulness practices strengthens focus and clarity. These techniques cultivate present-moment awareness, reduce mental clutter, and improve emotional regulation, fostering a more calm and centered leadership presence that can navigate complexities with greater precision. Neglecting self-reflection can lead to repeating mistakes, a lack of personal growth, reactive leadership, and an inability to adapt to new challenges with intentionality. By championing these introspective tools, you empower yourself to lead with greater self-awareness, learn continuously from your experiences, and cultivate a more resilient and effective leadership style, ultimately contributing to your sustained impact and the overall success of your organization

WHAT?	WHY?	HOW?	BEWARE!
GUIDED REFLECTION EXERCISES	**DEEPER LEARNING**	**PROMPT-BASED REFLECTION**	**SUPERFICIALITY**
Using frameworks to assess leadership experiences and personal growth.	Systematically extracts lessons from experiences.	Use specific questions or frameworks (e.g., "What went well?", "What could be improved?")	Reflection becoming a superficial exercise without true introspection.
JOURNALING FOR DEVELOPMENT	**AWARENESS & INSIGHT**	**CONSISTENT PRACTICE**	**CONSISTENCY**
Documenting insights to refine decision-making and emotional intelligence.	Builds a record of growth and reveals patterns in thinking/behavior.	Dedicate daily or weekly time to freely write about leadership experiences.	Maintaining a regular journaling habit amidst a busy schedule.
MEDITATION AND MINDFULNESS	**CALM & COMPOSURE**	**GUIDED SESSIONS**	**DISCIPLINE**
Strengthening focus and clarity for effective leadership presence.	Improves mental clarity, reduces stress, and enhances emotional regulation.	Practice daily mindfulness meditation or breathing exercises.	Finding quiet time and maintaining consistency in mindfulness practice.

For a business leader, actively employing feedback surveys is a critical strategic imperative for gathering structured insights and continuously refining leadership effectiveness. Beyond informal conversations, systematic surveys provide a valuable, quantifiable view of how your leadership is perceived by those you lead and serve, enabling data-driven improvement. Central to this is establishing robust employee and peer feedback loops. Encouraging anonymous and open communication through structured surveys allows for honest evaluation of your leadership impact, revealing strengths to leverage and areas needing development from the perspective of your direct reports and colleagues.

Furthermore, leveraging sentiment analysis tools is crucial. By using AI-driven insights to gauge stakeholder perceptions and leadership impact from text-based survey responses and comments, you can quickly identify emotional trends, understand underlying sentiment, and prioritize areas for intervention. Crucially, implementing continuous improvement surveys ensures ongoing leadership refinement. Establishing structured feedback mechanisms that are deployed regularly (e.g., pulse surveys) allows for agile adjustments to your leadership approach, demonstrating responsiveness and fostering a culture of continuous learning and adaptation. Neglecting structured feedback surveys can lead to blind spots, unaddressed issues, stagnant leadership styles, and a disconnect between leadership intent and actual impact. By championing a systematic, data-driven approach to feedback collection, you empower yourself to lead with greater awareness, build stronger relationships based on trust, and ultimately drive superior organizational performance through continuously evolving and effective leadership.

WHAT?	WHY?	HOW?	BEWARE!
EMPLOYEE + PEER FEEDBACK LOOPS	**HOLISTIC PERCEPTION**	**ANONYMOUS SURVEYS**	**TRUST**
Encouraging anonymous and open communication.	Provides diverse perspectives on leadership impact.	Deploy regular online surveys allowing for anonymous feedback from directs/ peers.	Ensuring psychological safety for honest and constructive feedback.
SENTIMENT ANALYSIS TOOLS	**DEEPER UNDERSTANDING**	**NLP INTEGRATION**	**NUANCE**
Using AI-driven insights to gauge stakeholder perceptions.	Reveals emotional tone and underlying sentiment in feedback.	Utilize AI/NLP tools to analyze open-ended survey comments for sentiment.	AI struggling with sarcasm, irony, or highly nuanced language.
CONTINUOUS IMPROVEMENT SURVEYS	**AGILE ADAPTATION**	**PULSE SURVEYS**	**SURVEY FATIGUE**
Establishing structured feedback mechanisms for ongoing leadership refinement.	Enables quick adjustments to leadership approaches based on real-time input.	Implement short, frequent surveys to track key metrics and collect ongoing feedback.	Overwhelming respondents with too many frequent surveys.

3. Retrospective sessions

For a business leader, rigorously conducting retrospective sessions is a critical strategic imperative for reviewing past experiences, extracting valuable lessons, and systematically enhancing future leadership decisions. In a dynamic and often unpredictable environment, learning from both successes and failures is paramount for continuous improvement and strategic agility. Central to this is meticulous success and failure analysis. Systematically identifying patterns from both wins and setbacks allows you to understand contributing factors, refine leadership strategies, and avoid repeating past mistakes. This objective assessment turns experience into actionable insight.

Furthermore, facilitating structured reflection meetings is crucial. Conducting dedicated discussions where teams and leaders evaluate collective leadership effectiveness and adaptation provides a safe space for open dialogue, shared learning, and constructive critique, fostering a culture of continuous improvement. Crucially, actively learning from past decisions is vital for future leadership improvements. Using retrospectives as a foundational practice ensures that insights gained are not just discussed but are deliberately integrated into future planning, decision-making frameworks, and leadership development initiatives, closing the loop on learning. Neglecting retrospective sessions can lead to repeating errors, a failure to capitalize on successes, a reactive approach to problem-solving, and a static leadership approach that struggles to adapt. By championing a disciplined, open, and forward-looking approach to reviewing past experiences, you empower your organization to become a learning powerhouse, accelerate leadership growth, and ultimately drive greater strategic success and resilience in a constantly evolving landscape.

WHAT?	WHY?	HOW?	BEWARE!
SUCCESS AND FAILURE ANALYSIS	**LEARNING & REFINEMENT**	**ROOT CAUSE ANALYSIS**	**BLAME AVOIDANCE**
Identifying patterns from wins and setbacks.	Enables leaders to understand contributing factors for better future strategies.	Apply techniques like Five Whys to understand the causes of outcomes.	Ensuring a blame-free environment for honest analysis of failures.
REFLECTION MEETINGS	**SHARED INSIGHTS**	**FACILITATED SESSIONS**	**TIME COMMITMENT**
Conducting structured discussions to evaluate leadership effectiveness and adaptation.	Fosters collective learning and collaborative problem-solving.	Organize regular meetings (e.g., quarterly) with clear agendas for reflection.	Allocating dedicated time for reflection amidst busy schedules.
LEARNING FROM PAST DECISIONS	**CONTINUOUS IMPROVEMENT**	**ACTIONABLE INSIGHTS**	**IMPLEMENTATION GAP**
Using retrospectives as a foundation for future leadership improvements.	Ensures insights gained are integrated into future strategic planning.	Translate lessons learned into concrete action items and incorporate them into future roadmaps.	Difficulty ensuring insights are actually applied and sustained over time.

For a business leader, rigorously conducting leadership evaluations is a critical strategic imperative for measuring competencies and performance, providing objective data that guides growth and ensures a high-performing leadership pipeline. Without systematic assessment, development efforts can be misdirected, and potential leadership gaps may go unnoticed. Central to this is implementing 360-degree leadership assessments. Gathering holistic feedback from multiple stakeholders—including superiors, peers, direct reports, and even external partners—provides a comprehensive view of your leadership effectiveness, highlighting strengths and identifying areas for targeted development from diverse perspectives.

Furthermore, employing competency-based evaluations is crucial. Analyzing specific technical (e.g., digital literacy), strategic (e.g., market foresight), and interpersonal leadership skills (e.g., empathy, communication) against defined benchmarks ensures that assessments are objective and align with the critical capabilities needed for success in your organization. Crucially, establishing robust leadership growth tracking is vital. Implementing structured reviews and performance metrics allows you to measure personal and professional development over time, ensuring that leadership development initiatives are effective and that leaders are continuously progressing towards their full potential. Neglecting systematic leadership evaluations can lead to stagnant leadership, unaddressed performance issues, inefficient development investments, and a struggle to identify and nurture future leaders. By championing a data-driven, multi-faceted approach to leadership assessment, you empower your organization to build a strong, capable, and continuously developing leadership team, ultimately driving a resilient future.

WHAT?	WHY?	HOW?	BEWARE!
360-DEGREE ASSESSMENTS	**COMPREHENSIVE VIEW**	**PLATFORM UTILIZATION**	**BIAS**
Gathering holistic feedback from multiple stakeholders.	Provides a well-rounded perspective on leadership impact.	Use 360-degree feedback platforms to collect anonymous input from various sources.	Potential for personal biases in feedback from various stakeholders.
COMPETENC-EVALUATIONS	**TARGETED DEVELOPMENT**	**ASSESSMENT MATRIX**	**OBJECTIVITY**
Analyzing technical, strategic, and interpersonal leadership skills.	Identifies specific skill gaps for focused growth plans.	Develop a matrix of required competencies and evaluate leaders against each.	Ensuring objective evaluation of soft skills and interpersonal competencies.
GROWTH TRACKING	**PROGRESS & ROI**	**REGULAR REVIEWS**	**ATTRIBUTION**
Establishing structured reviews to measure development.	Monitors individual development and validates effectiveness of growth initiatives.	Conduct periodic performance reviews and track progress against development goals.	Accurately attributing improvements to specific development interventions.

5. Behavioral tracking

For a business leader, systematically engaging in behavioral tracking is a critical strategic imperative for monitoring leadership habits and decision-making trends, enabling continuous optimization and enhanced effectiveness. Beyond subjective assessments, empirical data on your actions and their outcomes provides invaluable insights for growth. Central to this is leveraging performance analytics. Using data-driven methods to evaluate leadership engagement and effectiveness—such as team productivity metrics, project completion rates, or communication frequency—provides objective insights into the impact of your actions and helps identify areas for improvement.

Furthermore, integrating AI-driven behavioral insights is crucial. By leveraging machine learning to analyze patterns in your leadership behaviors, decision inputs, and communication styles, AI can provide adaptive leadership refinement, offering subtle nudges or insights that help you optimize your approach in real-time. Crucially, applying habit-building methodologies is vital for strengthening positive leadership behaviors through consistent tracking. By consciously monitoring and reinforcing desired actions (e.g., consistent delegation, proactive feedback), you can embed more effective habits into your daily leadership practice. Neglecting behavioral tracking can lead to unconscious biases, unaddressed leadership patterns that hinder team performance, and a reactive approach to personal development. By championing a data-driven and self-aware approach to monitoring your leadership habits, you empower yourself to lead with greater intentionality, continuously refine your impact, and ultimately drive superior organizational performance through optimized and adaptive leadership.

WHAT?	WHY?	HOW?	BEWARE!
PERFORMANCE ANALYTICS	**OBJECTIVE INSIGHTS**	**KPI DASHBOARDS**	**ATTRIBUTION**
Using data-driven methods to evaluate leadership engagement and effectiveness.	Provides empirical data on the impact of leadership actions.	Track team performance KPIs (e.g., productivity, project completion rates) linked to leadership.	Accurately attributing team performance directly to leadership behaviors.
AI-DRIVEN BEHAVIOR INSIGHTS	**PROACTIVE OPTIMIZATION**	**AI ASSISTANTS**	**PRIVACY**
Leveraging machine learning for adaptive leadership refinement.	Provides real-time nudges and insights for behavioral adjustment.	Utilize AI tools that analyze communication patterns, meeting effectiveness, or decision-making biases.	Addressing privacy concerns when using AI to track individual behaviors.
HABIT-BUILDING METHODOLOGIES	**SUSTAINED IMPROVEMENT**	**HABIT TRACKERS**	**DISCIPLINE**
Strengthening positive leadership behaviors through consistent tracking.	Embeds effective behaviors into daily leadership practice.	Use digital habit trackers to monitor desired leadership actions (e.g., daily check-ins, proactive feedback).	Maintaining consistency in tracking and reinforcing new habits.

For a business leader, establishing robust knowledge repositories is a critical strategic imperative for curating valuable leadership insights and best practices, ensuring long-term development for yourself and future leaders. In a rapidly evolving world, capturing and organizing institutional knowledge is vital for continuous learning and organizational resilience. Central to this is building comprehensive digital resource libraries. Collecting leadership frameworks, case studies, strategic guides, and relevant research papers in an easily accessible format provides a centralized hub for continuous learning and reference, empowering leaders at all levels.

Furthermore, creating shared learning platforms is crucial. Encouraging mentorship relationships and fostering active leadership communities (e.g., internal forums, discussion groups) enables continuous knowledge sharing, peer-to-peer learning, and collaborative problem-solving, building collective wisdom across the organization. Crucially, diligently pursuing thought leadership documentation is vital. Archiving experiences, reflections, and strategic approaches from senior leaders provides invaluable institutional memory, serving as a legacy of wisdom for future reference, succession planning, and continuous organizational learning. Neglecting knowledge repositories can lead to loss of valuable institutional memory when leaders depart, repetitive problem-solving, inefficient knowledge transfer, and a slower pace of leadership development. By championing a systematic approach to curating and sharing leadership knowledge, you empower your organization to learn continuously, build a smarter leadership culture, and ultimately ensure long-term effectiveness and competitive advantage through cumulative wisdom.

WHAT?	WHY?	HOW?	BEWARE!
DIGITAL RESOURCE LIBRARIES	**ACCESSIBLE LEARNING**	**KNOWLEDGE PLATFORM**	**CONTENT CURATION**
Collecting leadership frameworks, case studies, and strategy guides.	Provides a centralized hub for continuous leadership development.	Implement a digital platform (e.g., internal wiki, LMS) for organizing resources.	Ensuring resources are relevant, high-quality, and up-to-date.
SHARED LEARNING PLATFORMS	**COLLECTIVE WISDOM**	**COMMUNITY FORUMS**	**ENGAGEMENT**
Encouraging mentorship and leadership communities.	Fosters peer-to-peer learning and collaborative problem-solving.	Create internal forums, discussion groups, or mentorship programs for leaders.	Encouraging active participation and contribution from busy leaders.
THOUGHT LEADERSHIP DOCUMENTATION	**INSTITUTIONAL MEMORY**	**LEGACY CAPTURE**	**TIME COMMITMENT**
Archiving experiences, reflections, and strategies.	Preserves valuable insights for future leaders and organizational learning.	Implement a process for senior leaders to document key learnings and strategic decisions.	Leaders finding time to document their experiences comprehensively.

6x6™ Transformation Blueprint

AUTHORITY | Endurance

"Many of life's failures are people who did not realize how close they were to success when they gave up." - Thomas Edison.

An American inventor and businessman who developed numerous devices profoundly influencing modern life, including the phonograph and practical electric light bulb. He pioneered mass production and teamwork in invention, holding over 1,000 U.S. patents.

The Authority level is your Endurance, representing the final state where leadership becomes a legacy that withstands external shocks. This organizational resilience ensures that the transformation is not a temporary win, but a permanent standard.

The sustained capacity to navigate relentless challenges, maintain unwavering resilience in the face of adversity, and consistently inspire persistence and commitment over extended periods. This fundamental quality is paramount for achieving long-term impact, ensuring that leaders and their ambitious initiatives can withstand unforeseen obstacles and overcome setbacks.

"Endurance" ultimately achieves lasting, transformative change, embodying the grit and determination required to see complex visions through, fostering stability and trust within the organization, and ensuring sustained progress through difficult times.

The **"Endurance"** chapter is fundamental for business strategy maturity as it ensures the long-term effectiveness and resilience of leadership in implementing plans and solving problems. A mature business strategy requires leaders who demonstrate unwavering commitment and adaptability in the face of challenges and changing environments. This endurance ensures that strategic initiatives are not just launched but sustained through various phases, adapting to unforeseen circumstances. It's about building the fortitude necessary to see complex strategies through to completion, ensuring the organization can consistently achieve its long-term objectives and thrive amidst disruption. Here is how we approach our learning:

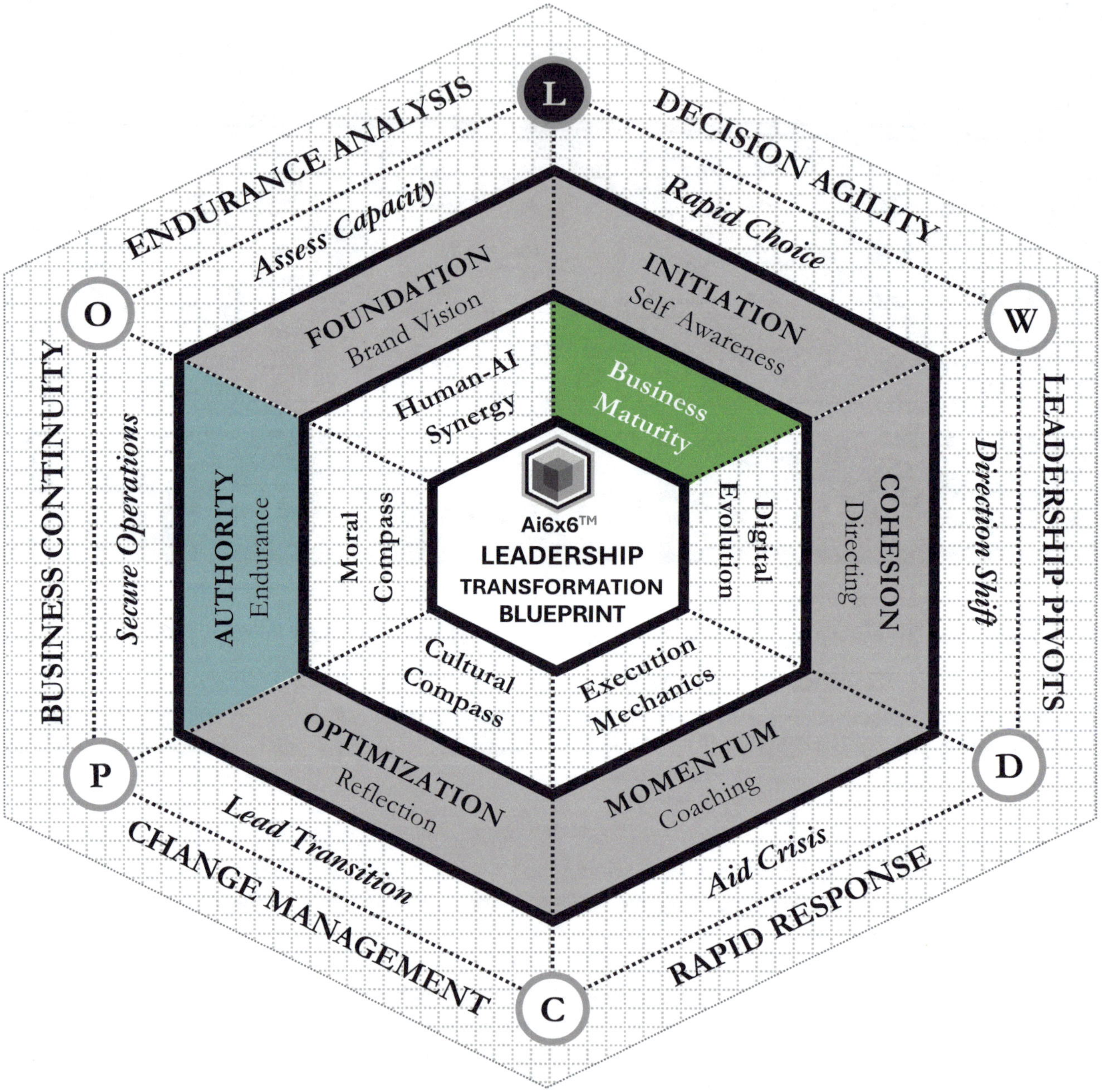

Ai6x6 BUSINESS MATURITY FOR ENDURANCE
LEADERSHIP TRANSFORMATION

1. Endurance analysis

For a business leader, rigorously conducting endurance analysis is a critical strategic imperative for evaluating both personal and organizational capacity to navigate setbacks, disruptions, and periods of prolonged stress. In today's volatile business landscape, the ability to withstand and recover from adversity is a hallmark of resilient leadership and a sustainable organization. Central to this is strengthening psychological endurance in leadership. Cultivating mental toughness, emotional resilience, and the ability to maintain composure and rational decision-making under sustained stress is paramount for effective crisis leadership and inspiring confidence in your teams.

Furthermore, meticulously measuring organizational endurance metrics is crucial. This involves assessing the adaptability, preparedness, and recovery strategies of your entire organization in the face of various shocks—from market downturns to supply chain disruptions—identifying vulnerabilities and building systemic resilience. Crucially, fostering a continuous improvement mindset is vital. Encouraging a culture where learning and refinement are the default response to challenges ensures that setbacks are viewed as opportunities for growth, rather than reasons for discouragement, driving ongoing adaptation and innovation. Neglecting endurance analysis can lead to burnout, brittle organizational structures, reactive crisis management, and a struggle to sustain performance through prolonged periods of adversity. By championing a proactive approach to assessing and building resilience, you empower yourself and your organization to not just survive but thrive amidst challenges, ultimately securing long-term stability, competitive advantage, and sustained success.

WHAT?	WHY?	HOW?	BEWARE!
PSYCHOLOGICAL ENDURANCE	**RESILIENT LEADERSHIP**	**STRESS MANAGEMENT**	**BURNOUT**
Strengthening mental toughness and decision-making under stress.	Ensures composure and rational decision-making during adversity.	Practice mindfulness, seek support, and prioritize self-care.	The demanding nature of leadership leading to mental fatigue.
ORGANIZATIONAL ENDURANCE METRICS	**SYSTEMIC RESILIENCE**	**SCENARIO PLANNING**	**MEASUREMENT**
Measuring adaptability, preparedness, and recovery strategies.	Identifies organizational vulnerabilities and strengthens overall adaptability.	Conduct stress tests and simulations to assess organizational response to various disruptions.	Quantifying abstract concepts like "adaptability" or "preparedness."
CONTINUOUS IMPROVEMENT MINDSET	**PROACTIVE GROWTH**	**RETROSPECTIVES**	**BLAME CULTURES**
Encouraging learning and refinement in response to challenges.	Transforms setbacks into opportunities for organizational learning and evolution.	Implement regular post-mortem reviews for all projects/initiatives (successes and failures).	Overcoming a culture of blaming that hinders honest learning from mistakes.

For a business leader, cultivating decision agility is a critical strategic imperative for making informed, fast-paced decisions in dynamic and often ambiguous environments. In today's rapidly changing world, slow decision-making is a significant liability; the ability to act swiftly and effectively is a key differentiator. Central to this is mastering real-time problem-solving. Adapting to shifting scenarios with data-driven insights ensures that your decisions are not based on outdated information but on the most current understanding of the situation, allowing for quick and precise responses.

Furthermore, rigorous scenario planning is crucial. Preparing contingency plans for various possible outcomes—from market shifts to technological disruptions—enables you to anticipate challenges and develop pre-emptive strategies, accelerating your response time when these scenarios materialize. Crucially, striking the right balance between intuitive vs. analytical decision-making is vital. Combining your experience and gut feeling with structured data analysis allows for rapid, yet robust, decision-making. This blend ensures that quick decisions are also well-informed, leading to more effective outcomes. Neglecting decision agility can lead to missed opportunities, reactive strategies, a struggle to keep pace with competitors, and a loss of market relevance. By championing a data-driven, foresight-driven, and balanced approach to decision-making, you empower your organization to respond swiftly and effectively to challenges, capitalize on fleeting opportunities, and ultimately secure your business's long-term success and competitive advantage in a volatile marketplace.

WHAT?	WHY?	HOW?	BEWARE!
REAL-TIME PROBLEM-SOLVING	**RAPID RESPONSE**	**LIVE ANALYTICS**	**DATA OVERLOAD**
Adapting to shifting scenarios with data-driven insights.	Enables quick and precise reactions to dynamic situations.	Integrate real-time data dashboards and AI-powered insights into decision-making processes.	Interpreting vast amounts of real-time data quickly and accurately.
SCENARIO PLANNING	**PROACTIVE READINESS**	**CONTINGENCY WORKSHOPS**	**COMPLEXITY**
Preparing contingency plans for various possible outcomes.	Anticipates challenges and accelerates response time.	Conduct workshops to develop "what if" scenarios and pre-defined responses.	Planning for every conceivable scenario can be overwhelming.
INTUITIVE VS. ANALYTICAL BALANCE	**ROBUST & FAST**	**DECISION FRAMEWORKS**	**BIAS**
Combining experience with structured analysis for rapid decision-making.	Ensures decisions are both well-informed and made swiftly.	Train leaders to integrate gut feeling with structured data analysis (e.g., decision matrix).	Over-reliance on intuition potentially leading to cognitive biases.

For a business leader, mastering leadership pivots is a critical strategic imperative for effectively adjusting your leadership direction in response to new challenges or emerging opportunities. In today's dynamic and often unpredictable business environment, the ability to recognize when a change in course is needed, and to execute that change decisively, is a hallmark of truly agile leadership. Central to this is strategic repositioning. Recognizing when to shift goals, priorities, or market strategies is vital for maintaining relevance and capturing new value. This involves a clear-eyed assessment of the current trajectory and a courageous decision to alter course when necessary.

Furthermore, fostering a culture of innovation and adaptability is crucial. Encouraging continuous evolution within teams and organizations ensures that the mindset for change is embedded, making pivots a natural part of your operational rhythm rather than a disruptive event. Crucially, meticulous stakeholder alignment is vital for successful pivots. Ensuring that strategic changes are communicated effectively, understood, and embraced by your teams and key stakeholders minimizes resistance, builds buy-in, and ensures a unified effort towards the new direction. Neglecting leadership pivots can lead to organizational inertia, missed market opportunities, prolonged struggles with outdated strategies, and a reactive stance that ultimately hinders growth. By championing a proactive, flexible, and communicative approach to adjusting leadership direction, you empower your organization to navigate change effectively, capitalize on new opportunities, and ultimately secure long-term relevance and sustained success in a constantly evolving marketplace.

WHAT?	WHY?	HOW?	BEWARE!
STRATEGIC REPOSITIONING	**MARKET RELEVANCE**	**MARKET ANALYSIS**	**RISK AVERSION**
Recognizing when to shift goals, priorities, or market strategies.	Ensures the business adapts to new market realities and opportunities.	Conduct regular market assessments to identify emerging challenges or opportunities that necessitate a pivot.	Leaders' reluctance to abandon established strategies or admit failure.
INNOVATION AND ADAPTABILITY	**ORGANIZATIONA L AGILITY**	**CULTURAL REINFORCEMENT**	**CULTURAL RESISTANCE**
Encouraging continuous evolution within teams and org.	Fosters a culture where change is embraced as a natural process.	Promote experimentation, learning from failure, and cross-functional collaboration.	Overcoming deeply ingrained resistance to change or new ideas.
STAKEHOLDER ALIGNMENT	**UNIFIED ACTION**	**COMMUNICATION PLAN**	**MISUNDERSTANDI NG**
Ensuring changes are communicated effectively and embraced by teams.	Minimizes resistance and ensures cohesive effort towards the new direction.	Develop a comprehensive communication strategy to articulate the "why" and "how" of the pivot.	Ensuring complex strategic shifts are understood by all stakeholders.

4. Rapid response

For a business leader, mastering rapid response is a critical strategic imperative for managing urgent situations with speed, precision, and minimal disruption. In today's fast-paced, interconnected world, crises can emerge unexpectedly, and your ability to react swiftly and decisively is paramount for mitigating damage, protecting reputation, and ensuring continuity. Central to this is adhering to crisis leadership principles. Maintaining composure, demonstrating decisiveness, and providing clear direction in high-pressure situations inspires confidence in your teams and allows for rational decision-making amidst chaos.

Furthermore, establishing clear emergency decision frameworks is crucial. Structuring responses to time-sensitive problems—defining escalation paths, pre-assigning roles, and outlining communication protocols—ensures that critical choices can be made efficiently, even when information is incomplete. Crucially, fostering robust cross-functional coordination is vital. Leveraging diverse expertise from across departments for fast, effective solutions ensures that all necessary resources and knowledge are brought to bear on the problem, accelerating resolution and minimizing negative impact. Neglecting rapid response capabilities can lead to prolonged crises, escalating damage, severe reputational harm, and a loss of customer and stakeholder trust. By championing a prepared, decisive, and collaborative approach to urgent situations, you empower your organization to react with agility, mitigate risks effectively, and ultimately secure your business's long-term resilience and stability in an unpredictable world.

WHAT?	WHY?	HOW?	BEWARE!
CRISIS LEADERSHIP PRINCIPLES	**EFFECTIVE NAVIGATION**	**TRAINING & SIMULATION**	**STRESS INOCULATION**
Maintaining composure and decisiveness in high-pressure situations.	Ensures calm and rational decision-making during crises.	Conduct regular crisis management training and simulated drills.	Replicating the true psychological stress of a real crisis during training.
EMERGENCY DECISION FRAMEWORKS	**ACCELERATED RESPONSE**	**PRE-DEFINED PROTOCOLS**	**FLEXIBILITY**
Structuring responses to time-sensitive problems.	Enables swift and organized decision-making under duress.	Develop clear decision trees, escalation paths, and communication plans.	Ensuring frameworks are adaptable enough for unforeseen types of crises.
CROSS-FUNCTIONAL COORDINATION	**HOLISTIC RESOLUTION**	**INCIDENT RESPONSE TEAMS**	**COMMUNICATION SILOS**
Leveraging diverse expertise for fast, effective solutions.	Ensures all necessary resources and knowledge are applied to the problem.	Establish pre-assigned, cross-functional teams for rapid crisis deployment.	Breaking down departmental silos during high-stress situations.

For a business leader, mastering change management is a critical strategic imperative for effectively leading teams through transitions with minimal disruption, fostering acceptance, and ensuring the successful adoption of new initiatives. In today's dynamic business environment, change is constant, and your ability to guide your organization through it gracefully is paramount for sustained success. Central to this is proactive resistance mitigation. Addressing concerns, engaging in open dialogue, and fostering acceptance of change initiatives minimizes pushback, builds buy-in, and ensures a smoother transition for all stakeholders.

Furthermore, implementing structured implementation strategies is crucial. Ensuring smooth adoption through clear planning, transparent communication, and meticulous execution of change initiatives reduces confusion, provides clear steps, and ensures that new processes or systems are embraced effectively. Crucially, prioritizing employee engagement during change is vital for maintaining morale and communication. Strengthening understanding, addressing fears, and fostering active participation through transformational shifts ensures that employees feel supported, valued, and connected throughout the process, preventing disengagement and burnout. Neglecting effective change management can lead to widespread employee resistance, project failures, low morale, increased turnover, and a struggle to realize the intended benefits of strategic initiatives. By championing a compassionate, systematic, and communicative approach to change, you empower your organization to adapt swiftly, embrace new directions, and ultimately secure its long-term resilience, innovation, and competitive advantage in an ever-evolving marketplace.

WHAT?	WHY?	HOW?	BEWARE!
RESISTANCE MITIGATION	**SMOOTHER TRANSITION**	**COMMUNICATION & EMPATHY**	**NEGATIVITY**
Addressing concerns and fostering acceptance in change initiatives.	Minimizes pushback and accelerates adoption of new processes.	Openly communicate the "why" of change, address concerns empathetically.	Overcoming ingrained negativity or cynicism towards change.
STRUCTURED IMPLEMENTATION STRATEGIES	**SUCCESSFUL ADOPTION**	**PHASED ROLLOUTS**	**COMPLEXITY**
Ensuring smooth adoption through clear planning.	Maximizes the likelihood of new initiatives being embraced effectively.	Implement changes in stages with clear milestones and communication.	Managing large-scale, intricate organizational changes.
EMPLOYEE ENGAGEMENT DURING CHANGE	**MORALE & PRODUCTIVITY**	**FEEDBACK CHANNELS**	**TRUST**
Strengthening morale and communication through transformational shifts.	Prevents disengagement and burnout, maintaining team effectiveness.	Establish open channels for employee feedback and support groups.	Rebuilding trust if past changes were poorly managed.

For a business leader, prioritizing business continuity is a critical strategic imperative for ensuring operational stability and safeguarding your business despite disruptions, challenges, or crises. In today's volatile global environment, unforeseen events are inevitable, and your ability to maintain essential functions is paramount to protecting assets, reputation, and profitability. Central to this is rigorous risk assessment and mitigation. Identifying vulnerabilities across your operations—from supply chain dependencies to IT infrastructure weaknesses—and developing proactive safeguards ensures that potential threats are minimized before they can escalate into major disruptions.

Furthermore, implementing sustainable contingency planning is crucial. Establishing robust backup protocols, redundant systems, and alternative operational sites prepares your organization for unforeseen events, guaranteeing that critical functions can quickly resume and minimize downtime, ensuring resilience. Crucially, fostering long-term adaptability involves creating flexible business models that can withstand significant industry changes. This strategic foresight ensures that your organization can pivot, innovate, and evolve in response to major market shifts, maintaining relevance and competitiveness over the long term. Neglecting business continuity can lead to catastrophic operational failures, severe financial losses, loss of customer trust, and a fundamental threat to your business's survival. By championing a proactive, systematic, and agile approach to business continuity, you empower your organization to withstand shocks, recover swiftly, and ultimately secure your business's long-term viability, resilience, and competitive advantage in an unpredictable world

WHAT?	WHY?	HOW?	BEWARE!
RISK ASSESSMENT AND MITIGATION	**PREVENTIVE MEASURES**	**RISK REGISTER**	**UNFORESEEN RISKS**
Identifying vulnerabilities and developing proactive safeguards.	Minimizes the likelihood and impact of potential disruptions.	Develop and maintain a comprehensive risk register, continuously updating threats and mitigation plans.	Difficulty anticipating every possible type of disruption.
SUSTAINABLE CONTINGENCY PLANNING	**RAPID RECOVERY**	**DR PLANS**	**COST & COMPLEXITY**
Establishing backup protocols for unforeseen events.	Ensures critical functions can quickly resume after a disruption.	Develop and regularly test disaster recovery and business continuity plans, including redundant systems.	High costs and logistical challenges of comprehensive contingency planning.
LONG-TERM ADAPTABILITY	**FUTURE-PROOFING**	**SCENARIO PLANNING**	**INERTIA**
Creating flexible business models that withstand industry changes.	Ensures the business remains relevant and competitive over time.	Develop flexible strategic plans and explore new business models adaptable to future disruptions.	Organizational resistance to fundamentally changing established business models.

BLUEPRINT SERIES

SECTION D

Digital Evolution

Ai6x6™ Transformation Blueprint

Story 1 — The Birth of COBOL and Business Computing (1959–1965): Early enterprises struggled to translate business logic into machine-readable form, limiting the usefulness of emerging computers. COBOL introduced a standardized language that mirrored human reasoning, enabling organizations to encode processes consistently. This shift allowed systems to scale beyond isolated experiments. Human analysts defined rules while machines executed them with precision. The partnership created the first bridge between business intent and digital execution.

Story 2 — ARPANET and Packet Switching (1969–1972): Communication networks were fragile, centralized, and prone to failure. Packet switching introduced a decentralized architecture that allowed information to flow even when parts of the network failed. Humans designed protocols while machines routed data autonomously. This resilience transformed digital communication from a technical novelty into a strategic infrastructure. ARPANET demonstrated how distributed systems could scale without collapsing. It laid the foundation for the modern internet.

Story 3 — Xerox PARC and the Graphical User Interface (1973–1983): Computers were powerful but inaccessible to most users, requiring specialized knowledge to operate. Xerox PARC introduced the graphical user interface, enabling humans to interact with machines visually rather than through code. This shift democratized computing and redefined usability. Machines handled complexity behind the scenes while humans engaged through intuitive metaphors. The interface became a catalyst for mass digital adoption. It showed that digital evolution accelerates when technology adapts to human cognition.

Story 4 — The Rise of Enterprise Resource Planning (1980s–1990s): Organizations operated in silos, with fragmented systems and inconsistent data. ERP platforms unified finance, operations, and supply chains into integrated digital environments. Machines enforced standardization while humans redesigned processes around shared information. This integration improved visibility, coordination, and control. ERP became the backbone of global enterprises. It demonstrated how digital evolution transforms not just tools but entire operating models.

Story 5 — The Commercial Internet and E-Commerce (1994–2005): The internet shifted from academic infrastructure to commercial ecosystem, enabling new forms of interaction, distribution, and value creation. Websites became digital storefronts, and payment systems matured into secure transactional platforms. Machines automated logistics while humans reimagined customer experience. The digital marketplace expanded global reach and lowered barriers to entry. It marked a turning point where digital capability became essential for competitiveness.

Story 6 — Cloud Computing and Elastic Infrastructure (2006–2015): Organizations struggled with rigid, capital-intensive infrastructure that limited experimentation and scale. Cloud computing introduced elastic resources that could expand or contract instantly. Machines automated provisioning while humans focused on architecture and innovation. This shift accelerated deployment cycles, reduced cost, and enabled global reach. Cloud became the foundation for modern digital ecosystems. It demonstrated how digital evolution thrives when infrastructure becomes fluid, on-demand, and programmable.

LESSONS FROM A LEADERSHIP TRANSFORMATION PERSPECTIVE

Leadership in digital evolution requires guiding organizations through successive waves of technological change while preserving clarity of purpose. As digital systems matured from isolated tools to global infrastructures, leaders learned to balance experimentation with governance and innovation with stability. Effective leadership recognizes that digital capability grows through layered progress—standardization, integration, democratization, and elasticity. Leaders must anticipate how each wave reshapes strategy, talent, and risk. Digital maturity emerges when leaders design environments where technology amplifies human intent, not distracts from it. The role of leadership becomes enabling transformation that is sustainable, coherent, and aligned with long-term institutional identity.

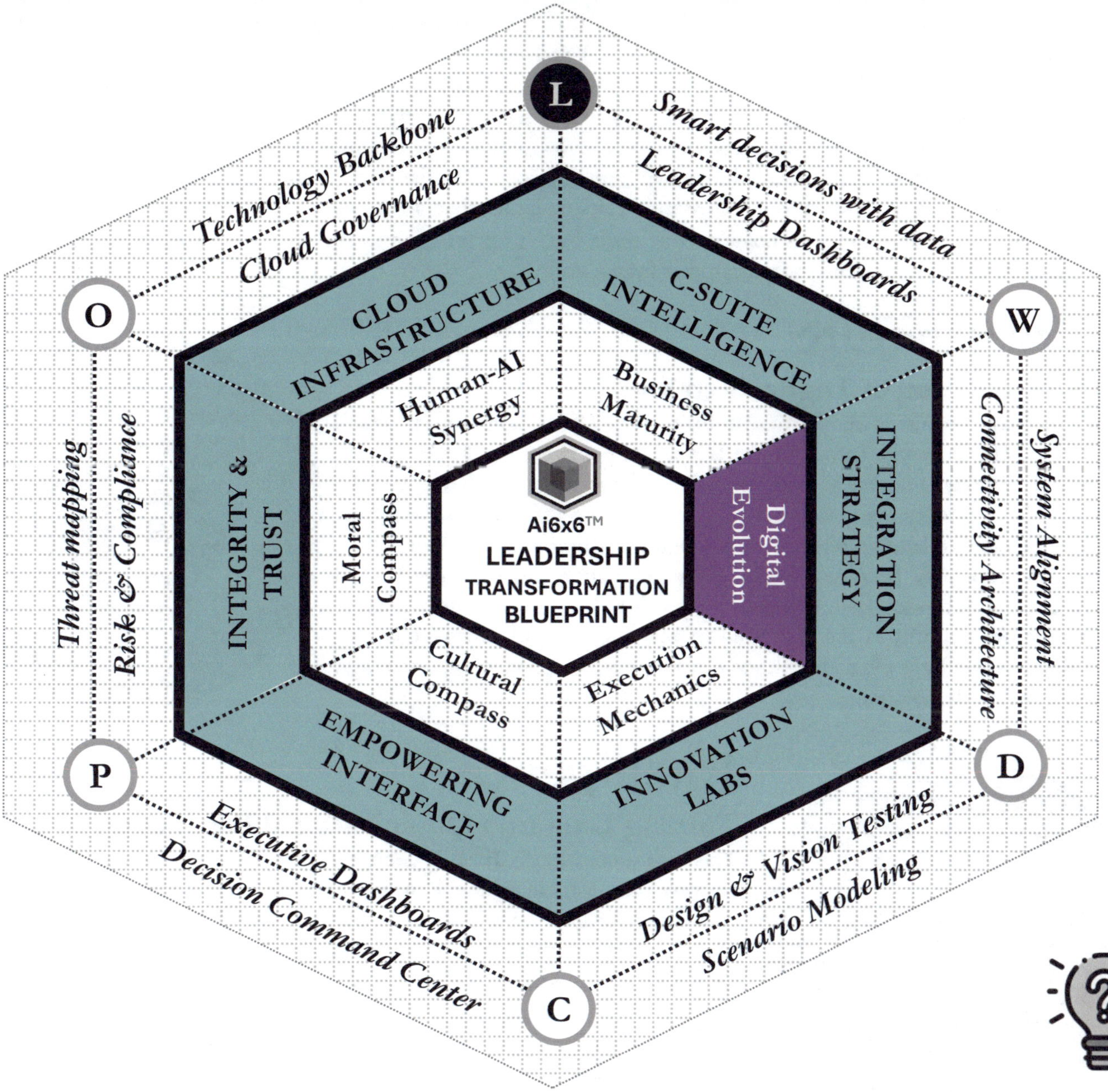

The **Ai6x6 Research Lab** chose this lens to solve the "mastery gap" that often leaves seasoned executives feeling like spectators in their own digital houses. We believe a leader must possess a quiet, unshakeable confidence in the systems they command—a digital gravitas that radiates from the mind into the organization. This isn't about learning to code; it is about the mental shift from being a passive consumer of technology to being its strategic master. It is vital because a leader cannot "visioneer" if they are constantly worried about the structural floor collapsing beneath them.

By mastering this evolution, the leader moves from technical anxiety to absolute strategic command. This lens ensures the C-suite is built on a foundation of trust, where "commanding the cloudscape" becomes a natural extension of executive intent. When a leader understands the integrity of their infrastructure, they gain the freedom to lead with an undistracted focus on the long-term horizon. This lens provides the psychological peace of mind that comes from knowing the backbone of the enterprise is secure. It is about the power that comes from clarity, ensuring the interface between human vision and digital execution is seamless and authoritative.

Digital Evolution is no longer a technical initiative. It is the structural rewiring of how an organization thinks, operates, and competes. For years, leaders treated digital transformation as a project—something with a start date, a roadmap, and a finish line. But the world has moved beyond that framing. Digital Evolution is not a destination; it is a permanent state of adaptation. It is the recognition that technology is no longer an enabler of strategy but a **shaper** of it. The organizations that thrive are those that treat digital capability as a living system—continuously upgraded, continuously integrated, and continuously aligned with the realities of the market. This lens reframes Digital Evolution as a leadership discipline rather than a technology program. It challenges leaders to see digital capability not as a collection of tools but as an interconnected ecosystem that influences every decision, every workflow, and every customer interaction.

Infrastructure

The Foundation of Digital Stability

The journey begins with **Infrastructure**, the foundation upon which everything else depends. Infrastructure is not just hardware or cloud capacity; it is the stability, scalability, and reliability of the systems that carry the organization's intelligence. Leaders who underestimate infrastructure discover that innovation collapses under its own weight. Without a strong foundation, every new capability becomes fragile, every integration becomes brittle, and every attempt at modernization becomes a patch rather than a transformation. Infrastructure maturity is the discipline of building for endurance—ensuring that the organization's digital backbone can support the speed, complexity, and volatility of modern markets.

Intelligence

Turning Data Into Organizational Foresight

The next dimension is **Intelligence**, representing the organization's ability to convert data into insight and insight into action. Intelligence is not about dashboards or reports; it is about building a system that learns. It requires data that flows freely, models that are trained responsibly, and decisions that are informed by evidence rather than assumption. Leaders who embrace intelligence understand that it is not a technical asset but a strategic one. Intelligence shapes priorities, exposes risks, and reveals opportunities that would otherwise remain invisible. It is the cognitive engine of Digital Evolution—the mechanism through which the organization perceives, interprets, and responds to its environment.

Integration

Making the Enterprise Whole

Integration is where Digital Evolution becomes real. Organizations often accumulate technologies faster than they can connect them. Systems become islands, data becomes siloed, and teams become fragmented. Integration is the discipline that ensures the enterprise operates as a coherent whole. It is the work of connecting platforms, harmonizing data, and designing workflows that move across boundaries. Without integration, digital investments become isolated experiments rather than enterprise capabilities. Integration maturity is the art of coherence—ensuring that the organization's digital landscape functions as a single, unified system rather than a patchwork of disconnected tools.

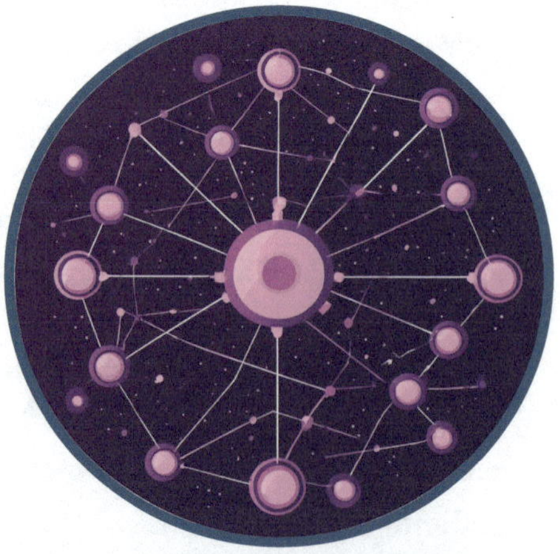

Innovation

Expanding the Field of Possibility

The fourth dimension, **Innovation**, expands the field of possibility. Innovation is not about chasing trends or adopting technology for its own sake. It is about using digital capability to reimagine how value is created, delivered, and experienced. Innovation demands that leaders challenge assumptions, explore alternatives, and test ideas quickly. It requires a culture where experimentation is encouraged, failure is reframed as learning, and progress is measured by insight rather than perfection. Innovation maturity is the discipline of curiosity—using digital tools not just to optimize the present but to invent the future.

Interface

Where Technology Meets Human Behavior

Interface represents the human experience of Digital Evolution. It is the point where technology meets behavior, where systems meet expectations, and where design meets trust. Leaders who ignore interface discover that even the most powerful systems fail if people cannot or will not use them. Interface is not cosmetic; it is strategic. It shapes adoption, accelerates learning, and determines whether digital capability becomes embedded or abandoned. Interface maturity is the discipline of empathy—designing digital experiences that are intuitive, accessible, and aligned with human needs.

Integrity

The Guardrail of Responsible Evolution

The final dimension, **Integrity**, grounds Digital Evolution in responsibility. Integrity ensures that systems are secure, data is protected, and decisions are ethical. It prevents speed from becoming recklessness and innovation from becoming harm. Leaders who prioritize integrity build organizations that are trusted, resilient, and prepared for the scrutiny that accompanies digital capability. Integrity maturity is the discipline of stewardship—ensuring that digital evolution strengthens the organization's moral and operational foundations rather than compromising them.

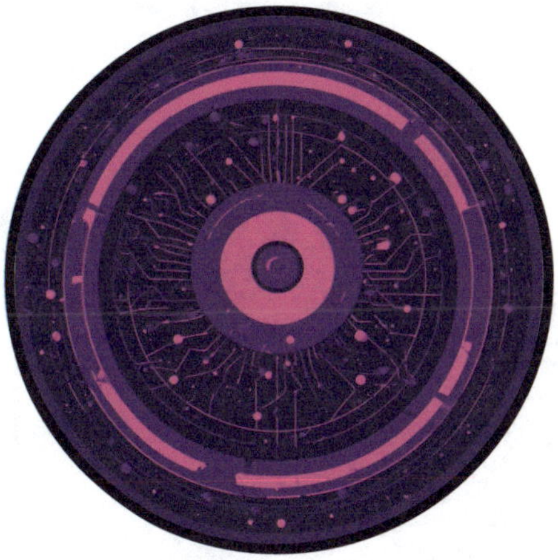

Digital Evolution is not a technology story. It is a leadership story. It is the recognition that the future belongs to organizations that treat digital capability as a core competency, not a side project. It is the shift from building tools to building systems, from adopting technology to evolving with it, and from reacting to change to shaping it. In this future, Digital Evolution is not a department, a project, or a budget line. It is the operating philosophy of the enterprise. It shapes how decisions are made, how systems are built, how teams collaborate, and how value is created. It turns technology from a source of disruption into a source of stability. It transforms the organization from a follower of change into a shaper of it.

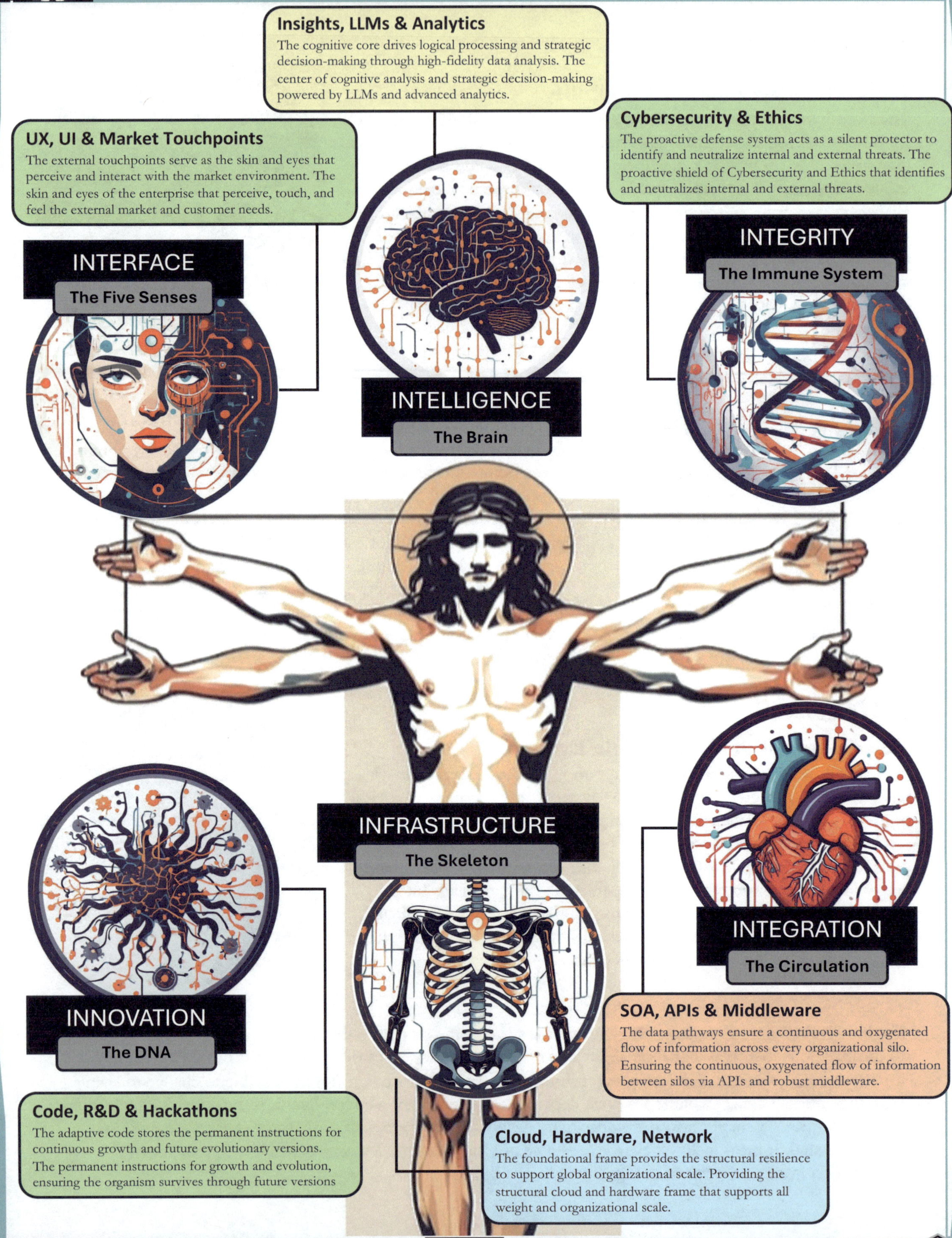
Insights, LLMs & Analytics
The cognitive core drives logical processing and strategic decision-making through high-fidelity data analysis. The center of cognitive analysis and strategic decision-making powered by LLMs and advanced analytics.
UX, UI & Market Touchpoints
The external touchpoints serve as the skin and eyes that perceive and interact with the market environment. The skin and eyes of the enterprise that perceive, touch, and feel the external market and customer needs.
Cybersecurity & Ethics
The proactive defense system acts as a silent protector to identify and neutralize internal and external threats. The proactive shield of Cybersecurity and Ethics that identifies and neutralizes internal and external threats.
INTERFACE
The Five Senses
INTELLIGENCE
The Brain
INTEGRITY
The Immune System
INFRASTRUCTURE
The Skeleton
INTEGRATION
The Circulation
INNOVATION
The DNA
SOA, APIs & Middleware
The data pathways ensure a continuous and oxygenated flow of information across every organizational silo. Ensuring the continuous, oxygenated flow of information between silos via APIs and robust middleware.
Code, R&D & Hackathons
The adaptive code stores the permanent instructions for continuous growth and future evolutionary versions. The permanent instructions for growth and evolution, ensuring the organism survives through future versions
Cloud, Hardware, Network
The foundational frame provides the structural resilience to support global organizational scale. Providing the structural cloud and hardware frame that supports all weight and organizational scale.

6x6™ Transformation Blueprint

Cloud 'Infrastructure'

"The cloud services we build will empower every organization to be a digital company." — Satya Nadella, CEO of Microsoft; led enterprise-wide cloud evolution with Azure as the transformation core.

Leadership's digital maturity begins not with flashy strategy but with quiet infrastructure authority. The "commanding cloudscape" is more than cloud adoption—it's about shaping a resilient foundation that supports scale, security, and speed. Executives don't inherit infrastructure—they craft it. Cloud platforms, API networks, and digital workspaces silently determine leadership agility. If vision is to reach velocity, this backbone must be modular, governed, and trusted. Transformative leaders orchestrate beneath the surface—designing infrastructures that flex for disruption, adapt for growth, and reinforce executive presence without visual clutter. Cloud isn't just a tool—it's the structural metaphor for leadership's invisible influence.

For leadership transformation, the "Cloud Infrastructure" chapter is foundational for digital maturity because it defines the robust technological backbone that empowers leaders to operate, collaborate, and strategize effectively in a complex, digital-first world. A digitally mature leadership team understands that a scalable, secure, and resilient infrastructure is not merely a support function but a strategic enabler. This includes intelligent cloud governance, adaptable hybrid cloud architectures, and robust cybersecurity, all crucial for scaling digital initiatives, ensuring data protection, and providing the reliable environment needed to drive organizational change, foster innovation, and maintain competitive advantage in an ever-evolving digital landscape.

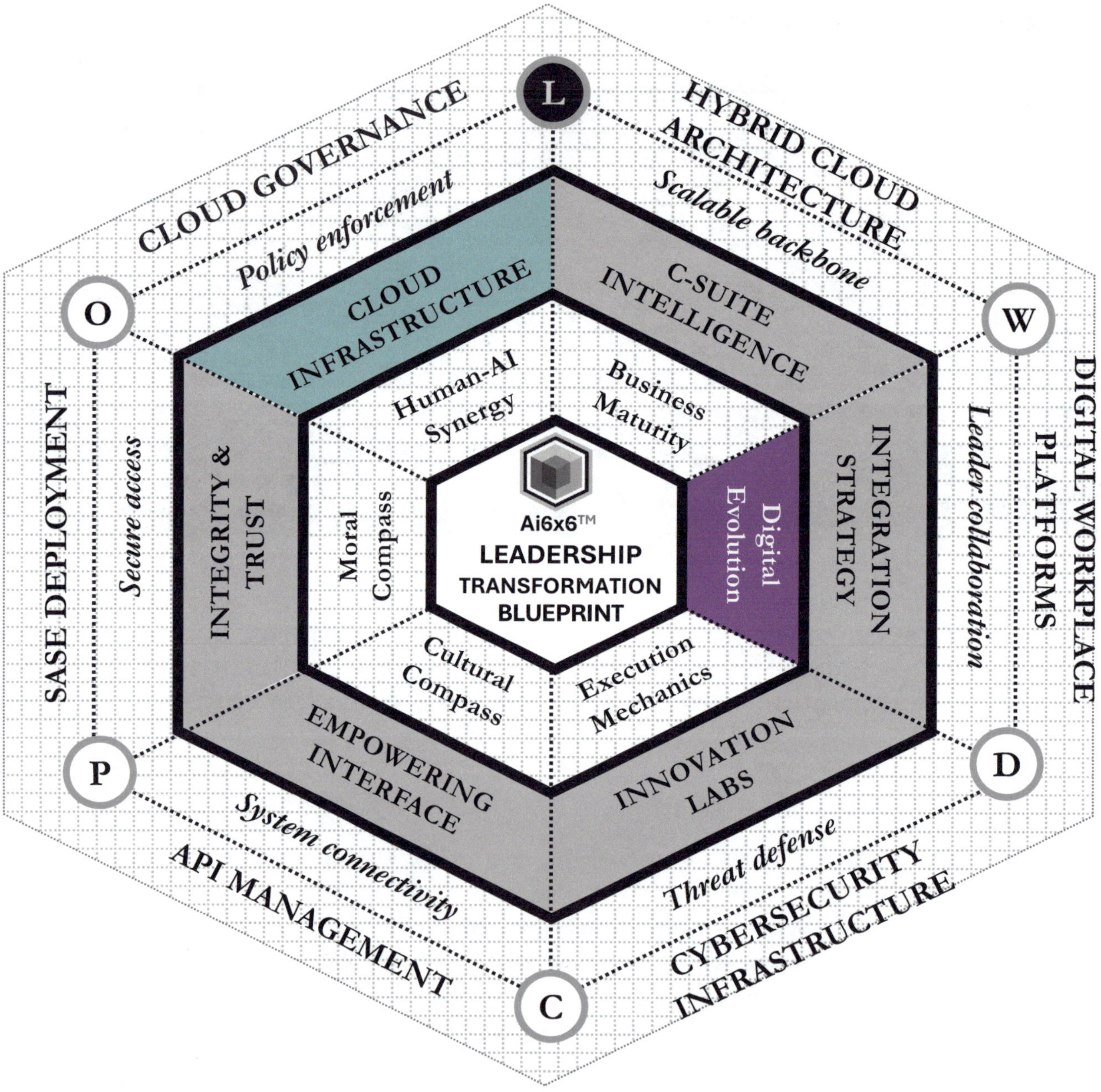

Ai6x6 DIGITAL EVOLUTION FOR INFRASTRUCTURE LEADERSHIP TRANSFORMATION

Cloud governance establishes a comprehensive framework of policies, procedures, and best practices for managing an organization's cloud resources and services across various cloud platforms. This structured approach ensures that cloud environments are operated securely, cost-effectively, and in compliance with regulatory requirements and internal standards. By defining clear roles, responsibilities, and automated guardrails, cloud governance provides leadership with the necessary control and visibility to mitigate risks, optimize resource utilization, and accelerate innovation confidently within a dynamic cloud landscape, preventing uncontrolled sprawl and ensuring strategic alignment.

Without effective cloud governance, organizations face significant risks and inefficiencies that can severely undermine their digital transformation efforts. Uncontrolled cloud resource provisioning often leads to spiraling costs due to unmanaged resources, substantial security vulnerabilities from misconfigurations, and non-compliance with data privacy regulations, resulting in hefty fines and reputational damage.

The absence of a clear governance framework hinders strategic agility, as leadership lacks the necessary insights and controls to make informed decisions about cloud investments and utilization, ultimately slowing down innovation and eroding trust in the digital infrastructure.

Implementation	Challenges
* Define clear policies & compliance standards * Establish roles, responsibilities & accountability * Implement automated guardrails & enforcement * Monitor usage, performance & expenditures * Conduct regular security & compliance audits * Continuously optimize resource allocation	* Lack of clear ownership & accountability * Rapidly evolving cloud services & features * Cultural resistance to new processes * Significant skills gap in cloud governance * Integrating with existing IT governance models * Complexity of multi-cloud environments

Products	Platform Cost	Project Cost	Duration
Cloud Custodian (Open Source)	$0-50K	$50-250K	3-6 m
AWS Config / Azure Policy / GCP Org Policy	$5-20K	$75-300K	4-8 m
CloudHealth by VMware	$10-50K	$100-400K	5-9 m
Flexera One (Cloud Management Platform)	$20-80K	$150-500K	6-12 m
Turbonomic (IBM)	$15-60K	$120-450K	5-10 m
Apptio Cloudability	$10-40K	$90-350K	4-8 m

Hybrid Cloud Architecture integrates on-premises infrastructure with public cloud services, creating a unified and flexible computing environment where data and applications can seamlessly move between environments. This strategic approach allows organizations to leverage the scalability, agility, and cost-effectiveness of public clouds for dynamic workloads, while retaining sensitive data or legacy applications within their private infrastructure. It provides unmatched flexibility, enhances disaster recovery capabilities, and enables seamless workload portability, empowering leadership to optimize resource allocation and accelerate digital initiatives with a balanced approach to cloud adoption.

Without embracing a hybrid cloud architecture, organizations often face significant limitations such as vendor lock-in, restricted scalability for fluctuating workloads, and inefficient resource utilization due to rigid infrastructure. This inability to burst capacity into the cloud or seamlessly integrate on-premises systems with modern cloud services results in suboptimal performance, higher operational costs for maintaining disparate environments, and challenges in managing diverse IT landscapes. The absence of a cohesive hybrid strategy hinders an organization's ability to innovate rapidly, adapt to evolving business demands, and leverage the full spectrum of cloud benefits, ultimately impacting agility and competitive responsiveness.

Implementation	Challenges
* Assess current infrastructure & business needs * Design network connectivity (VPN/Direct Connect) * Select appropriate public/private cloud services * Implement unified identity management * Deploy hybrid management and orchestration tools * Migrate/integrate initial applications/workloads	* Data synchronization & consistency across environments * Network latency between on-prem & cloud * Aligning security policies across hybrid infra * Increased management complexity for diverse IT * Significant skillset requirements for hybrid ops * Effective cost management across hybrid setup

Products	Platform Cost	Project Cost	Duration
VMware Cloud on AWS/Azure VMware Solution	$50-200K	$300-1000K	6-18 m
Azure Arc / Google Anthos	$20-100K	$200-800K	5-15 m
IBM Cloud Satellite	$30-150K	$250-900K	6-16 m
HPE GreenLake	$40-180K	$350-1200K	7-20 m
Red Hat OpenShift Container Platform	$25-120K	$200-750K	5-14 m
Nutanix Cloud Platform	$30-100K	$280-950K	6-18 m

Digital Workplace Platforms integrate a diverse suite of communication, collaboration, and productivity tools into a single, unified, and intuitive environment. These platforms enable employees to seamlessly connect, share information, co-create documents, and manage tasks from any location and device, fostering a dynamic and inclusive work culture. By streamlining workflows, enhancing real-time communication, and providing centralized access to necessary resources, they significantly improve employee experience, boost productivity, and support flexible work models (remote, hybrid, office), empowering leadership to cultivate a highly connected and agile workforce.

Without the adoption of robust digital workplace platforms, organizations struggle with fragmented toolsets, leading to communication silos, inefficient information sharing, and duplicated efforts across teams. Employees face constant context switching between disparate applications, resulting in frustration, reduced engagement, and hindered productivity, particularly in distributed workforces.

The absence of a cohesive digital environment impedes seamless collaboration, slows down decision-making, and limits an organization's ability to attract and retain top talent who expect modern work infrastructures, ultimately undermining operational efficiency and the pace of innovation.

Implementation	Challenges
* Assess current tools and user needs * Select platform based on integration needs * Plan data migration and content strategy * Customize workflows and user interfaces * Pilot deployment and comprehensive user training * Phased rollout and continuous feedback loop	* User adoption and managing change resistance * Data security and privacy concerns * Seamless integration with existing IT systems * Information overload and digital fatigue * Maintaining platform relevance and ongoing updates * Accurately measuring ROI and productivity gains

Products	Platform Cost	Project Cost	Duration
Microsoft 365 (Teams, SharePoint, Viva)	$5-30K	$50-300K	3-9 m
Google Workspace (Meet, Drive, Sites)	$3-25K	$40-250K	2-8 m
Slack Enterprise Grid	$10-40K	$60-200K	3-7 m
Salesforce Experience Cloud	$20-70K	$100-400K	4-10 m
Atlassian Confluence/Jira Service Mgmt	$5-20K	$30-150K	2-6 m
Workplace from Meta	$5-25K	$40-180K	3-8 m

Cybersecurity infrastructure encompasses the essential systems, tools, and processes designed to protect an organization's digital assets, networks, and data from a constantly evolving array of cyber threats. It forms the bedrock of digital trust, ensuring business continuity, safeguarding sensitive information, and maintaining regulatory compliance in an increasingly interconnected world. A robust cybersecurity infrastructure employs multi-layered defenses, proactive threat intelligence, and continuous monitoring to detect, prevent, and respond to attacks, allowing leadership to operate with confidence and protect the organization's reputation and financial stability.

Lacking a robust and up-to-date cybersecurity infrastructure exposes an organization to catastrophic risks, including debilitating data breaches, prolonged operational downtime, severe reputational damage, and significant financial losses from cyberattacks and regulatory fines. Without adequate protection, intellectual property can be stolen, customer trust can be irrevocably lost, and critical business operations can be disrupted, jeopardizing the very existence of the enterprise.

The absence of strong cybersecurity leadership leads to a vulnerable digital posture, hindering innovation due to fear of attack and ultimately eroding stakeholder confidence and long-term viability.

Implementation	Challenges
* Conduct comprehensive risk assessment * Deploy firewalls, IDS/IPS, EDR/XDR solutions * Implement Zero Trust principles & practices * Establish robust identity & access management * Implement continuous security awareness training * Regular vulnerability scanning/penetration testing	* Rapidly evolving and sophisticated threat landscape * Critical shortage of skilled cybersecurity talent * Managing complexity of diverse security toolsets * Balancing security with user experience/productivity * Constantly escalating budget constraints * Alert fatigue and false positives from systems

Products	Platform Cost	Project Cost	Duration
Palo Alto Networks Prisma SASE	$30-150K	$150-600K	4-12 m
CrowdStrike Falcon Platform (EDR/XDR)	$20-100K	$100-450K	3-9 m
Fortinet FortiGate Firewalls + Fabric	$25-120K	$120-500K	4-10 m
Okta (Identity and Access Management)	$15-80K	$80-350K	3-8 m
Splunk Enterprise Security (SIEM)	$50-200K	$300-1000K	6-18 m
Zscaler Zero Trust Exchange	$20-90K	$100-400K	4-9 m

API Management involves the comprehensive oversight of an organization's Application Programming Interfaces (APIs) throughout their entire lifecycle, from design and publication to versioning and retirement. This critical capability ensures that APIs are not only secure and performant but also easily discoverable, well-documented, and consumable for both internal and external developers, facilitating seamless digital interactions. By centralizing control over APIs, it enables organizations to expose valuable data and services securely, foster innovation through ecosystem partnerships, and accelerate the development of new digital products and experiences, transforming digital assets into strategic advantages.

Without robust API management, organizations risk exposing sensitive data through insecure endpoints, suffering from inconsistent API quality, and encountering significant friction for developers attempting to integrate with their services. The absence of a centralized management layer leads to API sprawl, versioning chaos, lack of visibility into API usage, and missed opportunities for data monetization or strategic partnerships. This ultimately hinders the ability to create new revenue streams, limits the speed of digital transformation, and erodes trust among partners and developers, severely impacting an organization's ability to innovate and compete in the digital economy.

Implementation	Challenges
* Define clear API strategy and standards * Select a comprehensive API management platform * Design, develop and publish APIs securely * Implement security policies (auth, rate limiting) * Monitor API usage, performance & health * Onboard developers & provide rich documentation	* Managing API sprawl and versioning complexity * Ensuring consistent API quality & performance * Effective developer onboarding & ongoing support * Scaling API infrastructure under high load * Governing API lifecycle (deprecating, retiring) * Balancing flexibility with tight security

Products	Platform Cost	Project Cost	Duration
Apigee (Google Cloud)	$20-100K	$150-500K	4-12 m
AWS API Gateway	$5-40K	$75-300K	3-9 m
Azure API Management	$10-50K	$90-350K	3-10 m
MuleSoft Anypoint Platform	$30-150K	$250-800K	6-18 m
Kong Enterprise	$15-70K	$120-400K	4-11 m
Postman Enterprise	$5-25K	$60-200K	2-7 m

SASE (Secure Access Service Edge) is a transformational cybersecurity framework that converges networking and security functions into a single, cloud-native service delivered from the edge. It integrates capabilities like SD-WAN, firewall as a service, secure web gateway, cloud access security broker (CASB), and Zero Trust Network Access (ZTNA) into a unified platform. This architecture provides secure, optimized access for users, devices, and applications anywhere, on any device, significantly simplifying network and security infrastructure, reducing operational complexity, and enhancing protection against modern threats by enforcing consistent policies from a single point of control.

Without adopting a SASE deployment strategy, organizations are often burdened with complex, fragmented security stacks that struggle to protect an increasingly distributed workforce and cloud-based applications. This leads to security gaps, inconsistent policy enforcement, poor network performance for remote users accessing cloud resources, and escalating operational costs from managing multiple disparate point solutions. The absence of SASE impedes agile digital transformation, increases the attack surface, and diminishes an organization's ability to provide secure, seamless connectivity for its evolving business needs, ultimately compromising both security posture and operational efficiency.

Implementation	Challenges
* Assess current network & security posture * Choose SASE vendor & deployment model * Plan phased migration strategy & timelines * Deploy SASE client software/edge connectors * Integrate with existing identity providers (IdP) * Monitor performance, security events & traffic	* Complex integration with existing network infrastructure * Potential for significant migration complexity & downtime * Managing user experience during the transition * Vendor lock-in considerations with single-platform SASE * Skillset requirements for SASE architecture & management * Ensuring data sovereignty & compliance across regions

Products	Platform Cost	Project Cost	Duration
Zscaler Zero Trust Exchange	$25-100K	$200-700K	6-15 m
Palo Alto Networks Prisma SASE	$30-120K	$250-800K	6-18 m
Fortinet FortiSASE	$20-90K	$180-600K	5-14 m
Cisco Umbrella + SD-WAN	$20-80K	$150-550K	5-13 m
Netskope Security Cloud	$25-110K	$220-750K	6-16 m
Versa SASE	$15-70K	$120-480K	4-12 m

6x6™ Transformation Blueprint

C-Suite 'Intelligence'

"Decisions should be based on data, not intuition." — Ginni Rometty, Former CEO of IBM; brought AI and analytics into boardroom strategy through Watson.

Insightful leadership isn't just about making informed choices—it's about shaping the environment in which choices emerge. Dashboards, predictive analytics, and sentiment engines represent the cognitive frontier of the executive suite. Yet tools alone don't deliver insight—leaders must sculpt decision ecosystems where real-time data meets strategic depth. It's not about replacing instinct with statistics—it's about refining executive judgment through calibrated intelligence. A mature C-suite doesn't merely view trends; it configures technologies that contextualize, clarify, and provoke action. In transformation, intelligence is not abstract—it's an ambient capacity shaped by the architecture of executive experience.

The "C-Suite Intelligence" chapter is crucial for leadership transformation's digital maturity as it focuses on empowering leaders with advanced data-driven insights for proactive and strategic decision-making. A digitally mature C-Suite moves beyond intuition and traditional reporting to leverage predictive analytics, AI-driven insights, and comprehensive leadership dashboards. This intelligence capability enables them to anticipate market shifts, optimize resource allocation, understand talent dynamics through people analytics, and gauge organizational sentiment. This fosters a culture of evidence-based leadership, essential for navigating complex digital landscapes and making agile, impactful choices that drive the organization forward.

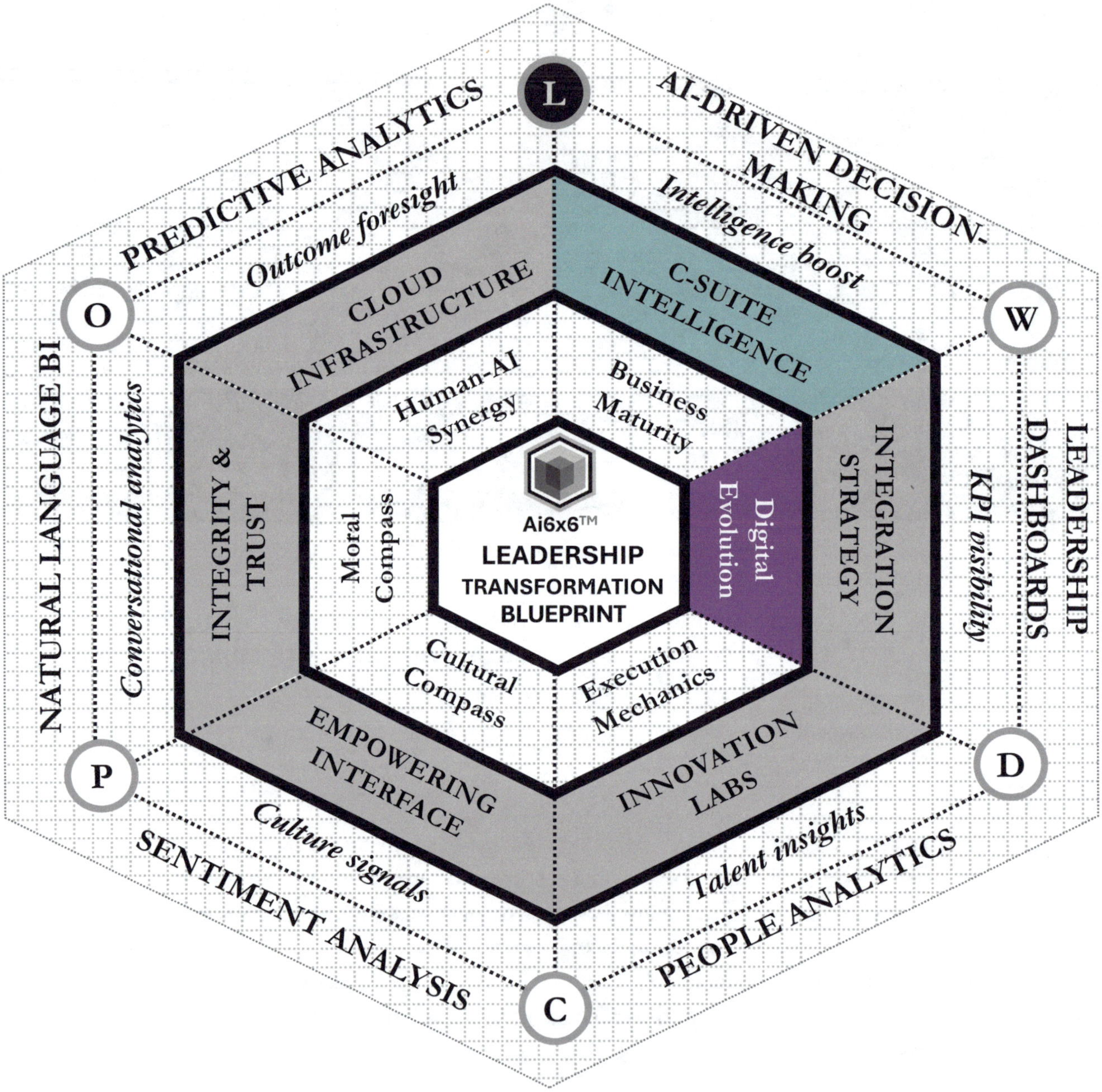

Ai6x6 DIGITAL EVOLUTION FOR INTELLIGENCE
LEADERSHIP TRANSFORMATION

1. PREDICTIVE ANALYTICS

Predictive analytics leverages historical data, statistical algorithms, and machine learning techniques to identify patterns and forecast future outcomes or behaviors. This advanced analytical capability enables leaders to move beyond reactive decision-making, anticipate market trends, predict customer churn, identify potential risks, and proactively capitalize on emerging opportunities. By transforming raw data into actionable foresight, predictive analytics empowers the C-Suite to make more informed and strategic choices, optimize resource allocation, and gain a significant competitive advantage in dynamic and uncertain business environments.

Without implementing predictive analytics, organizations remain largely reactive, basing crucial decisions on past performance or gut feelings rather than data-driven foresight. This reliance on historical data alone leads to missed opportunities, inefficient resource allocation, and a diminished capacity to anticipate challenges such as supply chain disruptions, customer attrition, or emerging market shifts. The absence of predictive capabilities results in slower response times to critical business events, a higher propensity for costly errors, and a general inability to proactively shape future outcomes, ultimately hindering agility and long-term sustainable growth.

Implementation	Challenges
* Define clear business problem & data needs * Collect & prepare diverse historical datasets * Select & train appropriate machine learning models * Validate model accuracy & performance rigorously * Integrate predictions into existing workflows * Continuously monitor & retrain models	* Ensuring high data quality & sufficient volume * Model complexity & interpretability challenges * Ethical considerations and potential data bias * Seamless integration with existing IT systems * Significant skills gap in data science & ML * Managing model drift over time

Products	Platform Cost	Project Cost	Duration
DataRobot	$50-200K	$200-800K	6-18 m
H2O.ai	$40-180K	$180-700K	5-16 m
IBM Watson Studio	$30-150K	$150-600K	5-15 m
SAS Viya	$60-250K	$300-1000K	7-20 m
Amazon SageMaker	$20-100K	$100-400K	4-12 m
Google Cloud AI Platform	$20-90K	$90-380K	4-11 m

AI-Driven Decision-Making involves leveraging artificial intelligence and machine learning models to analyze vast and complex datasets, identify nuanced patterns, and generate optimal recommendations or automate specific choices at scale. This capability augments human cognitive abilities, accelerates the decision-making process by providing rapid, data-backed insights, and reduces the impact of cognitive biases that can affect human judgment. By moving beyond traditional analytical approaches, AI empowers leaders to operate with unprecedented speed and precision, enabling more strategic, proactive, and resilient responses to complex business challenges and opportunities.

Without integrating AI into decision-making processes, organizations risk slower, less accurate, and potentially biased choices that fail to keep pace with dynamic market conditions and the sheer volume of available data. Relying solely on human intuition or traditional analytical methods means missing critical market signals, failing to identify subtle risks, and struggling to scale effective strategies across vast operations. The absence of AI's analytical power results in decreased competitive responsiveness, suboptimal resource allocation, and a missed opportunity to leverage advanced insights for sustained growth and innovation in an increasingly data-intensive world.

Implementation	Challenges
* Identify high-impact decision points * Integrate diverse data sources & define objectives * Develop/train AI models for actionable insights * Establish human-in-the-loop validation process * Integrate AI recommendations into workflows * Monitor performance & ethical implications	* Ensuring data privacy & addressing ethical concerns * Achieving explainability & transparency of AI * Cultural resistance to AI augmenting human roles * Strict data governance and quality assurance * Complex integration with legacy systems * Need for continuous model retraining & updates

Products	Platform Cost	Project Cost	Duration
Dataiku	$40-160K	$180-700K	5-16 m
C3 AI	$70-300K	$400-1500K	8-24 m
Palantir Foundry	$100-500K	$500-2000K	12-36 m
Microsoft Azure Machine Learning	$20-100K	$100-450K	4-12 m
Alteryx	$30-120K	$120-500K	4-13 m
IBM Watson Discovery	$25-100K	$100-400K	4-10 m

Leadership Dashboards are highly customized, interactive data visualization tools that provide senior executives with a real-time, consolidated, and intuitive overview of key performance indicators (KPIs), strategic metrics, and critical operational insights. These dashboards distill vast amounts of data into easily digestible visual formats, enabling quick comprehension of organizational health, performance against strategic goals, and emerging trends. By offering on-demand access to actionable intelligence, they empower leaders to monitor performance, identify areas for intervention, and align tactical decisions with overarching strategic objectives, significantly enhancing organizational agility and responsiveness.

Without well-designed leadership dashboards, executives often struggle with fragmented data, relying on outdated or manually compiled reports that provide delayed and inconsistent insights. This lack of a unified, real-time view of critical business information leads to reactive decision-making, an inability to quickly identify and address performance deviations, and difficulties in aligning various departmental efforts with overarching strategic goals. The absence of immediate, digestible insights results in inefficiencies, missed opportunities, and a diminished capacity for proactive strategic management, ultimately slowing down organizational responsiveness and hindering competitive advantage.

Implementation	Challenges
* Define critical KPIs aligned to strategy * Identify and integrate all relevant data sources * Design intuitive UI/UX for executive consumption * Develop robust data pipelines & visualizations * Implement stringent security & access controls * Pilot, gather feedback & iterate continuously	* Data fragmentation & inconsistency across systems * Defining truly impactful and actionable KPIs * Ensuring data freshness, accuracy & reliability * Balancing summary detail with drill-down capability * Tool selection and complex integration * User adoption and fostering data literacy

Products	Platform Cost	Project Cost	Duration
Tableau	$10-50K	$50-250K	3-9 m
Microsoft Power BI	$5-30K	$40-200K	2-8 m
Qlik Sense	$15-60K	$70-300K	3-10 m
Domo	$20-80K	$80-350K	4-11 m
Looker (Google Cloud)	$10-45K	$60-280K	3-9 m
MicroStrategy	$20-70K	$90-320K	4-10 m

People Analytics applies data science and statistical methods to human resources (HR) data to uncover insights into workforce trends, employee behavior, performance drivers, and engagement levels. By analyzing data from recruitment, retention, performance reviews, and other HR systems, it enables data-driven talent management, optimizes HR strategies, and helps forecast future workforce needs. This capability empowers leaders to make more objective decisions about hiring, development, retention, and organizational design, ensuring that human capital strategies are directly aligned with business objectives and maximizing employee potential.

Without leveraging people analytics, HR decisions are often based on intuition, anecdotal evidence, or outdated practices, leading to suboptimal outcomes and missed opportunities for workforce optimization. Organizations may suffer from high employee attrition without understanding its root causes, make ineffective hiring choices, fail to identify critical skill gaps, or implement engagement initiatives that miss the mark. The absence of data-driven HR insights results in inefficient talent acquisition, lower employee productivity, reduced morale, and an inability to proactively address workforce challenges, ultimately hindering organizational performance and competitive edge.

Implementation	Challenges
* Define HR data strategy & governance * Consolidate HR data from various sources * Implement analytics tools & platforms * Develop/apply statistical models to HR data * Interpret insights & recommend actionable strategies * Ensure strict data privacy & ethical use	* Navigating complex data privacy & anonymization rules * Integrating disparate HR data systems (HRIS, ATS) * Significant skillset gap in HR analytics & data science * Translating analytical insights into actionable HR policies * Overcoming resistance from traditional HR teams * Ensuring ethical considerations and preventing bias

Products	Platform Cost	Project Cost	Duration
Workday Prism Analytics	$30-100K	$150-500K	4-12 m
Visier	$40-150K	$180-600K	5-14 m
IBM Watson Talent Analytics	$25-90K	$120-400K	4-11 m
Oracle HCM Cloud Analytics	$20-80K	$100-380K	4-10 m
Tableau (with HR data integration)	$10-50K	$50-250K	3-9 m
Peakon (Workday)	$15-60K	$70-300K	3-9 m

Sentiment Analysis, a branch of Natural Language Processing (NLP), uses computational techniques to determine the emotional tone, attitude, or opinion expressed within text data, categorizing it as positive, negative, or neutral. This powerful capability allows organizations to extract real-time insights from unstructured text sources such as customer reviews, social media mentions, employee feedback, and market commentary, providing an immediate pulse on public or internal perception. By automating the analysis of vast volumes of qualitative data, sentiment analysis empowers leaders to understand feelings and opinions at scale, enabling proactive responses to address dissatisfaction, capitalize on positive trends, and protect brand reputation.

Without employing sentiment analysis, organizations are severely limited in their ability to understand and respond to the nuanced emotions and opinions of their customers, employees, and the broader market. Manually sifting through vast amounts of text data is impractical, leading to delayed insights, missed opportunities to address dissatisfaction proactively, and a failure to gauge public perception accurately. The absence of this capability means critical feedback goes unnoticed, brand reputation is vulnerable to unaddressed negative sentiment, and strategic decisions are made without a complete understanding of stakeholder feelings, ultimately hindering customer loyalty, employee engagement, and market responsiveness.

Implementation	Challenges
* Define relevant text data sources * Select NLP model/tool for sentiment extraction * Pre-process text data for optimal analysis * Train/fine-tune models for specific domain context * Visualize and report sentiment trends & changes * Integrate insights into decision workflows	* Handling ambiguity, sarcasm & subtle nuances in text * Domain-specific language variations & jargon * Processing vast data volumes efficiently * Achieving accuracy for multilingual content * Ethical concerns of monitoring private conversations * Translating aggregated sentiment into actionable steps

Products	Platform Cost	Project Cost	Duration
Brandwatch	$20-80K	$80-350K	3-9 m
Qualtrics (with Text iQ)	$15-70K	$70-300K	3-8 m
Clarabridge (Qualtrics XM Discover)	$25-90K	$100-400K	4-10 m
Medallia	$20-75K	$90-320K	3-9 m
AWS Comprehend	$5-30K	$40-150K	2-6 m
Google Cloud Natural Language API	$5-25K	$30-120K	2-5 m

Natural Language Business Intelligence (NLBI) empowers users to interact with data and extract insights using everyday conversational language, eliminating the need for complex queries, technical jargon, or specialized coding skills. This innovative approach allows non-technical business users, including senior leadership, to ask questions in plain English (or other natural languages) and receive immediate, relevant data visualizations or answers. By democratizing access to business intelligence, NLBI significantly speeds up insight discovery, fosters greater data literacy across the organization, and reduces reliance on data specialists, enabling faster, more intuitive, and widespread data-driven decision-making at all levels.

Without implementing Natural Language BI, data analysis typically remains a bottleneck, accessible primarily to a limited number of data specialists or IT personnel who are proficient in complex query languages or BI tools. This creates significant delays in obtaining critical insights, as business users must submit requests and wait for reports, hindering their ability to respond quickly to dynamic business needs. The absence of self-service, conversational data access leads to underutilized data assets, reduced agility in decision-making, and a limited ability for broader organizational engagement with valuable business intelligence, ultimately slowing down the pace of innovation and strategic execution.

Implementation	Challenges
* Define clear data models & business vocabulary * Select a suitable NLBI platform/tool * Integrate platform with all relevant data sources * Train/configure NL interface for accuracy * Conduct comprehensive user training & adoption programs * Continuously refine models based on user queries	* Accuracy of NLP interpretation and understanding * Handling ambiguity in natural language queries * Ensuring robust data governance and access controls * Maintaining optimal performance with large datasets * Training NL models for specific domain terminologies * Managing user expectations vs. system capabilities

Products	Platform Cost	Project Cost	Duration
Tableau Ask Data	$10-50K	$50-250K	3-9 m
Microsoft Power BI Q&A	$5-30K	$40-200K	2-8 m
Qlik Sense (Cognitive Engine)	$15-60K	$70-300K	3-10 m
ThoughtSpot	$25-100K	$100-450K	4-12 m
Sisense (Narrative Science integration)	$20-80K	$90-380K	3-10 m
	$10-40K	$60-280K	3-9 m

6x6™ Transformation Blueprint

'Integration' Strategy

"The systems should communicate, not compete." — Diane Greene
Co-founder of VMware; key figure in advancing cloud-native, interoperable architectures.

Integration begins with intent. Strategy synchronization isn't just system harmony—it's about making executive vision traceable across platforms. Data orchestration, event-driven architecture, and service bus logic convert leadership strategy into responsive infrastructure. Mature integration means each dashboard click, workflow automation, and interdepartmental signal reflects top-level ambition. Fragmented systems dilute transformation—aligned systems echo it. Leaders gain maturity when integration enables decisions to ripple outward without friction. Success lies not in tools but in choreography—an enterprise that dances in sync with executive foresight, not because it's told to, but because it's built that way.

"Integration Strategy" is vital for leadership transformation's digital maturity because it ensures that all organizational efforts, systems, and data flows are harmoniously aligned with strategic digital goals. A digitally mature leadership team champions seamless data and process integration through workflow automation, iPaaS platforms, and API orchestration, effectively breaking down silos. This creates a unified flow of information and operations, fostering cross-functional collaboration and real-time visibility. Such synchronization is critical for efficiently executing complex digital initiatives, accelerating agile builds through low-code integration, and adapting quickly to new opportunities or challenges in a dynamic market.

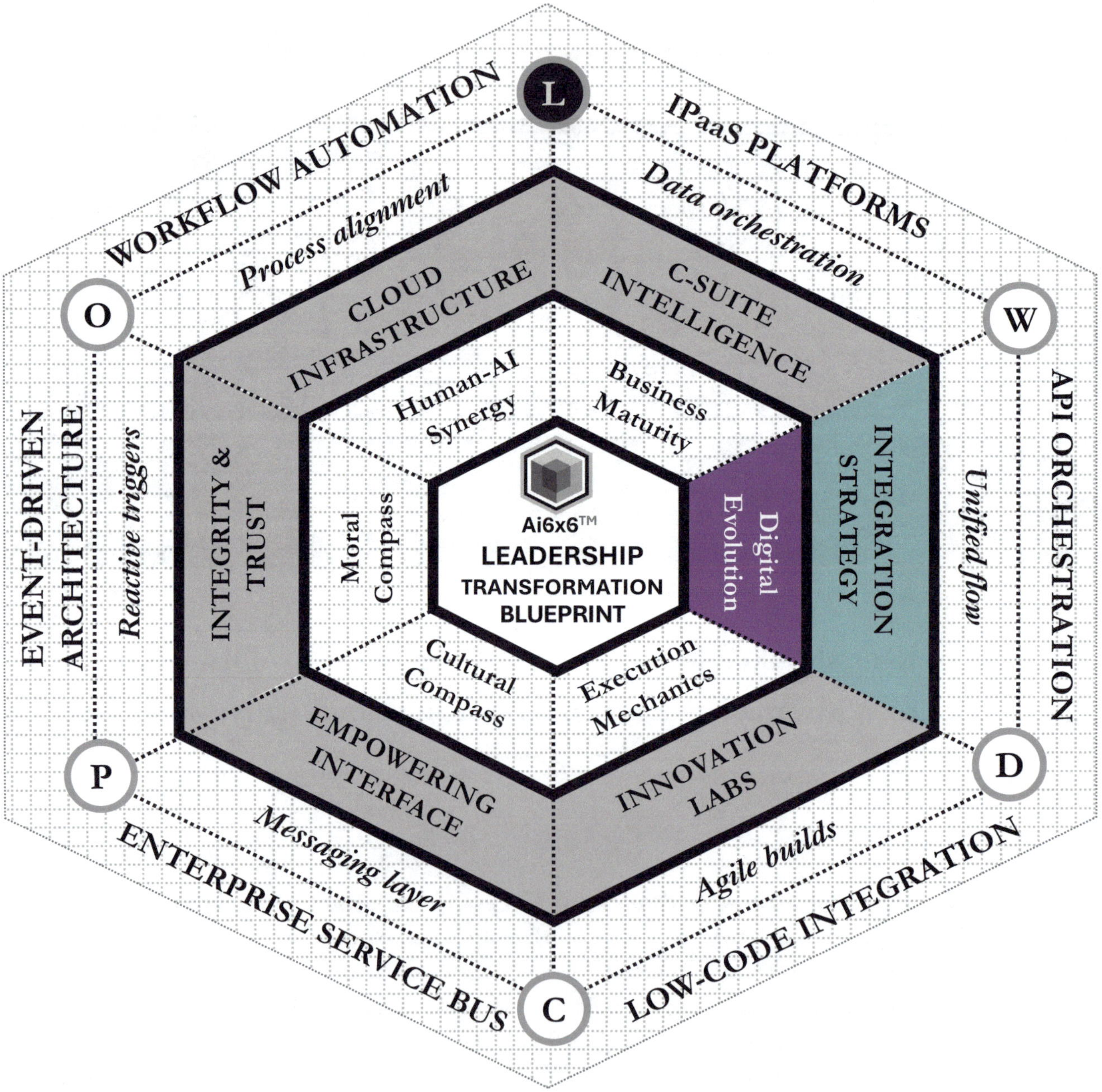

Ai6x6 DIGITAL EVOLUTION FOR INTEGRATION LEADERSHIP TRANSFORMATION

1. WORKFLOW AUTOMATION

Workflow automation streamlines repetitive, rule-based tasks and multi-step processes across an organization by using software and pre-defined logic to execute them automatically. This transformative shift from manual to automated execution significantly reduces human error, accelerates operational speed, and frees up valuable employee time from mundane, administrative activities, allowing them to focus on higher-value, strategic initiatives that drive growth and innovation. Effective automation ensures consistency in process execution, drastically improves data accuracy by minimizing manual data entry, and provides clear audit trails, all of which are critical for enhancing overall operational efficiency, ensuring compliance, and boosting employee satisfaction.

Failing to implement robust workflow automation leaves organizations burdened with inefficient, costly, and error-prone manual processes that drain resources and human capital. This perpetuates bottlenecks, leads to significant delays in critical operations, and inflates operational expenses due to the necessity of allocating valuable human resources to repetitive, low-value tasks. Furthermore, a lack of automation severely hinders an organization's scalability, as manual processes cannot easily adapt to increased volume or rapidly changing business demands, ultimately impeding its ability to respond quickly to market shifts, maintain a competitive edge, and achieve its full digital potential.

Implementation	Challenges
* Identify repetitive, rule-based processes * Map current workflows and identify bottlenecks * Select appropriate automation platform/tools * Design and configure automated workflows * Test, refine, and deploy automation solutions * Monitor performance and optimize continuously	* Resistance to change from employees * Accurately identifying suitable processes for automation * Complex integration with disparate legacy systems * Maintaining and updating complex automated workflows * Significant initial cost of tools and implementation * Risk of over-automating processes unnecessarily

Products	Platform Cost	Project Cost	Duration
UiPath	$30-150K	$150-700K	4-12 m
Microsoft Power Automate	$5-40K	$50-250K	2-8 m
Appian	$40-200K	$200-900K	6-18 m
Automation Anywhere	$30-140K	$140-650K	4-11 m
Pega Systems	$50-250K	$300-1200K	8-20 m
Zapier	$1-10K	$10-50K	1-3 m

An Integration Platform as a Service (iPaaS) provides a comprehensive, cloud-based suite of tools and services specifically designed for connecting disparate applications, data sources, and business processes, regardless of whether they reside on-premises or in the cloud. It offers a robust framework with pre-built connectors, intuitive data mapping capabilities, and powerful workflow orchestration engines, significantly simplifying complex enterprise-wide integrations that are vital for creating a unified and agile digital ecosystem. By abstracting the complexities traditionally associated with custom point-to-point integrations, iPaaS accelerates development cycles, ensures consistent data flow across all systems, and empowers IT teams to manage all integrations centrally, thereby enhancing overall organizational agility, reducing technical debt, and fostering real-time data accessibility.

Without an iPaaS, organizations typically rely on brittle, custom-coded point-to-point integrations that are inherently difficult to build, maintain, and scale as the IT landscape evolves. This fragmented approach invariably leads to pervasive data silos, inconsistent information across different departments and applications, increased operational overhead for troubleshooting and debugging, and significant delays in implementing new applications or services. The absence of a centralized and standardized integration strategy results in a chaotic and inflexible IT architecture, severely hindering critical strategic initiatives like comprehensive digital transformation, real-time data analysis, and seamless customer experiences, ultimately slowing down business innovation and eroding competitive advantage.

Implementation	Challenges
* Assess current integration needs & landscape * Select a suitable iPaaS vendor & platform * Plan detailed data flows and API strategies * Develop/configure integrations using connectors * Test, deploy, and monitor integrations * Establish governance & best practices for use	* Complexity of integrating with legacy systems * Intricate data mapping and transformation issues * Ensuring optimal performance and scalability * Managing security & compliance for data in transit * Specific skillset required for advanced iPaaS features * Potential for vendor lock-in and opaque pricing models

Products	Platform Cost	Project Cost	Duration
MuleSoft Anypoint Platform	$30-150K	$250-1000K	6-18 m
Boomi (Dell Technologies)	$20-100K	$180-800K	5-16 m
Workato	$25-110K	$200-900K	6-17 m
Tray.io	$10-60K	$80-400K	3-10 m
Microsoft Azure Integration Services	$15-70K	$120-600K	4-14 m
AWS AppFlow / Step Functions	$10-50K	$90-500K	4-12 m

API orchestration involves intelligently combining and sequencing calls to multiple individual Application Programming Interfaces (APIs) into a single, cohesive, and often more complex service or automated workflow, typically adding intricate business logic in between. This powerful capability allows for the creation of new, higher-value services and streamlined data retrieval without requiring alterations to underlying backend systems, significantly simplifying integration efforts for consuming applications and enhancing data consistency. By abstracting the complexity of multiple API interactions, orchestration dramatically boosts developer productivity, ensures consistent data presentation across diverse sources, and enables the rapid development of innovative digital products and experiences that rely on disparate data and functionalities.

Without effective API orchestration, developers would be compelled to make numerous individual API calls for each piece of data or function, manually stitching together responses, and painstakingly managing complex error handling for every interaction. This leads to greatly increased development time and costs, higher network traffic, a greater propensity for inconsistencies or errors due to manual data aggregation, and ultimately, a fragmented and inefficient user experience. The absence of this strategic capability severely limits an organization's ability to innovate quickly, build robust new applications, and leverage its existing digital assets effectively, thereby slowing down market responsiveness and impeding overall digital transformation efforts.

Implementation	Challenges
* Identify complex cross-system business processes * Design the logical flow for orchestrated APIs * Select an appropriate orchestration platform/tool * Implement business logic & detailed data mapping * Thoroughly test, secure, and deploy orchestrated APIs * Monitor performance and optimize continuously	* Managing the inherent complexity of chained APIs * Ensuring robust fault tolerance & comprehensive error handling * Addressing performance tuning & potential latency issues * Maintaining data consistency across multiple APIs * Implementing effective version control for orchestrated flows * Securing complex, multi-component API interactions

Products	Platform Cost	Project Cost	Duration
MuleSoft Anypoint Platform	$30-150K	$250-1000K	6-18 m
Apache Camel (Open Source)	$0-20K	$50-300K	3-9 m
WSO2 API Manager	$15-70K	$100-450K	4-12 m
IBM API Connect	$20-90K	$150-600K	5-14 m
Google Cloud Apigee (Mediation)	$20-100K	$150-500K	4-12 m
Azure Logic Apps / API Management	$5-40K	$75-300K	3-9 m

Low-code integration platforms empower a broader range of users, including business analysts and citizen developers, to connect applications, systems, and data sources with significantly reduced manual coding. This is primarily achieved through intuitive visual interfaces, drag-and-drop functionalities, and a rich library of pre-built components and connectors. This democratized approach to integration dramatically accelerates the development and deployment of new connections, empowering business users to create necessary linkages rapidly without requiring deep programming expertise. By reducing development backlogs and improving time-to-market for new functionalities, low-code integration fosters greater collaboration between IT and business teams, driving unprecedented agility and innovation across the entire organization.

The absence of low-code integration capabilities perpetuates a bottleneck where organizations remain heavily reliant on a limited pool of highly skilled developers for every single integration task. This dependency leads to prolonged development cycles, mounting IT backlogs, and an inability to swiftly respond to evolving business demands or integrate new technologies. Consequently, organizations become less responsive to critical market changes, accumulate significant technical debt from a proliferation of complex, custom-coded point-to-point integrations, and struggle to leverage their valuable data effectively across the enterprise, ultimately hindering their overall digital transformation journey and eroding their competitive standing.

Implementation	Challenges
* Identify integration pain points & opportunities * Select a suitable low-code integration platform * Design integration flows using visual interface * Configure connectors and perform data mapping * Rigorously test & deploy integrations * Establish governance and monitor solutions	* Limited flexibility for highly complex or unique logic * Potential for "shadow IT" creation without proper governance * Concerns regarding vendor lock-in and platform portability * Scalability limitations for extremely high-volume scenarios * Ensuring comprehensive governance and security oversight * Debugging and maintaining complex visual flows

Products	Platform Cost	Project Cost	Duration
Workato	$25-110K	$100-500K	3-10 m
Zapier	$1-10K	$10-50K	1-3 m
Microsoft Power Automate	$5-40K	$40-200K	2-8 m
Integrately	$1-5K	$5-30K	1-2 m
Appian	$40-200K	$150-700K	4-12 m
Retool	$5-30K	$30-150K	2-6 m

5. ENTERPRISE SERVICE BUS (ESB)

An Enterprise Service Bus (ESB) acts as a central communication backbone within an organization's IT architecture, facilitating standardized and managed interactions between disparate applications, services, and data sources. It provides robust capabilities for message routing, data transformation, protocol conversion, and request orchestration, enabling a loosely coupled integration paradigm that fosters service reuse and enhanced interoperability. By centralizing communication and mediation logic, an ESB significantly improves the manageability of complex enterprise landscapes, enhances the reliability and security of data exchange, and lays a robust foundational layer for Service-Oriented Architecture (SOA), which is crucial for operating large-scale, interconnected digital enterprises.

Without the implementation of an Enterprise Service Bus, organizations often find themselves struggling with a tangled, unsustainable web of point-to-point integrations, commonly referred to as "spaghetti architecture." This chaotic approach leads to immense complexity in managing system interactions, exorbitant maintenance costs, and extreme fragility where a change in one system can inadvertently break numerous others. The absence of a standardized and centralized integration layer severely hinders the ability to introduce new applications or services efficiently, results in pervasive data silos, creates inconsistent business logic across different applications, and drastically reduces an organization's agility to adapt to evolving business requirements or integrate with new technologies.

Implementation	Challenges
* Define clear integration strategy & scope * Select suitable ESB platform & design architecture * Develop connectors and message transformations * Implement robust routing and orchestration logic * Thoroughly test, deploy, and monitor services * Establish comprehensive governance & security policies	* High initial setup complexity and configuration effort * Potential performance bottlenecks if not scaled correctly * Requires specialized technical skills for implementation * Risk of becoming a single point of failure (if not clustered) * Can evolve into a monolithic integration bottleneck * Significant maintenance and upgrade complexities

Products	Platform Cost	Project Cost	Duration
MuleSoft Anypoint Platform	$30-150K	$250-1000K	6-18 m
Apache Camel (Open Source)	$0-20K	$50-300K	3-9 m
WSO2 Enterprise Integrator	$15-70K	$100-450K	4-12 m
IBM App Connect Enterprise	$20-90K	$150-600K	5-14 m
TIBCO BusinessWorks	$25-110K	$200-750K	6-16 m
Software AG webMethods	$20-100K	$180-700K	5-15 m

6. EVENT-DRIVEN ARCHITECTURE

Event-Driven Architecture (EDA) is a highly flexible and scalable software design pattern where loosely coupled services and applications communicate asynchronously by producing and consuming "events" through an event broker (e.g., a message queue or stream). This paradigm enables systems to react in real-time to changes or activities, fostering extreme scalability, resilience, and agility across the enterprise. By promoting true decoupling between services, components can operate and be updated independently without affecting others, which is fundamentally crucial for modern, dynamic applications and responsive business processes, significantly accelerating innovation cycles and improving system robustness.

Without adopting an event-driven architecture, systems typically rely on synchronous, tightly coupled interactions, often involving direct API calls or batch processing for data exchange. This creates rigid dependencies between components, making systems inherently brittle and exceptionally difficult to scale, especially under high load or when rapid changes are required across the distributed landscape. The absence of real-time reactivity means missed opportunities for immediate action on critical business events, slower data propagation across the enterprise, and reduced system resilience against individual component failures. This ultimately hinders the ability to build truly responsive, scalable, and adaptable digital products and services, slowing down competitive advantage and market leadership.

Implementation	Challenges
* Define events, their schemas, & business context * Select a robust event broker (e.g., Kafka, RabbitMQ) * Design and implement event producers and consumers * Implement robust error handling & idempotency * Monitor event streams, system health, & latency * Strategically refactor monolithic apps into microservices	* Ensuring consistent event delivery & ordering * Complexity in distributed tracing and debugging * Guaranteeing data delivery and processing reliability * Preventing "event sprawl" and maintaining strong governance * Significant skillset required for distributed systems design * Effectively handling dead letter queues and retries

Products	Platform Cost	Project Cost	Duration
Apache Kafka (Open Source)	$0-30K	$100-500K	4-12 m
Confluent Platform (Kafka Enterprise)	$20-100K	$150-700K	5-14 m
Amazon Kinesis / SNS / SQS	$5-40K	$75-300K	3-9 m
Azure Event Hubs / Service Bus	$5-35K	$60-280K	3-8 m
RabbitMQ (Open Source)	$0-15K	$40-200K	2-7 m
Google Cloud Pub/Sub	$5-30K	$50-250K	2-7 m

6x6™ Transformation Blueprint

'Innovation' Labs

"Innovation requires leadership that trusts its teams to challenge assumptions." — Satya Nadella , Microsoft CEO; architected decentralized innovation through trust and experimentation.

An Innovation Lab is not a playground—it's a pressure chamber for enterprise ambition. Innovation hubs push ideas to their limits, but in transformation, they must also protect coherence. Hackathons, digital twins, and portfolio governance offer structured chaos—allowing creativity without destabilizing strategic momentum. Mature leaders design labs where experimentation is both encouraged and accountable. Failure is accepted—but only if it fuels refinement. When leadership innovation is institutionalized yet adaptive, transformation becomes rhythmic: bold but measured, imaginative but disciplined. Labs no longer explore possibility—they rehearse future capability with precision and purpose.

The "Innovation Labs" chapter is paramount for leadership transformation's digital maturity as it emphasizes cultivating a dynamic environment where groundbreaking ideas are continuously generated, tested, and scaled. A digitally mature leadership fosters a culture of rapid prototyping, experimentation through hackathons, and human-centered design thinking. This enables the organization to proactively develop new digital products, services, and business models, leveraging tools like digital twin strategy and crowdsourced ideation. This commitment to agile innovation ensures sustained competitive advantage by constantly pushing boundaries and bringing transformative solutions to market.

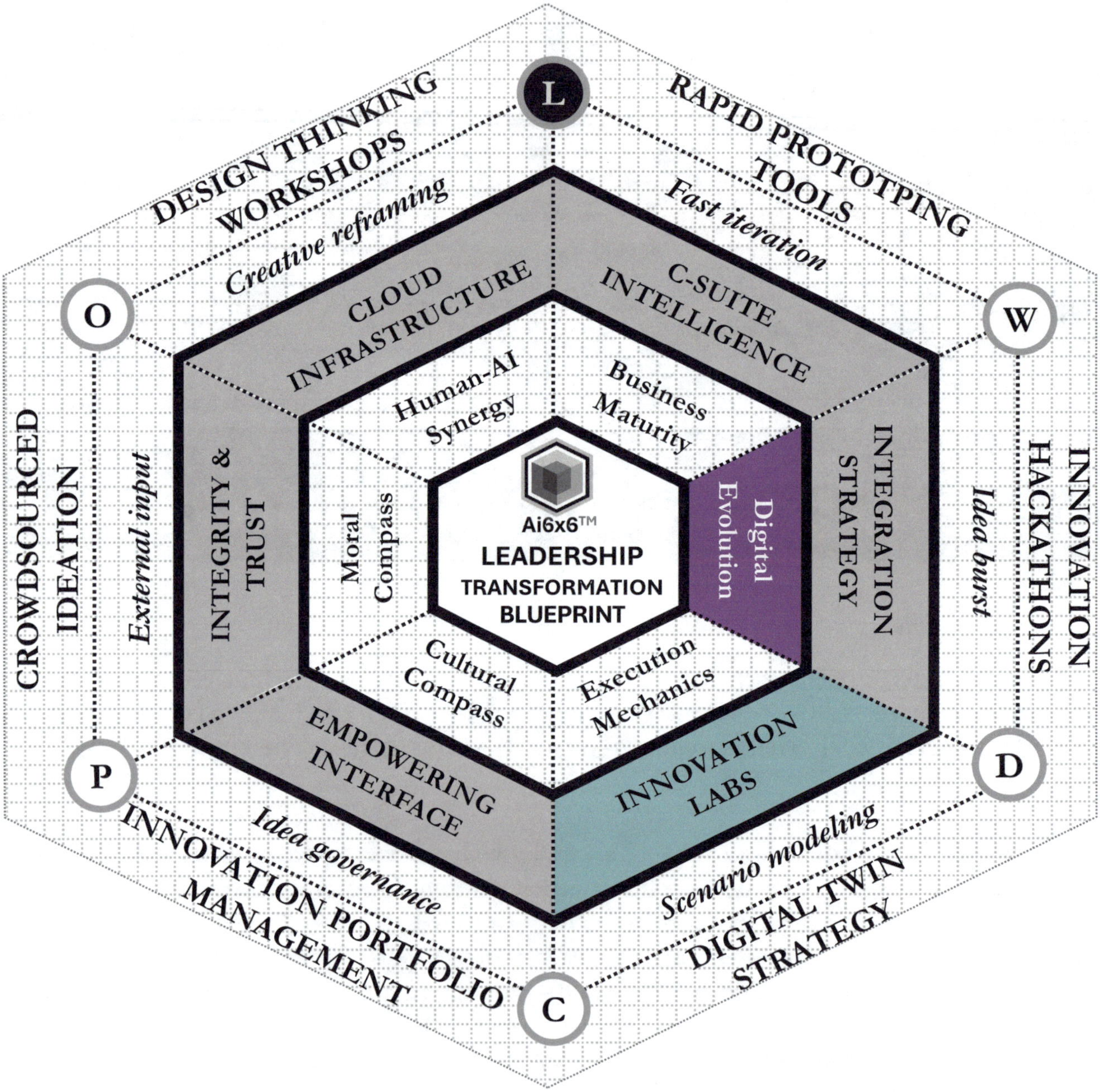

Ai6x6 DIGITAL EVOLUTION FOR INNOVATION
LEADERSHIP TRANSFORMATION

Design Thinking Workshops are highly collaborative, hands-on sessions that immerse participants in a human-centered problem-solving methodology, focusing intensely on understanding and addressing user needs. These structured workshops guide diverse teams through a cyclical process of empathy, ideation, rapid prototyping, and iterative testing, fostering a deep understanding of challenges before developing solutions. By promoting cross-functional collaboration and a culture of rapid experimentation, they empower teams to uncover profound insights, generate a wide array of creative ideas, and quickly validate concepts with real users, thereby significantly accelerating the innovation pipeline and ensuring that developed solutions truly resonate with their intended audience.

Without engaging in structured Design Thinking Workshops, organizations often develop products, services, or internal processes that fail to adequately address actual user needs or market demands, leading to low adoption rates, wasted resources, and ultimately, failed innovation initiatives. This traditional, often siloed approach can result in solutions that are technically sound but lack genuine market fit, user appeal, or practical usability. The absence of a systematic, empathetic innovation process means that problems are not thoroughly understood from the user's perspective, leading to reactive development cycles, missed opportunities for breakthrough ideas, and a culture that is less adept at creative problem-solving and agile, user-centric experimentation.

Implementation	Challenges
* Define clear problem statement & desired outcome * Assemble diverse, cross-functional team of participants * Facilitate empathy-building activities with users * Brainstorm and ideate a wide range of solutions * Rapidly prototype concepts and conduct user testing * Iterate on solutions based on user feedback	* Gaining full buy-in from all stakeholders * Requiring highly skilled facilitator expertise * Maintaining focus and managing scope creep * Translating workshop insights into actionable steps * Overcoming organizational resistance to failure/iteration * Sustaining momentum and implementing ideas post-workshop

Products	Platform Cost	Project Cost	Duration
Miro / Mural (Whiteboarding)	$1-10K	$5-50K	0.5-2 m
Figma (Prototyping)	$1-5K	$3-20K	0.5-1 m
InVision (Prototyping)	$1-5K	$3-20K	0.5-1 m
Google Jamboard (Digital Whiteboard)	$0-2K	$2-10K	0.5-1 m
Zoom/Teams (Collaboration)	$0-5K	$1-10K	0.5-1 m
Custom Workshop Facilitation	N/A	$10-100K	0.5-2 m

2. RAPID PROTOTYPING TOOLS

Rapid Prototyping Tools are specialized software applications and frameworks that enable the quick and cost-effective creation of functional models or simulations of products, services, or new features. These tools allow designers and developers to swiftly visualize abstract ideas, test underlying assumptions, and gather early, critical feedback from potential users or key stakeholders before committing significant resources to full-scale development. By drastically accelerating the iteration cycle, they effectively reduce overall development risks, identify usability issues at an early stage, and ensure that the final product aligns closely with both user expectations and evolving market demands, thereby fostering a more agile, efficient, and user-centric innovation process.

Without the strategic utilization of rapid prototyping tools, organizations are often compelled into lengthier, more expensive development cycles where product ideas are built out completely before any substantial user testing or validation occurs. This traditional "big bang" approach significantly increases the risk of delivering products that ultimately fail to meet user needs or market acceptance, frequently leading to costly reworks, extensive delays, or even complete abandonment of the project. The absence of quick and continuous feedback loops means that fundamental design flaws, usability issues, and market misalignment are discovered late in the development process, resulting in budget overruns, delayed market entry, and a diminished capacity for iterative improvement and responsive innovation, severely impacting competitive advantage.

Implementation	Challenges
* Define clear prototype scope & fidelity * Choose appropriate prototyping tool for context * Design and build prototype rapidly and iteratively * Conduct user testing and gather comprehensive feedback * Analyze feedback & iterate on prototype design * Seamlessly transition successful prototypes to development	* Balancing speed of creation with necessary prototype fidelity * Ensuring realistic and representative user testing scenarios<* Managing stakeholder expectations for unfinished prototypes * Smoothly translating prototype insights into production code * Overcoming potential "prototype fatigue" within teams * Balancing tool complexity versus ease of adoption for users

Products	Platform Cost	Project Cost	Duration
Figma	$1-5K	$3-20K	0.5-1 m
Adobe XD	$1-5K	$3-20K	0.5-1 m
Sketch	$1-3K	$2-15K	0.5-1 m
Proto.io	$1-3K	$2-15K	0.5-1 m
Axure RP	$1-4K	$3-18K	0.5-1 m
Framer	$1-3K	$2-15K	0.5-1 m

3. INNOVATION HACKATHONS

Innovation Hackathons are intense, time-bound events where diverse, cross-functional teams collaboratively work to develop innovative solutions to specific business challenges or open-ended problems within a compressed timeframe. These high-energy, creative sprints foster rapid ideation, hands-on prototype development, and interdisciplinary collaboration, frequently leading to the emergence of breakthrough concepts or minimum viable products (MVPs) in just a few days. By explicitly encouraging experimentation, managed risk-taking, and a playful yet focused approach to problem-solving, hackathons significantly energize employees, uncover hidden internal talent, and generate a vibrant pipeline of novel ideas that might not emerge through more traditional, rigid innovation processes.

Without regularly organizing Innovation Hackathons, organizations may struggle to effectively tap into the full creative potential of their vast internal workforce and broader external talent pools. Traditional hierarchical structures, daily operational demands, and rigid departmental silos can significantly stifle spontaneous ideation and inhibit interdisciplinary collaboration, often leading to a slower, more incremental, and less disruptive approach to innovation. The absence of such focused, high-energy events means fewer radical ideas are generated, a slower pace of internal experimentation, and a missed opportunity to cultivate an internal culture of rapid prototyping, creative problem-solving, and cross-pollination of ideas, ultimately impacting the organization's ability to disrupt markets or respond quickly to emerging changes

Implementation	Challenges
* Define clear & compelling challenge statement * Recruit diverse & skilled participants * Provide necessary resources (mentors, tools, data) * Facilitate intense ideation & rapid prototyping * Organize transparent judging & final presentations * Ensure structured follow-up on promising ideas	* Ensuring the problem definition is sufficiently clear * Sustaining high energy and participant engagement * Effectively managing diverse team dynamics * Translating raw hackathon ideas into actionable projects * Providing adequate resources and expert mentorship * Crucial post-hackathon follow-through and funding

Products	Platform Cost	Project Cost	Duration
Devpost (Hackathon Platform)	$2-10K	$10-50K	0.5-1 m
HackerEarth (Hackathon Platform)	$2-10K	$10-50K	0.5-1 m
Zoom/Teams (Virtual Collaboration)	$0-5K	$1-10K	0.5-1 m
Miro/Mural (Whiteboarding)	$1-10K	$5-30K	0.5-1 m
AWS/Azure/GCP Credits (Dev Resources)	$0-10K	$5-20K	0.5-1 m
Event Management Software	$1-5K	$5-25K	0.5-1 m

A Digital Twin Strategy involves the sophisticated creation of virtual replicas, or "digital twins," of physical assets, intricate processes, or entire operational systems, which are continuously updated with real-time data from embedded sensors and other data sources. These dynamic digital models allow for comprehensive, granular monitoring, sophisticated simulation, and advanced predictive analysis of their physical counterparts without impacting the real-world system. By providing unprecedented real-time visibility, predictive capabilities, and controlled experimentation environments, a robust digital twin strategy empowers organizations to make profoundly data-driven decisions, anticipate operational issues before they occur, optimize performance across the board, and innovate more safely and efficiently across their products, operations, and customer experiences, leading to significant competitive advantage.

Without establishing a comprehensive Digital Twin Strategy, organizations operate with critically limited visibility into the real-time performance, operational health, and potential issues of their physical assets and complex processes. This reliance on fragmented or historical data leads to reactive maintenance practices, inefficient operations due to a lack of predictive insights, and a critical inability to simulate "what-if" scenarios or test innovations without disrupting live systems, which is both costly and risky. The absence of a dynamic digital replica means missed opportunities for continuous performance optimization, significant downtime due to unforeseen failures, and a constrained capacity for developing new, data-driven services, ultimately hindering efficiency, stifling innovation, and diminishing competitive advantage in an increasingly data-intensive and interconnected world

Implementation	Challenges
* Identify critical assets/processes for twinning * Select appropriate digital twin platform & sensors * Integrate diverse real-time data streams * Build accurate virtual models & simulation logic * Develop insightful analytics & visualization dashboards * Pilot, validate accuracy, and scale deployment	* High initial investment in sensors, hardware, & platform * Complex data integration and ensuring data quality * Achieving high model accuracy and continuous calibration * Robust cybersecurity for IoT and twin data * Significant skill gaps in modeling & data science * Scalability and managing a large number of twins

Products	Platform Cost	Project Cost	Duration
Siemens Digital Twin	$50-200K	$300-1500K	9-24 m
GE Digital Predix	$40-180K	$250-1200K	8-20 m
AWS IoT TwinMaker	$20-100K	$100-500K	5-15 m
Azure Digital Twins	$20-90K	$90-450K	5-14 m
PTC ThingWorx	$30-120K	$150-700K	6-18 m
Dassault Systèmes 3DEXPERIENCE	$60-250K	$400-2000K	12-36 m

Innovation Portfolio Management involves the strategic selection, rigorous prioritization, and disciplined oversight of a diverse collection of innovation initiatives, ensuring their alignment with overarching organizational goals and maximizing the potential return on investment. It provides a structured, data-driven framework for systematically evaluating new ideas, allocating critical resources effectively across various projects, and tracking progress and performance across a spectrum of innovation efforts, from incremental improvements to potentially disruptive breakthroughs. By maintaining a balanced and dynamic portfolio, organizations can intelligently mitigate risks, optimize resource utilization, and ensure a continuous, robust pipeline of innovations that directly support long-term strategic objectives, fostering sustainable growth and maintaining competitive relevance in rapidly evolving markets.

Without a robust and disciplined Innovation Portfolio Management approach, organizations often suffer from fragmented, uncoordinated innovation efforts that critically lack strategic alignment and suffer from diffuse focus. This typically leads to significant resource waste on low-impact or ill-conceived ideas, a detrimental lack of concentrated focus on high-potential projects, and a critical inability to track the overall health, progress, and return of the innovation pipeline. The absence of a clear, centralized framework results in reactive decision-making based on intuition rather than data, missed opportunities for synergy between disparate projects, and a diminished organizational capacity to consistently deliver market-leading innovations, ultimately hindering sustainable growth, adaptability, and long-term market leadership.

Implementation	Challenges
* Define clear innovation strategy & criteria * Establish structured idea submission & evaluation process * Implement specialized portfolio management tools * Prioritize projects based on value, risk, & alignment * Allocate resources and track progress diligently * Regularly review & adjust portfolio composition	* Inherent subjectivity in early-stage idea evaluation * Balancing short-term incremental vs. long-term disruptive goals * Intense resource contention across multiple projects * Overcoming organizational resistance to killing "pet projects" * Accurately forecasting complex innovation ROI * Integrating seamlessly with existing project management systems

Products	Platform Cost	Project Cost	Duration
Planview ProjectPlace	$10-50K	$50-250K	3-9 m
Aha! (Roadmap & Product Mgmt)	$5-20K	$30-100K	2-6 m
ServiceNow Strategic Portfolio Mgmt	$20-80K	$100-400K	4-12 m
monday.com (Customizable)	$1-10K	$5-50K	1-3 m
Asana (Project Management)	$1-5K	$3-20K	1-2 m
Spigit (MindManager)	$15-60K	$70-300K	3-9 m

Crowdsourced Ideation is a powerful innovation approach that strategically leverages the collective intelligence, diverse perspectives, and varied experiences of a large group—which can include employees, customers, partners, or the broader public—to generate, refine, and select novel ideas for specific challenges. This open innovation methodology taps into a vast pool of creativity far beyond internal R&D teams, often leading to a greater quantity and significantly higher quality of ideas than traditional, insular brainstorming sessions. By actively engaging a broad community in the problem-solving process, it fosters a strong sense of ownership, dramatically increases the likelihood of identifying truly unique and breakthrough solutions, and cultivates a dynamic culture of open innovation, thereby enriching the organization's innovation pipeline and strengthening its engagement with its entire ecosystem.

Without actively implementing crowdsourced ideation strategies, organizations severely limit their innovation potential, restricting it primarily to internal teams or a select few individuals. This insular approach often leads to "groupthink," a narrow range of predictable ideas, and a significantly slower pace of innovation, as potential solutions are developed in isolation from the broader community that might ultimately benefit from them or contribute valuable insights. The absence of an open, accessible ideation platform means missed opportunities for truly groundbreaking ideas to emerge, reduced employee engagement in solving complex problems, and a diminished capacity to quickly adapt to evolving market needs or effectively leverage external expertise, ultimately hindering competitive advantage and long-term growth

Implementation	Challenges
* Define clear and inspiring challenge statement * Select a suitable crowdsourcing platform * Promote challenge effectively to target crowd * Facilitate idea submission & collaborative refinement * Systematically evaluate, refine, and select best ideas * Provide transparent feedback & appropriate rewards	* Ensuring high quality and relevance of submitted ideas * Efficiently managing and filtering a large volume of submissions * Effectively engaging and motivating the participating crowd * Navigating intellectual property (IP) and ownership considerations * Smoothly integrating selected ideas into existing projects * Avoiding "idea fatigue" and maintaining sustained crowd interest

Products	Platform Cost	Project Cost	Duration
Ideascale	$10-50K	$50-200K	2-6 m
Spigit (MindManager)	$15-60K	$70-300K	3-9 m
Brightidea	$20-70K	$90-350K	3-10 m
Innocentive	$10-40K	$60-250K	2-7 m
Crowdicity	$5-30K	$40-150K	2-6 m
Yammer/Teams (Internal Crowdsourcing)	$0-5K	$10-50K	1-3 m

6x6™ Transformation Blueprint

Empowering 'Interface'

"Interfaces shouldn't be barriers; they should be invitations." — Aza Raskin

Inventor of the infinite scroll; known for ethical tech and accessible design principles.

Empowering Interface isn't about sleekness—it's about presence. The executive interface must deliver clarity at a glance and control with a gesture. Dashboards, voice interfaces, and immersive briefings should amplify decision confidence—not bury it. Mature transformation means designing platforms that don't distract leaders—they anchor them. Real-time insights must be intuitive, not intrusive. Leadership doesn't scale through complexity—it scales through refinement. When executives interact seamlessly with organizational data, performance, and sentiment, their digital footprint mirrors their strategic influence. These interfaces aren't surfaces—they're power translators, turning cognition into command.

"Empowering Interface" is critical for leadership transformation's digital maturity because it focuses on empowering leaders with intuitive, accessible, and highly effective digital tools for communication, decision-making, and influence. A digitally mature leadership team leverages advanced interfaces, from interactive dashboards and executive chatbots to voice-enabled systems and mobile leadership apps. These interfaces ensure clarity, efficiency, and broad reach in their interactions, allowing leaders to command effectively across distributed environments, gain immersive insights through AR/VR briefings, and inspire widespread digital adoption throughout the organization, enhancing their overall digital presence and impact.

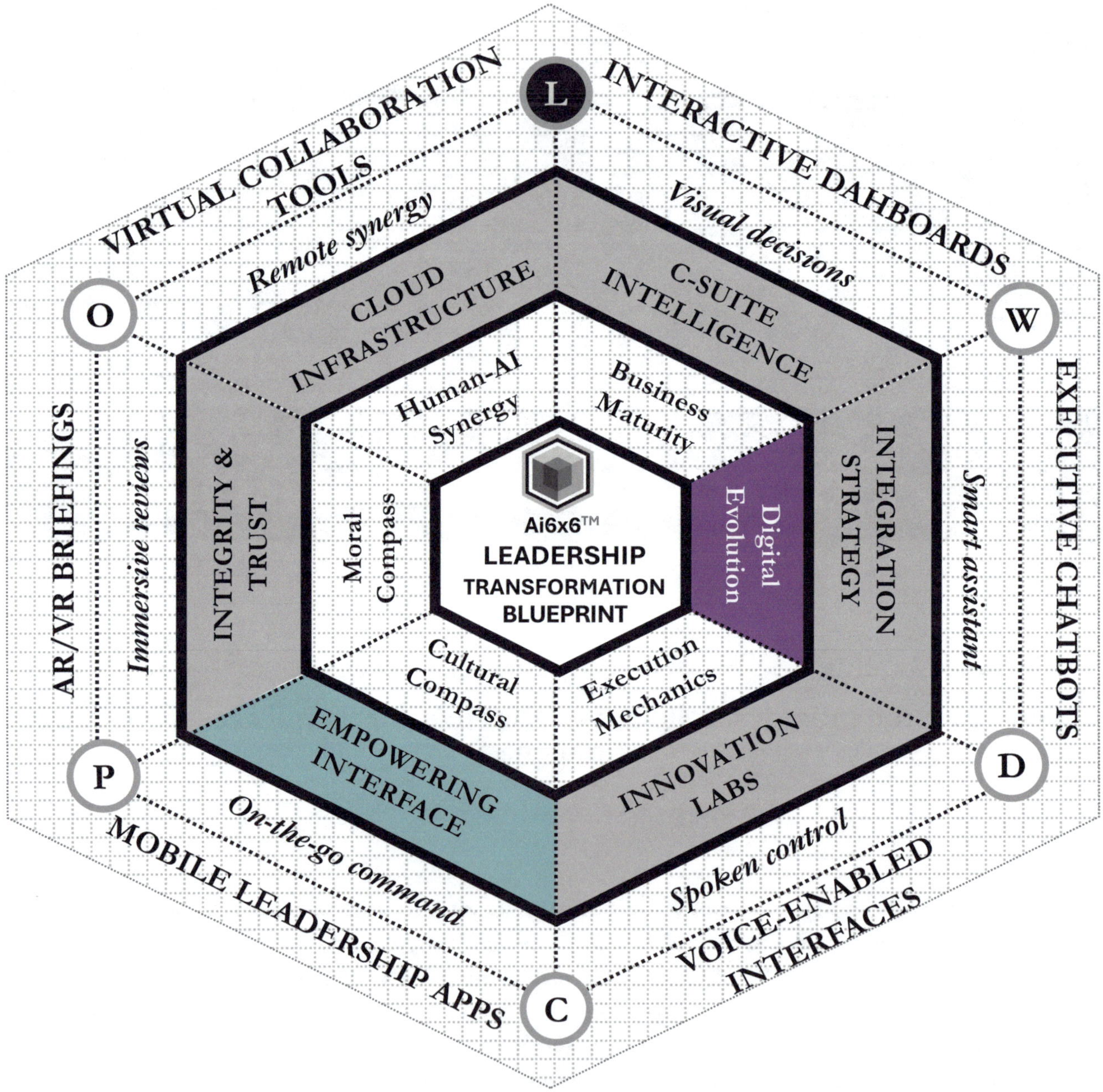

Ai6x6 DIGITAL EVOLUTION FRAMEWORK FOR INTERFACE LEADERSHIP TRANSFORMATION

Virtual Collaboration Tools are sophisticated software applications designed to enable distributed teams to communicate seamlessly, share information efficiently, and work together effectively, regardless of their physical location or time zone. These comprehensive tools often integrate a rich suite of features, including high-definition video conferencing, instant messaging, real-time shared document editing, and robust project management functionalities, all meticulously engineered to mimic and significantly enhance in-person interactions. By facilitating fluid communication, providing centralized shared workspaces, and enabling asynchronous teamwork, they ensure robust business continuity, dramatically boost overall team productivity, and foster a connected, inclusive organizational culture, which is absolutely paramount for the success of modern hybrid and fully remote work models, as well as global operations.

Without the strategic adoption of robust virtual collaboration tools, organizations face significant and escalating challenges in maintaining cohesive, engaged, and productive remote or hybrid teams. Communication silos inevitably emerge, leading to frequent misunderstandings, delayed decision-making processes, and costly duplicated efforts due to fragmented information sharing across disparate channels. The absence of a unified, intuitive platform can also negatively impact employee engagement and morale, as individuals may feel isolated, disconnected, and struggle to effectively align with team and organizational objectives. Ultimately, a lack of effective virtual collaboration severely hinders an organization's ability to operate efficiently in a distributed environment, stifles the very innovation that relies on collective brainstorming, and significantly limits its talent acquisition capabilities to narrow constraints.

Implementation	Challenges
* Assess comprehensive communication & collaboration needs * Select suitable platform(s) with robust integrations * Plan phased rollout with extensive user training * Establish clear usage guidelines & best practices * Monitor adoption rates & gather continuous user feedback * Provide dedicated and ongoing technical support	* Significant user adoption hurdles and resistance to change * Managing information overload and notification fatigue * Complex integration with existing legacy systems * Addressing critical data security and privacy concerns * Ensuring equitable participation in virtual meetings * Resolving technical issues (bandwidth, device compatibility)

Products	Platform Cost	Project Cost	Duration
Microsoft Teams	$0-30K	$30-150K	2-6 m
Zoom Enterprise	$5-25K	$20-100K	1-4 m
Slack Enterprise Grid	$10-40K	$40-200K	2-7 m
Google Workspace (Meet, Chat)	$3-25K	$20-120K	2-5 m
Webex by Cisco	$5-20K	$25-110K	2-5 m
Mural / Miro (Whiteboarding)	$1-10K	$5-50K	0.5-2 m

Interactive Dashboards are dynamic and highly intuitive data visualization tools that empower users to explore complex datasets, apply various filters, and drill down into granular details to uncover deeper insights without requiring specialized technical skills or assistance. Unlike static reports that provide fixed views, these dashboards enable leaders and decision-makers to directly manipulate data, personalize their views based on specific needs, and instantly identify trends, anomalies, or performance shifts in real-time. By providing immediate, intuitive, and self-service access to critical business intelligence across various functions, interactive dashboards accelerate data-driven decision-making, enhance understanding of intricate information, and foster a pervasive culture of analytical self-service throughout the entire organization, leading to greater agility.

Without the strategic implementation of interactive dashboards, organizations are often stuck with outdated, static reports that provide extremely limited, pre-defined views of data, requiring constant and time-consuming requests to IT or data analysts for any new or deeper insights. This creates significant bottlenecks in the crucial decision-making process, as leaders lack immediate, granular access to the nuanced information needed to respond swiftly and effectively to rapidly changing market conditions or emerging operational issues. The absence of self-service data exploration capabilities leads to delayed insights, less informed strategic choices, and a general underutilization of valuable organizational data assets, ultimately hindering agility, competitive responsiveness, and the organization's capacity for rapid adaptation.

Implementation	Challenges
* Define critical key metrics & data sources * Choose optimal BI/dashboarding platform * Design intuitive layouts & effective visualizations * Develop robust data pipelines for real-time feeds * Implement stringent security & access controls * Train users & gather feedback for continuous iteration	* Ensuring high data quality and consistency issues * Complex integration of disparate data sources * Designing effective, unbiased, and non-misleading visuals * Ensuring timely data refresh rates and high performance * Overcoming user adoption hurdles and fostering data literacy * Avoiding "dashboard sprawl" and maintaining clarity

Products	Platform Cost	Project Cost	Duration
Tableau	$10-50K	$50-250K	3-9 m
Microsoft Power BI	$5-30K	$40-200K	2-8 m
Qlik Sense	$15-60K	$70-300K	3-10 m
Domo	$20-80K	$80-350K	4-11 m
Looker (Google Cloud)	$10-45K	$60-280K	3-9 m
MicroStrategy	$20-70K	$90-320K	4-10 m

3. EXECUTIVE CHATBOTS

Executive Chatbots are advanced AI-powered conversational agents specifically engineered to provide high-level insights, concise summaries, and immediate answers to specific queries for senior leadership using natural language interactions. These intelligent chatbots seamlessly integrate with various core business systems, such as CRM, ERP, and Business Intelligence tools, to rapidly pull relevant, aggregated data and present it in an easily digestible format, often in real-time. By offering immediate, on-demand access to crucial business information and freeing up valuable executive time from manual data gathering and report generation, executive chatbots significantly enhance leadership efficiency, accelerate strategic decision-making by providing quick answers to pressing questions, and allow leaders to focus more intently on high-level strategic initiatives.

Without the strategic implementation of executive chatbots, leaders often spend an inordinate amount of time submitting requests for reports, manually searching for specific data points, or waiting for data to be aggregated by their teams, creating significant information bottlenecks. This heavy reliance on traditional, slower methods creates critical delays in the decision-making process and significantly reduces the agility of leadership to respond swiftly to dynamic business environments, market shifts, or emerging crises. The absence of immediate, self-service access to aggregated and contextualized insights can lead to less informed decisions, reduced overall operational efficiency at the leadership level, and a profound missed opportunity to leverage cutting-edge AI for augmenting executive intelligence and enhancing real-time responsiveness.

Implementation	Challenges
* Define precise scope and common executive queries * Select robust AI/chatbot platform with NLU * Integrate securely with all relevant data sources * Train robust NLP models for executive vocabulary * Pilot with select leaders & iterate on responses * Ensure stringent data security & access controls	* Ensuring extreme accuracy of AI interpretation & responses * Addressing critical data security and access permissions * Complex integration with diverse back-end systems * Ensuring high quality of training data for NLP * Effectively managing diverse executive expectations * Continuous refinement of chatbot knowledge base & capabilities

Products	Platform Cost	Project Cost	Duration
IBM Watson Assistant	$10-50K	$50-250K	3-9 m
Google Dialogflow	$5-30K	$40-180K	2-7 m
Microsoft Bot Framework + Azure AI	$5-25K	$30-150K	2-6 m
Kore.ai	$15-70K	$70-300K	3-10 m
Amelia (IPsoft)	$20-90K	$100-400K	4-12 m
Custom Development (using ML libs)	$0-10K	$80-500K	4-12 m

4. VOICE-ENABLED INTERFACES

Voice-Enabled Interfaces represent a transformative technological leap, allowing users to interact with enterprise systems, applications, and critical data using intuitive spoken commands rather than traditional manual inputs. This hands-free, natural interaction method significantly streamlines operational workflows, dramatically accelerates access to vital insights, and enhances overall productivity, particularly benefiting executives who are frequently mobile, multitasking, or require immediate, efficient information retrieval without visual interaction. By moving beyond conventional graphical user interfaces, voice integration simplifies complex tasks, greatly improves accessibility for diverse users, and opens entirely new avenues for interacting with sophisticated business intelligence and operational control systems, making technology more natural, pervasive, and responsive to human command.

Without the strategic adoption of voice-enabled interfaces, leaders and employees remain predominantly reliant on traditional input methods such as keyboards, mice, and touchscreens, which can be considerably less efficient, especially when performing multiple tasks concurrently or operating on the go. This technological limitation fundamentally slows down access to immediate information, impedes rapid query resolution, and significantly reduces the potential for hands-free operational control in specific environments, such as smart offices, specialized manufacturing floors, or during dynamic meetings. The absence of sophisticated voice interaction means missed opportunities for profound enhancements in executive efficiency, limited technological accessibility for certain tasks, and a slower pace in adopting intuitive, next-generation interaction paradigms, ultimately impacting overall digital sophistication and competitive agility.

Implementation	Challenges
* Define precise voice commands & functionalities * Select robust voice AI platform/SDK * Integrate seamlessly with backend systems & data * Train speech recognition & NLP models * Rigorously test for accuracy and user experience * Deploy & iterate on continuous usage feedback	* Achieving accuracy of speech recognition (accents, noise) * Accurately understanding natural language context * Addressing data security and privacy in voice commands * Complex integration with diverse enterprise systems * Managing user adoption, comfort with voice interfaces * Designing intuitive, effective voice user interfaces (VUI/VUX)

Products	Platform Cost	Project Cost	Duration
Amazon Alexa for Business	$5-30K	$30-150K	2-6 m
Google Assistant (Enterprise)	$5-25K	$25-120K	2-5 m
Microsoft Cortana (Enterprise features)	$5-20K	$20-100K	2-5 m
Nuance Conversational AI	$15-70K	$70-300K	3-9 m
Custom voice AI (e.g., using AWS Polly/Transcribe)	$5-40K	$50-250K	3-8 m
Houndify (SoundHound)	$10-50K	$40-200K	2-7 m

Mobile Leadership Apps are purpose-built or meticulously configured applications specifically designed to provide senior executives with immediate, secure, and intuitive on-the-go access to mission-critical business data, real-time performance metrics, and essential communication tools. These highly optimized apps offer a streamlined, visually appealing interface tailored for mobile devices, enabling leaders to effortlessly monitor key performance indicators (KPIs), review comprehensive reports, approve critical requests, and communicate securely from virtually any location worldwide. By ensuring constant connectivity and delivering real-time, actionable insights directly to their fingertips, mobile apps significantly enhance executive agility, facilitate quicker and more informed decision-making, and dramatically improve responsiveness to urgent matters, aligning perfectly with the demanding and dynamic schedules of modern leadership.

Without dedicated and highly optimized mobile leadership apps, executives are frequently confined to their desktops or forced to struggle with non-optimized, cumbersome web interfaces when attempting to access critical information on mobile devices, leading to substantial delays in accessing crucial data and making timely decisions. This pervasive lack of convenient, secure, and fast mobile access can result in missed strategic opportunities, slower responses to rapidly evolving market changes or unexpected crises, and reduced personal productivity for leaders who are frequently traveling, working remotely, or away from their primary workspace. The absence of tailored mobile solutions for leadership fundamentally limits executive flexibility and responsiveness, ultimately impacting the pace, effectiveness, and overall strategic execution across the entire organization in a fast-moving global economy.

Implementation	Challenges
* Define precise executive needs & desired features * Design intuitive and efficient mobile UI/UX * Integrate securely with core business systems * Develop robust mobile app (iOS/Android native/hybrid) * Implement stringent security protocols & MFA * Pilot, deploy, and gather continuous feedback	* Ensuring robust data security on mobile devices * Complex integration with diverse back-end enterprise systems * Maintaining optimal performance and responsiveness * Managing app distribution, updates, and version control * User adoption challenges and providing comprehensive training * Implementing effective offline capabilities and data sync

Products	Platform Cost	Project Cost	Duration
Salesforce Mobile App Cloud	$20-80K	$100-400K	4-12 m
Microsoft Power Apps	$5-30K	$50-250K	2-8 m
SAP Mobile Platform	$15-70K	$80-350K	3-10 m
OutSystems (Low-code Mobile)	$20-90K	$120-450K	3-10 m
Custom Native App Development	N/A	$200-800K	6-18 m
Oracle Mobile Cloud Service	$10-50K	$70-300K	3-9 m

AR/VR Briefings strategically leverage advanced Augmented Reality (AR) and Virtual Reality (VR) technologies to deliver profoundly immersive and interactive presentations of complex data, intricate scenarios, or critical strategic plans for senior leadership. Instead of relying on static slides or traditional two-dimensional visualizations, executives can actively explore high-fidelity 3D models, visualize vast datasets within spatial environments, or participate in realistic virtual simulations of complex operations or future product designs. This immersive approach dramatically enhances comprehension, enables deeper and more intuitive engagement with multifaceted information, and facilitates significantly more informed decision-making by offering richer context and a more natural understanding of intricate details, revolutionizing how strategic insights are consumed and discussed at the highest organizational levels.

Without adopting cutting-edge AR/VR briefings, leaders remain confined to traditional, often passive, and inherently two-dimensional methods of information consumption, which frequently struggle to adequately convey the true complexity, scale, and spatial relationships inherent in modern business challenges. This pervasive limitation can lead to a less intuitive understanding of critical data, reduced engagement during crucial strategic discussions, and a slower, less comprehensive grasp of complex operational processes or future product designs that are inherently better visualized immersively. The absence of AR/VR means missed opportunities for truly groundbreaking strategic analysis, less impactful executive communication, and a diminished capacity for leaders to genuinely "experience" and internalize complex information, thereby hindering future-forward decision-making and limiting the organization's ability to innovate and compete effectively in an increasingly visual and data-rich world

Implementation	Challenges
* Identify suitable use cases for immersive content * Select AR/VR hardware & software platforms * Develop high-fidelity 3D models & visualizations * Create interactive scenarios or detailed simulations * Conduct user training for AR/VR devices * Ensure stringent data security & content mgmt.	* High initial investment in specialized hardware & content development * Significant technical complexity and specialized skill requirements * Addressing user comfort and potential for motion sickness * Complex data integration and real-time updates for models * Substantial content creation time and resource allocation * Scalability for widespread executive and team adoption

Products	Platform Cost	Project Cost	Duration
Unity 3D / Unreal Engine (Development)	$0-20K	$100-500K	4-12 m
Microsoft HoloLens (Hardware/SDK)	$3-10K	$50-250K	3-9 m
Oculus for Business (Hardware/SDK)	$1-5K	$30-150K	2-7 m
Vuforia (PTC) (AR SDK)	$1-5K	$20-100K	2-6 m
Varjo (High-end VR/AR Hardware)	$5-15K	$70-300K	3-9 m
Custom VR/AR Content Dev Agencies	N/A	$150-700K	6-18 m

6x6™ Transformation Blueprint

'Integrity' & Trust

"You can't innovate without trust. Security isn't separate—it's foundational."
— Padmasree Warrior , Former CTO of Cisco; advocate for embedding security into the heart of innovation.

Integrity & Trust isn't paperwork—it's architecture. Mature leadership embeds trust not through compliance but through code: identity governance, blockchain audits, and zero-trust models. These aren't reactive layers—they're proactive frameworks. In transformation, ethical posture becomes infrastructural. When leadership designs for integrity—from privacy workflows to risk analytics—trust becomes ambient. Employees don't just follow rules—they operate inside them. Mature governance isn't visible until tested—and when it is, resilience shows. The enterprise doesn't pause to verify trust—it accelerates because it's already there. That's trust by design, not declaration.

The "Integrity & Trust" chapter is fundamental for leadership transformation's digital maturity as it underscores the paramount importance of embedding ethics, security, and compliance into every digital initiative from the outset. A digitally mature leadership ensures robust governance through Zero Trust Architecture, strong identity management, and automated compliance. This commitment to data privacy, risk management, and transparent blockchain audit trails builds unwavering trust with all stakeholders. Such integrity is essential for maintaining brand reputation, mitigating escalating digital risks, and achieving long-term sustainability and credibility in the increasingly complex digital age.

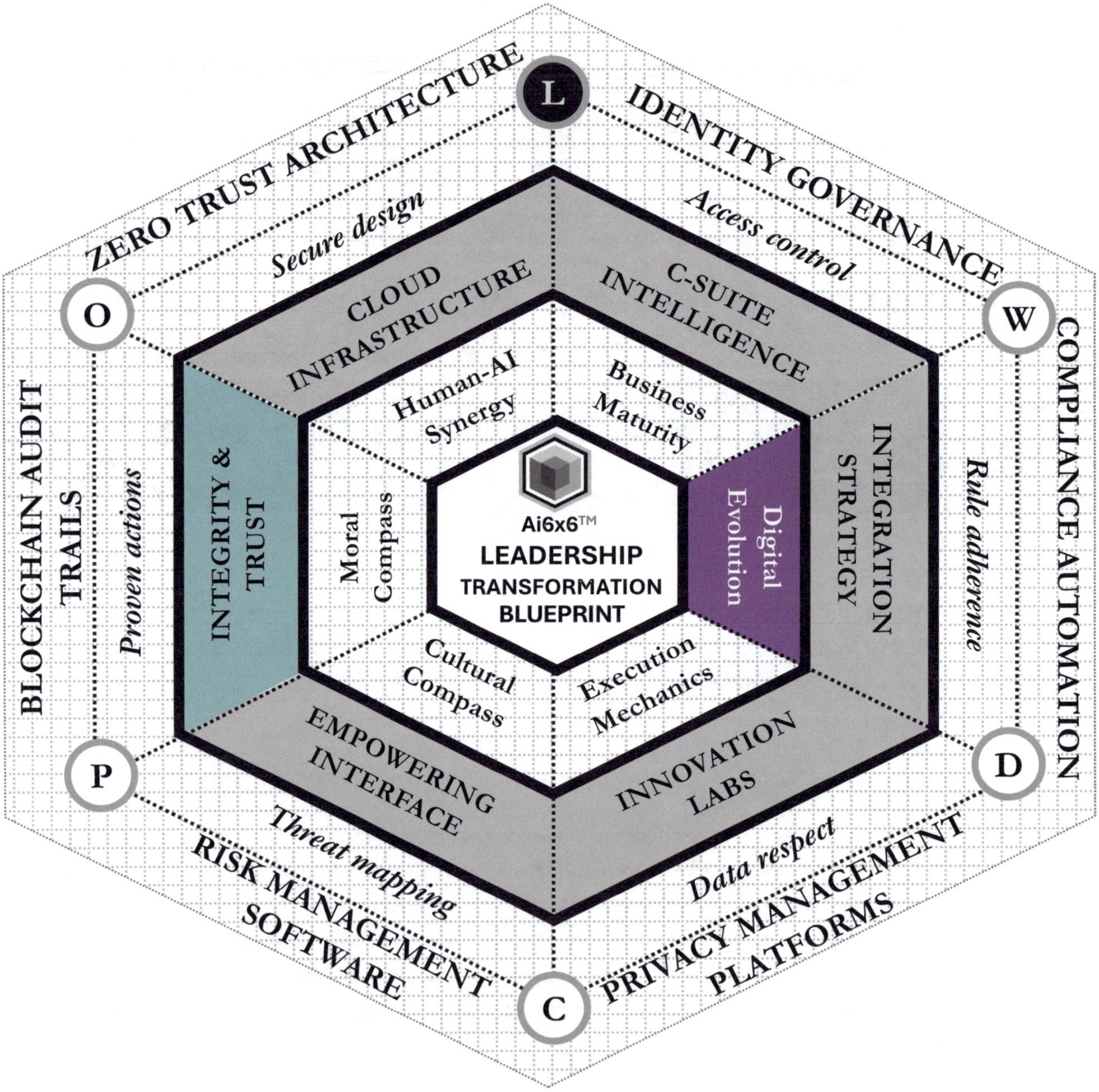

Ai6x6 DIGITAL EVOLUTION FOR INTEGRITY | LEADERSHIP TRANSFORMATION

1. DATA ETHICS FRAMEWORK

A Data Ethics Framework establishes a structured set of principles, guidelines, and practices that govern the responsible and fair collection, storage, analysis, and use of data within an organization. This framework extends beyond mere legal compliance, focusing on moral considerations such as privacy, bias, transparency, and accountability in all data-driven activities, especially those involving AI and personal information. By embedding ethical considerations into every stage of the data lifecycle, a robust data ethics framework builds public trust, mitigates reputational and regulatory risks, and ensures that data is leveraged not just for profit, but also for societal benefit, allowing leadership to make decisions that are both effective and morally sound.

Without a clearly defined and actively enforced Data Ethics Framework, organizations face significant risks including legal challenges, consumer backlash, and severe reputational damage stemming from data misuse, privacy breaches, or algorithmic bias. Relying solely on legal compliance often proves insufficient as societal expectations for responsible data handling rapidly evolve, potentially leading to a loss of trust among customers, employees, and stakeholders. The absence of ethical guardrails can result in unintended negative societal impacts from data-driven initiatives, hinder innovation by fostering a culture of fear rather than responsible experimentation and ultimately erode the long-term sustainability and legitimacy of the organization's digital initiatives in an increasingly scrutinized data landscape.

Implementation	Challenges
* Define core ethical principles for data * Establish clear data governance and privacy policies * Implement internal training on data ethics * Conduct ethical impact assessments for new projects * Create a transparent reporting & accountability mechanism * Regularly review & adapt framework to new tech	* Ensuring universal adoption across all departments * Rapidly evolving data privacy regulations (e.g., GDPR, CCPA) * Addressing inherent bias in historical datasets * Balancing innovation with ethical constraints * Measuring the impact of ethical practices * Obtaining executive-level buy-in and consistent enforcement

Products	Platform Cost	Project Cost	Duration
OneTrust (Privacy & Ethics Management)	$20-80K	$100-400K	4-12 m
TrustArc (Privacy Management)	$15-70K	$80-350K	3-10 m
BigID (Data Discovery & Privacy)	$25-100K	$120-500K	4-13 m
IBM Watson OpenScale (AI Explainability/Bias)	$10-50K	$60-250K	3-9 m
Ethyca (Data Privacy Ops)	$5-30K	$40-180K	2-7 m
Custom Policy & Governance Solutions	N/A	$50-300K	3-9 m

Digital Identity Management (DIM) encompasses the entire lifecycle of an individual's digital identity within an organization, from creation and authentication to authorization and de-provisioning, ensuring secure and controlled access to digital resources. It involves robust systems for verifying user identities, managing access privileges across various applications and data sources, and monitoring user activity to prevent unauthorized access. By implementing a comprehensive DIM strategy, organizations enhance security, streamline user experience, ensure regulatory compliance, and reduce administrative overhead, providing a critical foundation for a Zero Trust security model and empowering leadership to confidently manage access in an increasingly interconnected and digital ecosystem.

Without a robust Digital Identity Management system, organizations are highly vulnerable to security breaches, unauthorized data access, and compliance violations, as user identities and permissions become fragmented and difficult to control. This lack of centralized identity governance leads to weak authentication processes, excessive access privileges, and a complex, inefficient user onboarding/offboarding experience, increasing operational costs and security risks. The absence of a unified DIM approach hinders the implementation of essential security frameworks like Zero Trust, complicates audits, and ultimately erodes trust by exposing the organization to heightened risks of insider threats and external attacks, severely impacting operational integrity and regulatory standing.

Implementation	Challenges
* Conduct identity & access risk assessment * Select IAM/IDaaS platform for unified control * Integrate with existing applications & directories * Implement strong authentication (MFA) & SSO * Establish roles, policies & least privilege access * Continuous monitoring & access review processes	* Complex integration with diverse legacy systems * User resistance to new authentication methods * Managing vast number of identities & entitlements * Ensuring consistent policy enforcement across all apps * Balancing security with user experience * Staying compliant with evolving data privacy regulations

Products	Platform Cost	Project Cost	Duration
Okta (IDaaS)	$15-80K	$80-350K	3-8 m
Azure Active Directory Premium	$10-60K	$60-280K	3-7 m
Ping Identity	$20-90K	$100-400K	4-9 m
SailPoint (Identity Governance)	$25-110K	$120-500K	4-12 m
Auth0 (Okta)	$5-30K	$40-180K	2-6 m
IBM Security Verify	$10-50K	$70-300K	3-9 m

Automated Compliance utilizes specialized software and intelligent systems to continuously monitor, assess, and report on an organization's adherence to various regulatory requirements, industry standards, and internal policies. This proactive approach moves beyond manual, periodic checks by embedding compliance rules directly into operational workflows and IT infrastructure, leveraging real-time data analysis to identify and remediate deviations swiftly. By providing continuous visibility into compliance posture, automated compliance significantly reduces the risk of fines and legal penalties, streamlines audit processes, and frees up compliance teams to focus on higher-level strategic risk management, allowing leadership to operate with greater assurance and maintain unwavering trust with regulators and stakeholders.

Without the implementation of robust automated compliance solutions, organizations are forced to rely heavily on manual, time-consuming, and often error-prone processes for monitoring regulatory adherence. This outdated approach leads to significant delays in identifying compliance gaps, increased susceptibility to human error in reporting, and a reactive posture that often results in hefty fines, legal challenges, and severe reputational damage due to non-compliance. The absence of continuous, real-time compliance monitoring hinders an organization's agility in adapting to evolving regulations, creates bottlenecks in audit readiness, and ultimately exposes leadership to unpredictable risks, eroding trust and undermining the integrity of its operations in an increasingly complex regulatory landscape

Implementation	Challenges
* Identify all relevant regulations & standards * Map compliance requirements to IT controls * Select & configure automated compliance platform * Integrate with existing IT & security tools * Define automated alerts & reporting dashboards * Conduct regular audits & refine rules continuously	* High complexity of integrating disparate systems * Keeping pace with rapidly evolving regulatory changes * Defining accurate, actionable compliance rules * Managing false positives and alert fatigue * Securing sensitive compliance data itself * Gaining buy-in from various compliance stakeholders

Products	Platform Cost	Project Cost	Duration
MetricStream	$30-120K	$150-600K	4-12 m
GRC Platforms (e.g., ServiceNow GRC)	$20-90K	$100-450K	4-10 m
LogicManager (ERM/GRC)	$15-70K	$80-350K	3-9 m
AuditBoard (Audit & Risk Management)	$10-50K	$60-280K	3-8 m
Compliance Automation via Cloud Native	$5-30K	$50-200K	2-7 m
OneTrust (GRC & Compliance)	$20-80K	$100-400K	4-12 m

Ethical AI Development establishes a proactive and systematic approach to designing, building, and deploying artificial intelligence systems that are fair, transparent, accountable, and respectful of human values and rights. This involves implementing robust practices to identify and mitigate biases in data and algorithms, ensuring explainability of AI decisions, safeguarding privacy, and establishing clear lines of responsibility for AI system outcomes. By embedding ethics throughout the AI lifecycle, organizations can build trusted AI solutions that avoid unintended discriminatory impacts, comply with emerging regulations, and maintain public confidence, allowing leadership to harness the full potential of AI responsibly and sustainably.

Without a dedicated focus on Ethical AI Development, organizations risk deploying AI systems that perpetuate or even amplify societal biases, make unfair or discriminatory decisions, and operate with a lack of transparency that erodes public trust. This oversight can lead to significant reputational damage, costly legal challenges, regulatory penalties, and widespread public backlash, severely undermining the benefits of AI adoption. The absence of ethical guidelines and controls also fosters a culture of irresponsible AI innovation, limiting the organization's ability to achieve long-term value from its AI investments and potentially facing outright rejection of its AI-powered products or services in an increasingly ethics-conscious market.

Implementation	Challenges
* Define organizational AI ethics principles * Implement bias detection & mitigation tools * Develop AI explainability (XAI) capabilities * Ensure data privacy & security in AI models * Establish clear human oversight & accountability * Conduct continuous ethical AI audits & reviews	* Identifying and mitigating subtle biases in data & models * Achieving true explainability in complex AI systems * Balancing ethical constraints with performance goals * Lack of standardized ethical AI regulations * Significant skillset gap in AI ethics expertise * Integrating ethical considerations into agile dev cycles

Products	Platform Cost	Project Cost	Duration
IBM Watson OpenScale	$10-50K	$70-300K	3-9 m
Google Cloud Explainable AI	$5-30K	$40-180K	2-7 m
Microsoft Azure Responsible AI	$5-25K	$30-150K	2-6 m
Arthur AI	$15-70K	$80-350K	3-10 m
Fiddler AI	$10-60K	$60-280K	3-8 m
Custom Framework & Tools (Open Source ML)	$0-20K	$50-250K	3-9 m

Blockchain technology, by its decentralized and immutable ledger nature, provides an unprecedented level of trust and transparency across various business processes by creating verifiable and tamper-proof records of transactions or data exchanges. This distributed ledger technology ensures data integrity, enhances security through cryptographic principles, and eliminates the need for central intermediaries, significantly reducing fraud and increasing auditability. For leadership, implementing blockchain can enable transparent supply chains, secure digital identity verification, trustworthy data sharing among partners, and verifiable compliance records, fundamentally strengthening stakeholder confidence, streamlining complex operations, and establishing a new paradigm of verifiable trust in digital interactions.

Without leveraging blockchain for trust and transparency, organizations often grapple with opaque processes, fragmented data records, and a persistent lack of verifiable trust among multiple parties, leading to inefficiencies, disputes, and heightened risks of fraud or data manipulation. This reliance on traditional, centralized record-keeping systems can make auditing complex, introduce single points of failure, and limit the ability to share data securely and immutably across ecosystems. The absence of blockchain's unique attributes means missed opportunities to enhance supply chain visibility, strengthen digital asset provenance, and build robust, auditable multi-party trust networks, ultimately hindering strategic collaboration, increasing operational friction, and diminishing overall integrity in digital commerce and data exchange.

Implementation	Challenges
* Identify specific trust/transparency pain points * Select appropriate blockchain platform (public/private) * Design smart contracts & data models * Integrate with existing enterprise systems * Pilot with key stakeholders & validate immutable data * Scale deployment & establish governance framework	* High initial complexity and development cost * Scalability limitations for high transaction volumes * Significant energy consumption for some public chains * Regulatory uncertainty and legal frameworks * Lack of skilled blockchain developers * Interoperability with non-blockchain systems

Products	Platform Cost	Project Cost	Duration
Hyperledger Fabric (IBM Blockchain)	$20-100K	$150-600K	6-18 m
Ethereum Enterprise (Quorum, Besu)	$15-70K	$120-500K	5-16 m
Azure Blockchain Service (Deprecated but examples)	$10-50K	$80-350K	4-12 m
Amazon Managed Blockchain	$10-40K	$70-300K	3-10 m
R3 Corda Enterprise	$25-120K	$200-800K	7-20 m
Custom Blockchain Development	N/A	$300-1M+	9-24 m

6. THIRD-PARTY RISK MANAGEMENT (TPRM)

Third-Party Risk Management (TPRM) is a critical discipline focused on identifying, assessing, and mitigating the risks associated with engaging external vendors, suppliers, and service providers. This comprehensive process evaluates potential risks across various domains, including cybersecurity, data privacy, regulatory compliance, operational resilience, and financial stability, ensuring that an organization's extended ecosystem does not introduce vulnerabilities that could impact its own integrity or operations. By systematically vetting and continuously monitoring third parties, TPRM protects the organization's reputation, safeguards sensitive data, ensures business continuity, and maintains regulatory adherence, empowering leadership to confidently expand partnerships while mitigating the complex risks inherent in modern supply chains and digital ecosystems.

Without a robust Third-Party Risk Management program, organizations are highly susceptible to significant disruptions, data breaches, and reputational damage originating from vulnerabilities within their extended network of vendors, suppliers, and partners. A lack of systematic vetting and continuous monitoring of third parties exposes critical assets to unknown security gaps, compliance failures, and operational dependencies that can lead to severe financial penalties and customer distrust. The absence of a strong TPRM framework means missed opportunities to proactively identify and mitigate supply chain risks, hinders the ability to scale secure partnerships, and ultimately leaves the organization exposed to unpredictable and potentially catastrophic external threats, compromising its overall integrity and resilience.

Implementation	Challenges
* Define clear TPRM policy & risk appetite * Establish a centralized vendor assessment process * Implement TPRM platform for automation * Conduct due diligence and continuous monitoring * Develop incident response plans for third-party breaches * Regularly review and update vendor risk profiles	* Managing a vast and complex ecosystem of third parties * Obtaining accurate and timely risk data from vendors * Ensuring consistent assessment across diverse vendor types * Integrating TPRM with existing GRC/security tools * Skillset challenges in risk assessment & contract negotiation * Balancing thoroughness with efficient vendor onboarding

Products	Platform Cost	Project Cost	Duration
Prevalent	$20-80K	$100-400K	4-12 m
LogicManager (ERM/GRC)	$15-70K	$80-350K	3-10 m
OneTrust (Third-Party Risk Management)	$20-90K	$90-450K	3-11 m
Archer (RSA)	$30-120K	$150-600K	5-14 m
ServiceNow Vendor Risk Management	$25-100K	$120-500K	4-12 m
CyberGRX	$15-60K	$70-300K	3-9 m

BLUEPRINT SERIES

SECTION E

Execution Mechanics

Ai6x6™ Transformation Blueprint

Story 1 — The Manhattan Project Coordination Model (1942–1946): The Manhattan Project united scientists, engineers, and military leaders across dispersed sites under extreme secrecy and urgency. Workstreams were interdependent, requiring precise coordination without full visibility. Structured reporting, controlled interfaces, and disciplined compartmentalization enabled progress despite complexity. Human judgment guided trade-offs while machines and labs executed specialized tasks. The project demonstrated how execution succeeds when governance, clarity, and cadence align. It became a template for managing high-stakes, multi-disciplinary programs.

Story 2 — The Berlin Airlift Logistics System (1948–1949): With land routes blocked, Allied forces had to supply an entire city solely by air. Aircraft schedules, loading patterns, and turnaround times were optimized to the minute. Machines provided lift and speed, but humans orchestrated sequencing, prioritization, and contingency planning. Continuous feedback loops refined operations daily. The airlift proved that disciplined execution can overcome structural constraints. It showed how precision, rhythm, and adaptability create operational breakthroughs.

Story 3 — The Interstate Highway Construction Program (1956–1970s): The United States embarked on building a nationwide highway network requiring coordination across states, contractors, and engineering disciplines. Standardized designs ensured consistency while local teams adapted to terrain and conditions. Machines handled heavy construction, but humans managed sequencing, safety, and quality. The program relied on predictable funding, clear specifications, and phased execution. It demonstrated how large-scale delivery succeeds when structure and flexibility coexist. The highways became a symbol of disciplined national execution.

Story 4 — Toyota's Just-in-Time Logistics (1960s–1980s): As Toyota refined its production system, logistics became a critical execution layer. Parts arrived exactly when needed, minimizing inventory and waste. Machines tracked flow, but humans monitored signals, resolved disruptions, and maintained rhythm. Kanban created transparency across the value chain. The system required precision, trust, and continuous calibration. It showed how execution excellence emerges when every component of the workflow reinforces timing, quality, and flow.

Story 5 — FedEx's Overnight Delivery Network (1973–1990s): FedEx built a hub-and-spoke system capable of moving packages across the country overnight. Aircraft, sorting machines, and routing algorithms provided speed, but humans managed exceptions, weather disruptions, and customer commitments. The network operated on tight cycles that demanded flawless coordination. Data informed decisions while people ensured reliability. The system demonstrated how execution mechanics scale when tech and human oversight operate in synchronized cadence.

Story 6 — The Rise of Agile Software Delivery (2001–2015): Software teams struggled with long cycles, rigid plans, and late-stage failures. Agile introduced iterative delivery, cross-functional collaboration, and continuous feedback. Machines automated testing and deployment, while humans prioritized work, clarified requirements, and adapted to change. Execution shifted from linear phases to dynamic cycles. The approach improved predictability and responsiveness. It became a global standard for modern execution, proving that disciplined iteration outperforms rigid planning.

LESSONS FROM A LEADERSHIP TRANSFORMATION PERSPECTIVE

Leadership in execution mechanics evolves from directing tasks to designing systems that deliver reliably under pressure. Across decades, leaders learned that execution succeeds when clarity, cadence, and coordination are institutionalized. Effective leaders create environments where teams understand priorities, interfaces are well-defined, and decisions flow without friction. They balance structure with adaptability, ensuring that processes remain stable while accommodating uncertainty. Leadership maturity emerges when the organization can execute predictably without constant intervention. The leader's role becomes enabling disciplined rhythm—aligning people, processes, and technology so execution remains consistent, resilient, and aligned with strategic intent.

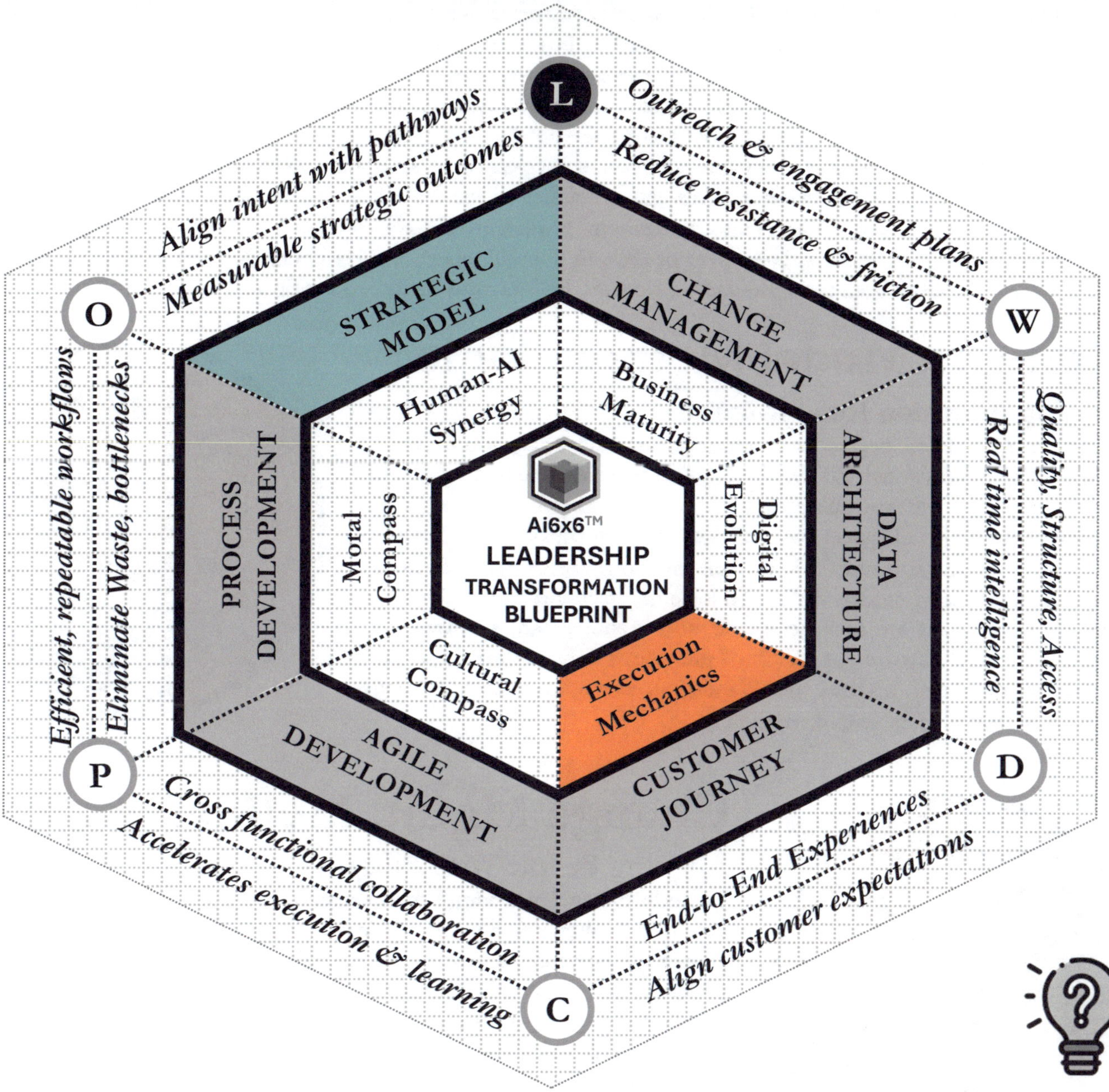

The **Ai6x6 Research Lab** identified this lens as the primary antidote to executive "imposter syndrome" during periods of rapid change. We chose it because the human mind requires a disciplined, repeatable system to make high-stakes decisions without the friction of hesitation. This lens isn't merely about management tools; it is about the psychological certainty and mental clarity that come from a proven methodology. It is vital because it provides a mental scaffolding, turning the chaos of transformation into a series of surgical, confident strikes. This is the "grit" required to ground ambition.

This lens allows the leader's heart to remain calm while the organization moves at terminal velocity. By relying on structured mapping and evaluation, the leader replaces intuition-based guesswork with evidence-based conviction. We believe execution is the ultimate form of leadership mindfulness—the practice of being fully present in the "how" so the "what" is achieved with excellence. This lens ensures every initiative has a defined purpose, allowing the leader to communicate a clear pathway. It is the discipline that turns a fragile hope for growth into a robust engine for value creation, providing the steady footing necessary to leap into new markets.

Execution has always been the crucible where strategy proves its worth. Leaders can articulate bold visions, design elegant frameworks, and outline ambitious roadmaps, but none of it matters if the organization cannot execute with precision, coherence, and resilience. In the age of AI and accelerated change, **Execution Mechanics** becomes the lens that determines whether transformation becomes real or remains theoretical. It forces leaders to confront the uncomfortable truth that most failures are not failures of strategy but failures of execution. The gap between intention and impact is rarely philosophical; it is operational. Execution Mechanics exposes that gap and provides the architecture to close it. This lens reframes execution as a system—an interconnected set of methodologies that convert ambition into momentum. It recognizes that execution is not the final step of strategy but the environment in which strategy lives or dies. In a world where AI accelerates decision cycles, compresses timelines, and amplifies both strengths and weaknesses, leaders must design execution as they design vision.

Strategic Model Methodologies

Turning Ambition Into Structure

Execution Mechanics begins with **strategic model methodologies**, the structured approaches that translate ambition into actionable pathways. These methodologies are not academic exercises; they are the scaffolding that ensures strategy is grounded. Leaders who neglect this dimension find themselves improvising under pressure, relying on intuition rather than disciplined frameworks. Strategic models define the logic of execution—how choices are made, how priorities are sequenced, and how resources are allocated. They ensure that every initiative has a defined purpose, a coherent structure, and a measurable outcome. Without this foundation, execution becomes reactive, fragmented, and vulnerable to disruption.

Change Mgmt. Methodologies

Aligning People With Progress

The next dimension, **change management methodologies**, recognizes that execution is ultimately a human endeavor. No transformation succeeds unless people understand, accept, and commit to the change. Leaders who underestimate this reality discover that resistance is not a barrier but a predictable response to uncertainty. Effective change management acknowledges the emotional, cultural, and psychological dimensions of transformation. It ensures that communication is transparent, expectations are clear, and support is available. It turns change from something imposed into something shared. In this dimension, execution becomes relational—rooted in trust, empathy, and the ability to guide people toward possibility.

Data Architecture Methodologies

Building the Backbone of Execution

Data architecture provide the backbone for modern execution. In a world where decisions must be informed by evidence, organizations cannot afford fragmented data, inconsistent definitions, or unreliable pipelines. Data architecture is not a technical concern; it is a strategic one. It determines whether insights are trustworthy, whether systems can scale, and whether AI can operate responsibly. Leaders who invest in strong data architecture create the conditions for intelligence to flow freely across the enterprise. In this dimension, execution becomes intelligent—powered by clarity, consistency, and evidence.

Customer Journey Methodologies

Anchoring Execution in Value

Customer journey ensure that execution remains anchored in value creation. Too many organizations design processes around internal convenience rather than customer experience. Customer journey thinking forces leaders to see the organization from the outside in, identifying friction, gaps, and opportunities that internal perspectives overlook. It ensures that execution is not only efficient but meaningful. It aligns operations with the expectations, needs, and emotions of the people the organization serves. In this dimension, execution becomes empathetic—shaped by the reality of customers rather than the assumptions of leaders.

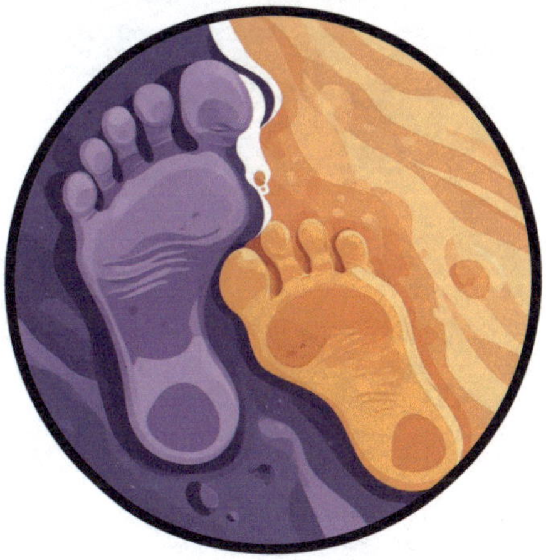

Agile Development Methodologies

Enabling Adaptation at Speed

Agile development introduce the adaptability required in a world where conditions shift rapidly. Traditional execution models assume stability, predictability, and linear progress. Agile assumes the opposite. It embraces iteration, experimentation, and continuous learning. It allows teams to adjust quickly, respond to feedback, and deliver value incrementally. Leaders who adopt agile mindsets create organizations that move with speed without sacrificing quality. In this dimension, execution becomes dynamic—capable of evolving in real time as new information emerges.

Process Improvement Methodologies

Sustaining Excellence Over Time

The final dimension, **process improvement methodologies**, completes the system by ensuring that execution evolves rather than stagnates. They provide the discipline to identify inefficiencies, eliminate waste, and refine workflows. They ensure that execution is not a one-time effort but a continuous practice. Leaders who embrace process improvement build organizations that get better over time, not simply bigger or busier. In this dimension, execution becomes self-correcting—driven by curiosity, discipline, and a commitment to excellence.

Execution Mechanics is the lens that transforms strategy into reality by ensuring that ambition is supported by structure, that change is supported by people, and that progress is supported by systems. It reveals that execution is not the final step of leadership but the continuous arena in which leadership is tested. In a world where AI accelerates both opportunity and exposure, leaders must design execution as deliberately as they design vision, building the operational muscle that allows the organization to move with coherence under pressure. This lens makes clear that execution is not a matter of effort but of engineering—of constructing the models, architectures, and behaviors that convert intention into repeatable performance. Execution becomes a living capability rather than a periodic push, a discipline that strengthens with use rather than erodes under stress.

STRATEGIC MODEL

Flight Schedule

Establishing a universal cadence ensures every department operates with synchronized precision. The master timeline that dictates when every resource must be in place to ensure a coherent, global operation.

DATA ARCHITECTURE

Air Traffic Control

Maintaining a high-fidelity signaling system eliminates the risk of operational collisions and blind spots. The invisible infrastructure that provides the real-time telemetry needed for confident, evidence-based decisions at high velocity.

CUSTOMER JOURNEY

The Terminal Flow

Eliminating friction at every touchpoint transforms the user experience into a seamless journey of value. Redesigning the "Gate" flow to remove bottlenecks and demonstrate respect for the customer's time and needs.

AGILE DEVELOPMENT

Quick Turnaround

Compressing execution cycles minimizes the cost of delay and keeps strategic assets in constant motion. The ability to de-plane, refuel, and re-load (iterate) with extreme speed, ensuring the organization never "idles" on the tarmac.

CHANGE MANAGEMENT

Ground Crew Synergy

Cultivating organizational elasticity allows the workforce to pivot instantly without losing mechanical momentum. Leaders create psychological safety for the "crew" to adapt during disruptions, synchronized under pressure.

PROCESS IMPROVEMENT

Logistics Optimization

Embedding a habit of marginal gains compounds small refinements into a structural advantage. Identifying every millisecond lost in baggage handling or fueling to maximize the total capacity and throughput of the hub.

COMMON FRAMEWORK - TOGAF (The Open Group Architecture Framework)

Book 1: Leadership Transformation—Strategic Direction : The frameworks in this pillar ensure that all efforts are strategically aligned and that leaders are focused on the right outcomes for value creation.

1. **McKinsey 7-S Framework:** Aligns the organization's strategy with its structure, systems, staff, and skills to ensure all components support the desired transformation.
2. **Business Model Canvas:** Creates a shared understanding of how value will be created, delivered, and captured, providing a clear map for all teams.
3. **McKinsey Three Horizons of Growth:** Structures innovation efforts across short-term improvements (Horizon 1), scaling existing successes (Horizon 2), and future-proofing the business (Horizon 3), ensuring a sustainable pipeline of value.
4. **Balanced Scorecard:** Translates the strategic vision into measurable performance objectives, connecting operational activities to high-level financial and non-financial goals.
5. **SWOT Analysis:** Provides a structured way to evaluate the strategic context, enabling leaders to leverage strengths, mitigate weaknesses, seize opportunities, and counter threats.
6. **Blue Ocean Strategy:** Drives leaders to look for uncontested market space, shifting the focus from competing for value to creating new value.

Book 2: Workforce Transformation—Adoption and Engagement : This pillar addresses the human element of change, ensuring that employees are prepared, engaged, and able to sustain new ways of working.

1. **Kotter's 8-Step Change Model:** Provides a step-by-step process for leading major change, from creating a sense of urgency to solidifying new approaches in the corporate culture.
2. **ADKAR Model:** Guides change managers in ensuring that individual employees have the Awareness, Desire, Knowledge, Ability, and Reinforcement needed for successful change.
3. **Lewin's Change Management Model:** Conceptualizes change as a three-stage process: "Unfreeze" existing habits, implement the "Change," and "Refreeze" the new processes to make them stick.
4. **Bridges' Transition Model:** Focuses on the psychological journey of individuals, helping them navigate the emotional stages of an ending, a neutral zone, and a new beginning.
5. **Prosci Change Management Methodology:** Provides a structured, research-based methodology for applying change management to any project or initiative.
6. **The Satir Change Model:** Helps leaders understand and anticipate the emotional and performance impacts of change on individuals and teams.

Book 3: Data Transformation—Measurement and Insights - This pillar provides the tools and architectures to manage data as a strategic asset, ensuring that value can be reliably measured and tracked.

1. **Zachman Framework:** Offers a comprehensive, logical structure for classifying and organizing enterprise architecture, ensuring data management is part of the larger business strategy.
2. **DAMA-DMBOK:** Provides a set of best practices for managing data throughout its lifecycle, ensuring the quality and integrity of data used for value measurement.
3. **Data Mesh:** Promotes a decentralized, domain-driven approach to data ownership, empowering teams to create and manage data as a product.
4. **Lambda Architecture (or Kappa Architecture):** Establishes a technical architecture for processing large volumes of data efficiently, supporting real-time value tracking and analytics.
5. **Master Data Management (MDM) Frameworks:** Ensure that an organization's critical business data is consistent, accurate, and provides a "single source of truth" for reliable reporting.
6. **Data Fabric:** Creates a flexible, interconnected data architecture that supports seamless access and integration of data, regardless of its location.

Book 4: Customer Transformation—External Value Delivery: This pillar focuses on defining and delivering value from the customer's perspective, ensuring that transformation efforts translate to measurable customer outcomes.

1. **Customer Journey Mapping:** Visualizes the customer's end-to-end experience, identifying pain points and opportunities for value creation.
2. **Service Blueprinting:** Expands on journey mapping by detailing the front-stage and backstage processes, people, and systems that support a customer experience.
3. **Jobs to be Done (JTBD) Framework:** Helps define "value" from the customer's perspective by focusing on the functional and emotional outcomes they seek.
4. **Design Thinking Methodology:** Provides a human-centered approach to innovation that focuses on empathizing with customers to design meaningful solutions.
5. **CX (Customer Experience) Maturity Models:** Provides a benchmark for tracking and improving customer experience over time, linking strategic initiatives to better customer outcomes.
6. **Persona Development Frameworks:** Creates detailed profiles of target users, ensuring that transformation efforts are grounded in a deep understanding of customer needs.

Book 5: Product Transformation—Internal Value Delivery : This pillar establishes an agile, iterative process for continuously building and improving the products and services that deliver customer value.

1. **Scrum:** Enables cross-functional teams to deliver value incrementally in short sprints, allowing for continuous feedback and adaptation.
2. **Kanban:** Helps manage and visualize workflow, limiting work-in-progress and improving the flow of value to the customer.
3. **Lean Startup:** Applies a "build-measure-learn" feedback loop to product development, ensuring teams are constantly validating their assumptions about value creation.
4. **Scaled Agile Framework (SAFe):** Provides a system for coordinating and aligning multiple agile teams across the enterprise to deliver large-scale solutions.
5. **Feature Flags / Toggles Architecture:** Enables teams to deploy new features and capabilities quickly and safely, allowing for real-time testing of value propositions.
6. **Continuous Integration / Continuous Delivery (CI/CD) Pipeline Architecture:** Automates the software delivery process, allowing teams to deliver new value to customers efficiently and reliably.

Book 6: Operations Transformation—Efficiency and Sustainability
This pillar focuses on optimizing the internal processes and systems that support the entire value chain, ensuring efficiency and sustainability.

1. **Lean Management:** Aims to eliminate waste and optimize the flow of value through the organization's processes, generating cost savings and efficiency gains.
2. **Six Sigma (DMAIC Methodology):** Uses a data-driven approach to improve processes by defining, measuring, analyzing, improving, and controlling them.
3. **BPM (Business Process Management) Lifecycle:** Provides a structured approach to continuously improving business processes, ensuring sustained value delivery over the long term.
4. **SCOR Model (Supply Chain Operations Reference):** Offers a standard framework for managing and improving supply chain efficiency, a key driver of operational value.
5. **Digital Twin Architecture:** Creates virtual models of physical assets or processes, enabling simulation and optimization before real-world changes are made.
6. **Robotic Process Automation (RPA) Frameworks/Methodologies:** Automates repetitive, rule-based tasks, freeing up human resources for higher-value activities.

TOGAF (The Open Group Architecture Framework) is a widely recognized and comprehensive enterprise architecture methodology that provides a structured approach for developing and managing enterprise architecture. It helps organizations design, plan, implement, and govern an enterprise information architecture. At its core is the Architecture Development Method (ADM), a detailed, iterative process for developing various architectural domains: Business, Data, Application, and Technology. For leadership transformation, particularly in achieving digital maturity, TOGAF is indispensable. Digital transformation is not just about implementing new technologies; it requires aligning business strategy with IT capabilities across the entire enterprise. Without a coherent architecture, digital initiatives can become fragmented, leading to integration challenges, redundant systems, security vulnerabilities, and a lack of scalability. TOGAF provides the necessary rigor to ensure that digital investments are strategically aligned, well-integrated, and support the organization's long-term digital vision. It helps leaders understand the current state of their digital landscape, define the desired future state (e.g., cloud-native infrastructure, API-driven integration, AI-enabled applications), and create a roadmap to get there. By fostering a common language and framework for architects and stakeholders, TOGAF enables effective communication and collaboration across complex digital initiatives. It ensures that the digital transformation is not just a series of isolated projects but a cohesive, well-governed evolution of the entire enterprise, a critical component of achieving and sustaining comprehensive digital maturity. It provides the blueprint for how an organization's digital capabilities (applications, data, infrastructure) support its strategic business goals, ensuring that every technological investment contributes to a unified, adaptable, and future-ready digital enterprise. This structured approach helps avoid costly misalignments and ensures that digital initiatives are built on a solid, scalable foundation, allowing the organization to truly harness the power of technology for competitive advantage.

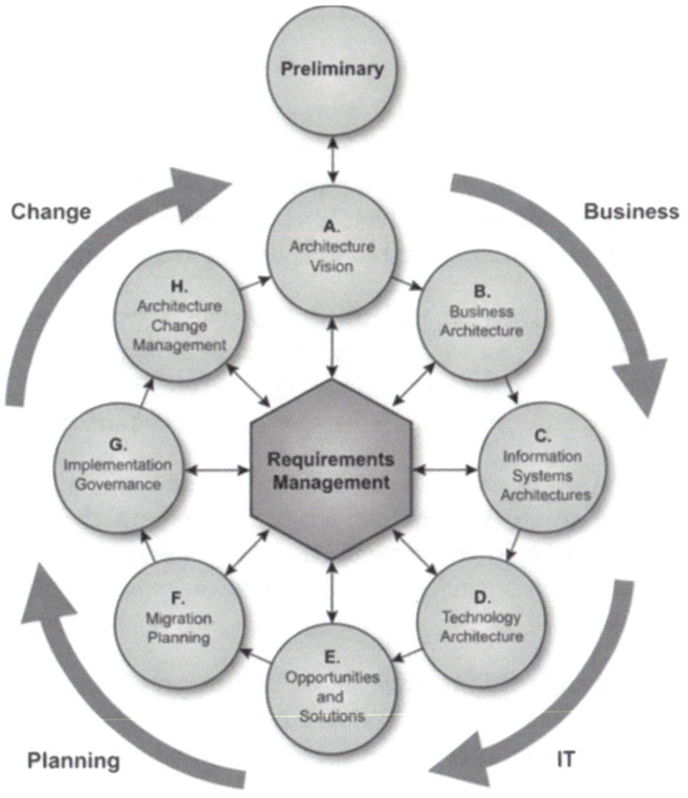

1. **Preliminary Phase: Framework and Principles:** Establish the architecture capability for the enterprise. This involves defining the organizational context for architecture work, identifying key stakeholders, establishing the architecture principles (e.g., "cloud-first," "API-driven," "data-driven decision making"), and defining the scope and governance of the architecture effort.

2. **Phase A: Architecture Vision:** Develop a high-level vision for the target architecture that aligns with business strategy and digital transformation goals. This involves identifying key stakeholders, defining the business value of the architecture work, and securing approval for the architecture development cycle.

3. **Phase B: Business Architecture:** Develop a baseline (current state) and target (future state) Business Architecture that describes the organization's processes, functions, and organizational structure. This ensures that the digital transformation efforts are aligned with and support the core business capabilities and strategic objectives, identifying where digital enablement can optimize business value, streamline workflows, and enhance customer interactions through digital means.

4. **Phase C: Information Systems Architectures (Data & Application):** Develop baseline and target Data Architecture (logical and physical data models, data governance, data flows) and Application Architecture (applications, their interactions, and how they support business functions). This phase is critical for designing integrated digital systems, ensuring data flows seamlessly across digital platforms, and supporting new digital applications that drive business value and customer engagement.

5. **Phase D: Technology Architecture:** Develop baseline and target Technology Architecture that describes the logical software and hardware capabilities required to support the Business, Data, and Application components. This includes defining infrastructure (e.g., cloud platforms, edge computing), platforms (e.g., microservices, AI/ML platforms), and security standards necessary for a robust, scalable, and resilient digital enterprise.

6. **Phase E: Opportunities & Solutions:** Identify and evaluate various implementation options and opportunities to deliver the target architecture. This involves assessing gaps between baseline and target architectures, identifying potential solutions (e.g., Commercial Off-The-Shelf (COTS), custom development, cloud migration, SaaS adoption), and defining high-level implementation projects. For digital maturity, this prioritizes digital solutions that offer the greatest business value and strategic alignment.

7. **Phase F: Migration Planning:** Develop a detailed implementation and migration plan. This involves prioritizing projects, estimating costs, defining a timeline, and outlining the sequence of activities required to transition from the baseline to the target architecture. This ensures a structured and managed approach to digital transformation implementation, minimizing disruption and maximizing efficiency.

8. **Phase G: Implementation Governance:** Oversee the implementation of the architecture. This involves establishing governance processes to ensure that projects adhere to the defined architecture, managing architectural contracts, and providing architectural guidance during development and deployment. This continuous oversight is vital for maintaining architectural integrity and consistency throughout the digital transformation.

9. **Phase H: Architecture Change Management:** Establish processes for managing changes to the architecture. This ensures that the architecture remains relevant and responsive to evolving business needs, technological advancements (e.g., new AI capabilities, quantum computing), and market shifts.

10. **Requirements Management:** This phase runs concurrently throughout all ADM phases. It involves managing the architectural requirements, ensuring they are consistent, complete, and aligned with stakeholder needs. For digital maturity, this ensures that evolving digital requirements (e.g., real-time data access, scalability for new digital services) are captured and integrated into the architectural design and implementation.

Strategic Model Frameworks

"Strategy without execution is hallucination."

— *Thomas Edison. This emphasizes that a strategic model framework is the essential bridge that connects abstract strategy to concrete, measurable execution.*

Strategic Model Frameworks are critical for transforming strategic ambitions into a controlled, executable reality. Tools like the **Target Operating Model (TOM)** and **Enterprise Architecture (EA)** provide the essential **structural language** that ensures consistency across the entire enterprise, aligning IT, HR, and business functions under a unified blueprint. This structure is paramount for **risk mitigation**, compelling leaders to establish robust governance and make decisive choices on data standards, ethical AI compliance, and resource allocation *before* significant capital expenditure. By explicitly defining the desired future state across **People, Process, Technology, and Governance**, the framework prevents organizational fragmentation. It enables effective **value mapping**, helping leaders systematically identify which core capabilities are best augmented by AI, ensuring transformation spending is deliberate, measurable, and contributes to a cohesive, future-ready business.

The **Strategic Model Framework** pillar is foundational for ensuring that all organizational efforts are decisively aligned with high-value outcomes and the overarching strategic direction. This pillar utilizes proven frameworks to establish clarity, measure progress, and drive innovation. For instance, the McKinsey 7-S Framework aligns internal components—like structure and systems—to support a unified strategy, while the Business Model Canvas visually articulates how the organization creates, delivers, and captures value. Strategic innovation is structured by the McKinsey Three Horizons of Growth, which manages investment across short-term gains, scaling, and future development. The Balanced Scorecard translates the abstract strategic vision into tangible, measurable performance objectives, linking daily operations to financial and non-financial goals. Contextual evaluation is managed via SWOT Analysis (Strengths, Weaknesses, Opportunities, Threats), and a competitive edge is often sought using the Blue Ocean Strategy, which guides leaders to create new, uncontested market space rather than fighting for existing value. Collectively, these tools ensure leaders are focused on the right areas for sustainable value creation and execution excellence.

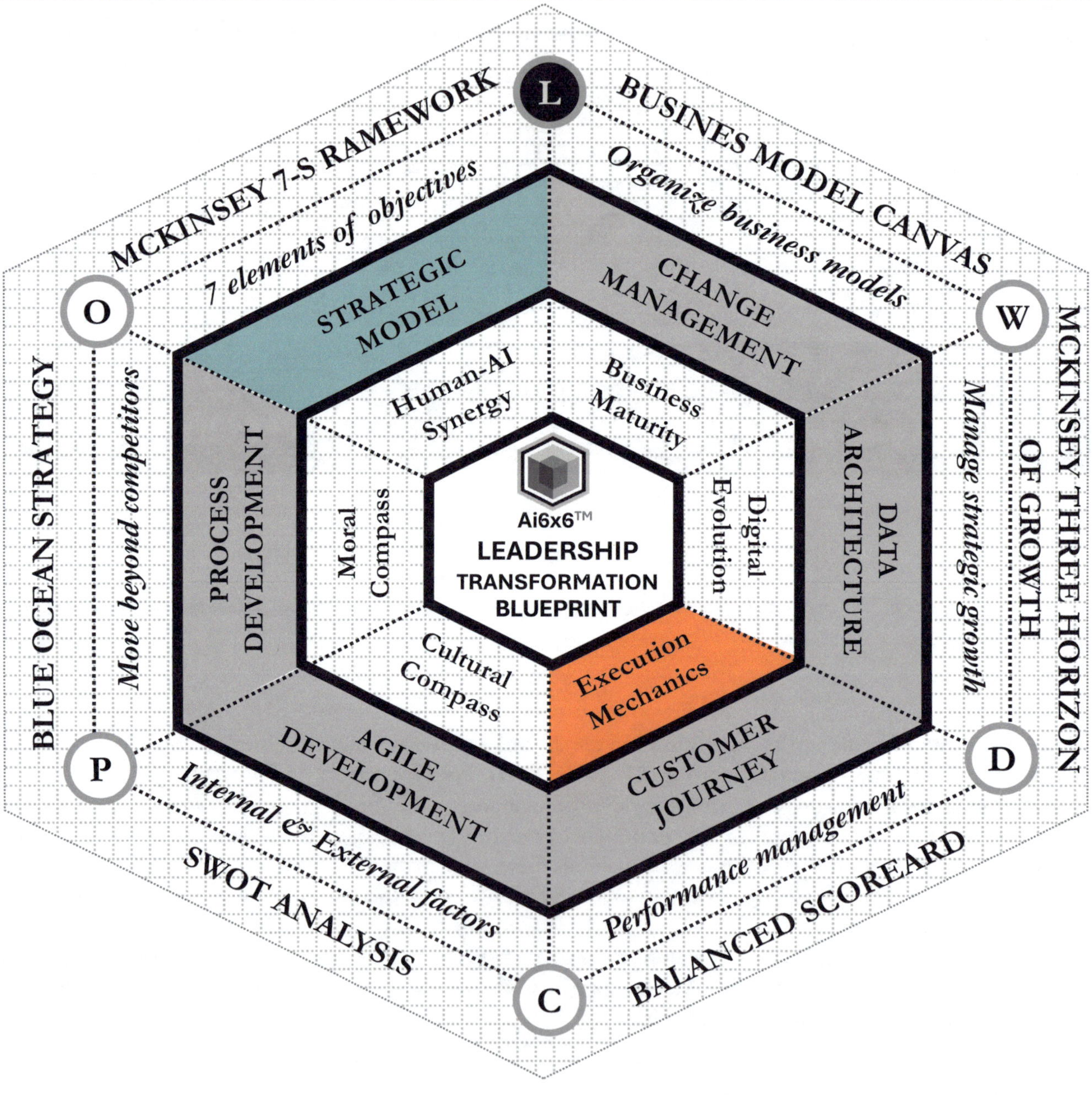

Ai6x6 STRATEGIC MODEL FRAMEWORK | LEADERSHIP TRANSFORMATION

The McKinsey 7-S Framework is a powerful analytical tool that helps organizations understand how seven key internal elements interact and must align to achieve strategic objectives. Developed by McKinsey & Company consultants, it posits that effective organizational change, particularly large-scale transformations like digital maturity initiatives, requires a holistic view, not just focusing on one or two areas. The "hard" elements—Strategy, Structure, and Systems—are relatively easy to identify and change. However, the "soft" elements—Shared Values, Skills, Staff, and Style—are often more challenging to define and shift, yet they are crucial for sustainable transformation. For leadership transformation, this framework is invaluable because it forces leaders to look beyond superficial changes and consider the deep interdependencies within their organization. A new digital strategy, for instance, will fail if the organizational structure remains siloed, if the systems don't support agility, if the workforce lacks the necessary digital skills, or if the leadership style doesn't foster innovation and collaboration. The 7-S model provides a diagnostic lens to identify misalignments that hinder digital adoption and agility. By analyzing how each element currently supports or impedes the desired digital future, leaders can pinpoint specific areas for intervention. For example, a traditional, hierarchical structure (Structure) might stifle cross-functional digital initiatives, or a risk-averse culture (Shared Values) might prevent experimentation with new digital technologies. The framework helps leaders design a coherent transformation roadmap where changes in leadership approach (Style) are congruent with shifts in digital strategy, talent development, and technological infrastructure. Ultimately, mastering the alignment of these seven elements is a hallmark of a digitally mature organization, where leadership effectively orchestrates all internal facets to leverage digital capabilities for competitive advantage and rapid response to market shifts. It moves beyond simply implementing new tech to truly embedding a digital-first operating model.

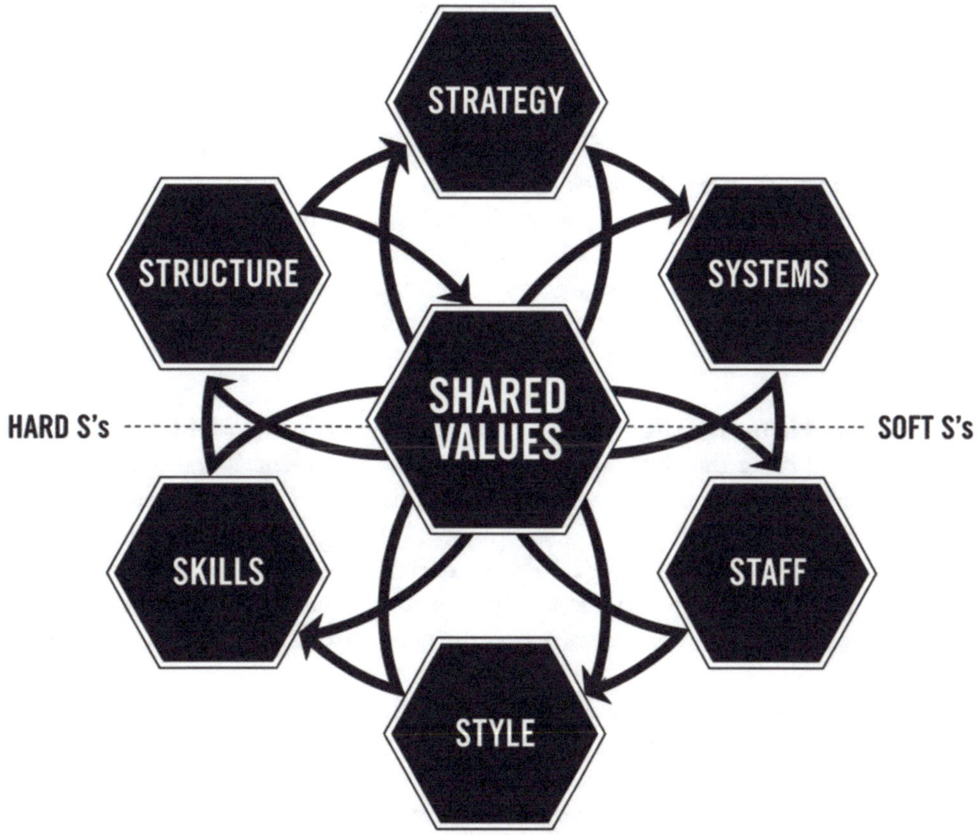

McKinsey 7-S Frameworks

1. **Identify 7 Elements:** Begin by thoroughly listing the current state of each of the seven interconnected elements: Strategy (digital strategy), Structure (organizational design), Systems (IT systems, digital workflows), Shared Values (digital-first mindset), Skills (digital literacy, data fluency), Staff (digital talent acquisition), and Style (agile leadership). This initial inventory provides a baseline understanding of the organization's current operational and cultural landscape, highlighting existing strengths and weaknesses before any transformation initiatives begin. For digital maturity, this step reveals how well current elements support or hinder digital adoption.

2. **Analyze Current State:** Conduct a deep assessment of how these seven elements currently function individually and, more importantly, how they interact and influence one another. This involves examining formal reporting lines, informal communication flows, decision-making processes, and the prevailing organizational culture. The goal is to pinpoint areas of misalignment or inefficiency that may hinder strategic objectives, providing a comprehensive diagnostic view of the organization's operational reality. In a digital context, this analysis might reveal how legacy systems impede new digital strategies, or how a siloed structure prevents cross-functional digital initiatives, highlighting critical areas for leadership intervention.

3. **Define Desired Future State:** Based on the strategic vision for leadership transformation, articulate precisely what each of the 7 elements *should* look like in the future. For instance, if the vision is for a more agile and innovative organization that leverages digital technologies, leaders must define how the structure will become more networked, how systems will integrate digitally, how skills will evolve to embrace data analytics, and how leadership style will foster digital experimentation. This step sets clear targets for the transformation, ensuring all efforts are directed towards a coherent and desirable future, where digital capabilities are seamlessly integrated.

4. **Identify Gaps & Inconsistencies:** Perform a rigorous comparison between the current state (as-is) and the desired future state (to-be) for each of the seven elements. This critical step reveals specific misalignments, missing capabilities, or areas where current practices directly conflict with future aspirations, particularly concerning digital readiness. For example, a gap might be a traditional leadership style that stifles digital innovation, or a lack of digital skills among staff that prevents adoption of new systems. Identifying these gaps allows leaders to prioritize which aspects of the organization require the most significant attention and resources during the transformation journey, specifically targeting digital enablement.

5. **Develop Action Plans for Alignment:** Create specific, actionable initiatives designed to bridge the identified gaps and bring the elements into harmonious alignment. These plans should detail the steps required to change structures (e.g., forming cross-functional digital squads), update systems (e.g., implementing cloud-based platforms), develop new skills (e.g., digital upskilling programs), and foster a supportive culture (e.g., promoting data-driven decision-making). It's vital to ensure that leadership development and changes in leadership style are integrated with broader organizational shifts, reinforcing new behaviors and processes across the enterprise to drive digital maturity.

6. **Monitor and Adjust:** Implement a continuous monitoring system to track the progress of alignment efforts and the impact of transformation initiatives on each of the seven elements. Regularly collect feedback, analyze performance metrics (including digital adoption rates and ROI of digital investments), and periodic reviews to assess effectiveness. Be prepared to make necessary adjustments to strategies and action plans as the transformation unfolds, responding to unforeseen challenges or emerging opportunities with agility and adaptability, ensuring the organization remains digitally responsive and continuously improves its strategic alignment.

The Business Model Canvas (BMC) is a strategic management tool that allows organizations to describe, design, challenge, invent, and pivot their business models. Created by Alexander Osterwalder, it provides a visual chart with nine blocks, outlining a company's value proposition, infrastructure, customers, and finances.

For leadership transformation, especially in the context of digital maturity, the BMC is indispensable. Traditional business models may not be sustainable in a rapidly evolving digital landscape. Leaders need a clear, shared understanding of how their organization creates, delivers, and captures value, and how digital technologies can fundamentally reshape these processes. The BMC compels leaders to think systematically about new digital value propositions (e.g., personalized services enabled by AI), innovative digital channels (e.g., direct-to-consumer e-commerce, mobile apps), and novel digital revenue streams (e.g., subscription models, data monetization). It encourages experimentation with digital business models without committing to full-scale development. By mapping out current and future digital business models, leaders can identify opportunities for digital disruption, assess the digital capabilities required (Key Resources like data platforms, AI models), and understand the digital activities (Key Activities like platform management, digital marketing) necessary for success. It also highlights potential digital partnerships (e.g., with tech vendors, digital agencies) that are crucial for expanding digital reach. Ultimately, the BMC provides a common language and framework for leaders to collaboratively design and iterate on digitally-enabled business models, ensuring that the organization's strategic direction is not just about technology adoption, but about fundamentally reinventing how it does business in a digital world, a core indicator of digital maturity.

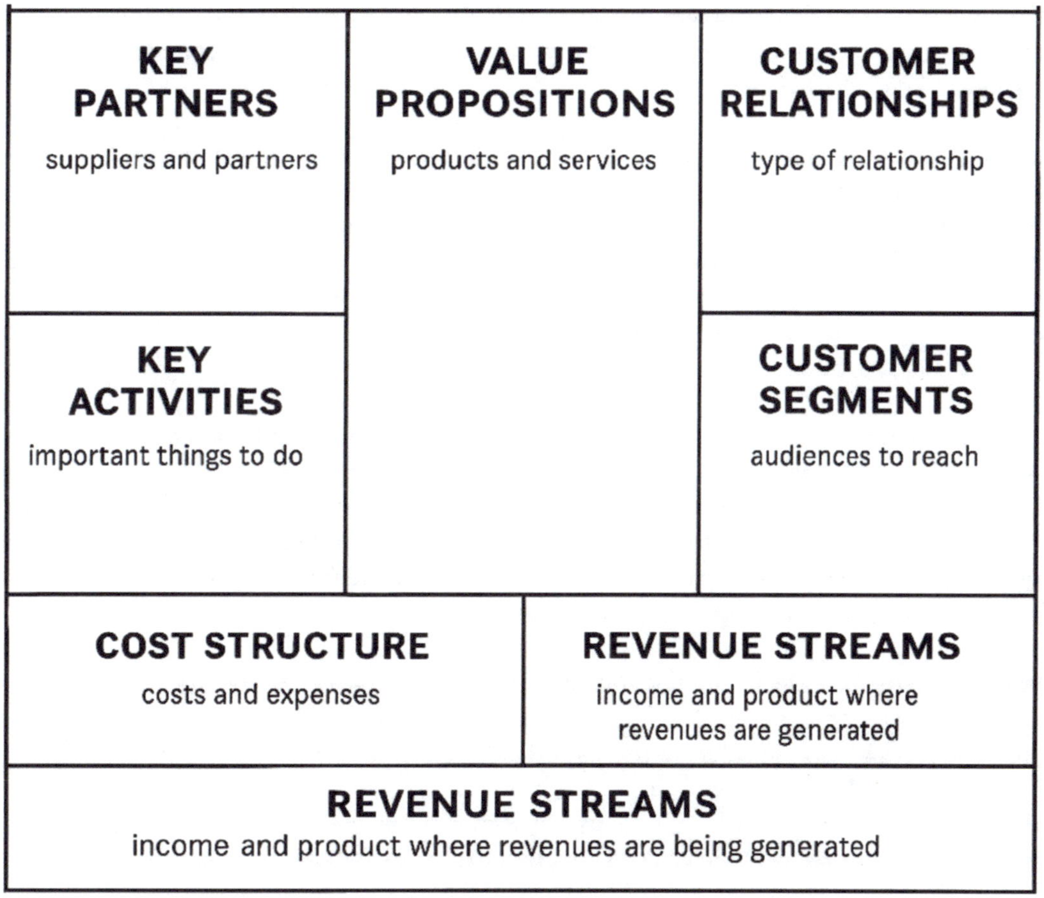

Business Model Canvas

1. **Define Customer Segments & Value Propositions:** Start by identifying the specific groups of people or organizations the business aims to serve (Customer Segments), considering how digital channels might enable reaching new segments. Then, clearly articulate the unique value (products, services, solutions) offered to these segments that solves their problems or satisfies their needs (Value Propositions). This foundational step ensures a clear understanding of who the business serves and what compelling benefits it provides. For digital maturity, this often involves defining value propositions that are enhanced or enabled by digital technologies, such as personalized services or data-driven insights.

2. **Determine Channels & Customer Relationships:** Specify how the company communicates with, reaches, and delivers its value propositions to its customer segments (Channels). This includes traditional channels but critically emphasizes digital channels like e-commerce platforms, mobile apps, social media, and online communities. Simultaneously, describe the types of relationships the company establishes and maintains with specific customer segments (Customer Relationships), which can range from personal assistance to automated self-service, influencing customer loyalty and retention. Digital maturity means optimizing these channels and relationships for seamless digital engagement.

3. **Specify Revenue Streams & Cost Structure:** Outline precisely how the company generates cash from each customer segment, detailing the pricing mechanisms and payment methods (Revenue Streams). This is where leaders can explore new digitally-enabled revenue models like subscriptions, freemium, platform fees, or data monetization. Concurrently, describe all the most significant costs incurred to operate the business model, including fixed and variable costs, to understand the financial viability and profitability (Cost Structure). Digital maturity often involves reducing operational costs through automation and digital efficiency.

4. **Identify Key Resources & Activities:** List the most important assets required to offer and deliver the value proposition. These can be physical (e.g., factories), intellectual (e.g., patents, proprietary algorithms), human (e.g., skilled employees, data scientists), or financial (e.g., cash reserves) (Key Resources). For digital maturity, this increasingly includes digital assets like data platforms, AI models, and cloud infrastructure. Then, outline the most important things the company must do to operate successfully, such as production, problem-solving, or platform management (Key Activities), often leveraging digital processes.

5. **Define Key Partnerships:** Name the network of suppliers, alliances, and non-competitive partners that are essential for the business model to function efficiently. These partnerships are often crucial for acquiring key resources or performing key activities that the company chooses not to do in-house, optimizing costs and reducing risks while gaining access to specific resources or capabilities. In a digitally mature context, these partnerships often involve technology providers, data partners, or platform collaborators, extending the digital ecosystem.

6. **Analyze and Iterate:** Once all nine building blocks are filled, review the complete canvas for coherence, consistency, and viability. Identify areas for innovation, improvement, or potential pivots. This iterative process allows leaders to test hypotheses about their business model, gather feedback, and continuously refine their approach to ensure it remains competitive and relevant in a dynamic market. For digital maturity, this means continuous experimentation with digitally-enabled business models, learning rapidly from market feedback, and adapting to new digital opportunities.

The McKinsey Three Horizons of Growth framework provides a powerful lens for leaders to manage strategic growth in a dynamic environment, particularly crucial for navigating digital transformation. It helps organizations balance the demands of running the current business with the need to innovate for the future. Horizon 1 represents the core business, requiring optimization and defense. Horizon 2 comprises emerging opportunities that are likely to generate substantial new revenue in the medium term, often building on existing capabilities. Horizon 3 involves speculative, high-potential ideas that could become future core businesses, often requiring disruptive technologies and new business models. For leadership transformation, this framework is vital because it addresses the inherent tension between short-term performance and long-term innovation. Leaders must strategically allocate attention, funding, and talent across these three horizons. Over-investing in Horizon 1 can lead to complacency and eventual decline in a digitally disrupted market, while neglecting Horizon 1 can jeopardize current profitability. Similarly, investing too much too soon in Horizon 3 without proper validation can be wasteful. In the context of digital maturity, this framework helps leaders consciously fund and manage digital initiatives at all levels: optimizing existing digital channels and processes (H1), scaling new digital services or platforms (H2), and exploring truly disruptive technologies like advanced AI or quantum computing for future competitive advantage (H3). It ensures that the organization has a continuous pipeline of digital innovation, preventing stagnation and fostering a culture of future-oriented leadership. This balanced approach to digital investment and strategy is a hallmark of truly digitally mature organizations.

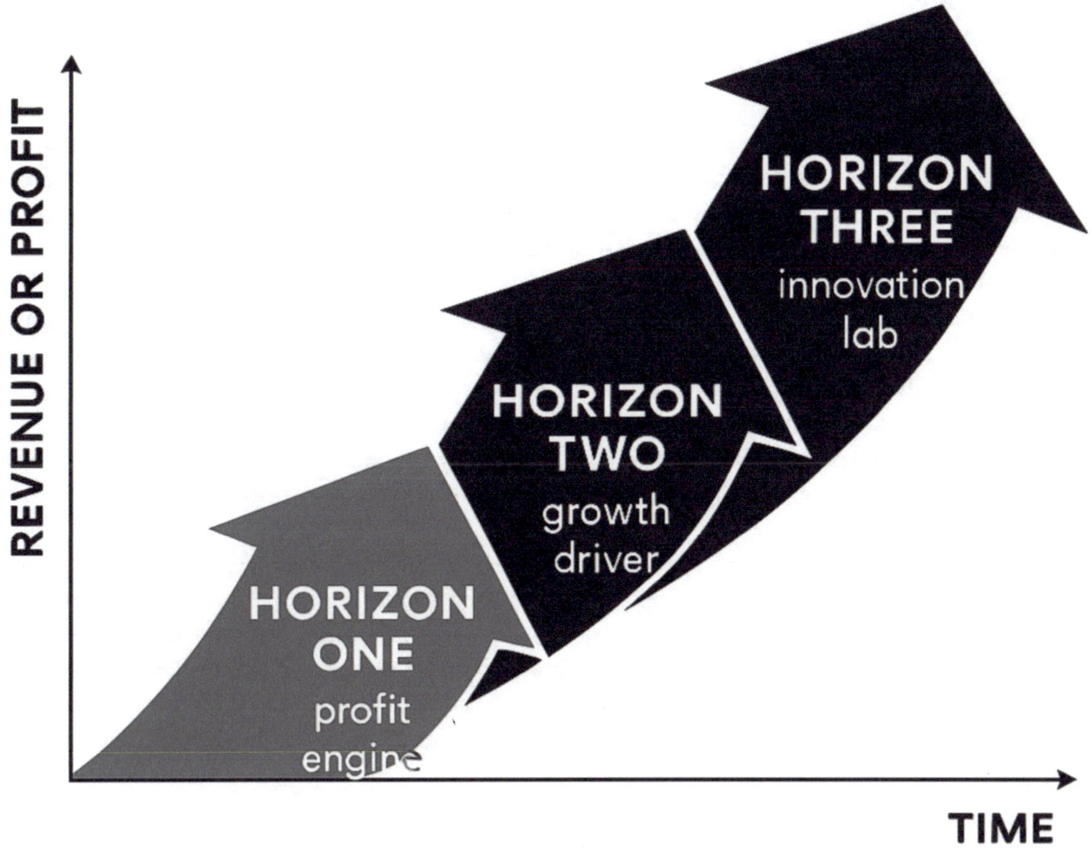

McKinsey Three Horizons of Growth

1. **Identify Horizon 1 (Core Business):** Define the existing businesses and initiatives that currently generate the most revenue and profit. These are the mature, established operations that require continuous optimization and defense against competition. Leaders should focus on improving efficiency and maintaining market share within this horizon, ensuring its ongoing health as the primary revenue driver. In a digitally mature context, this involves optimizing existing digital channels, improving core digital products, and enhancing operational efficiency through digital automation of current processes.

2. **Identify Horizon 2 (Emerging Opportunities):** Pinpoint initiatives that are likely to generate substantial new revenue in the medium term (typically 2-3 years). These opportunities often build on existing capabilities or target adjacent markets. They require significant investment to scale up and move beyond initial experimentation, becoming the "next big thing" for the organization. For digital maturity, Horizon 2 might include scaling a new digital service, expanding into a new digital market segment, or integrating advanced analytics into a core offering to create new value.

3. **Identify Horizon 3 (Future Growth Engines):** Explore speculative, high-potential ideas that could become future core businesses in the long term (typically 5+ years). These often involve disruptive technologies, new business models, or entirely new markets. While highly uncertain, they are crucial for long-term survival and require early-stage investment in research, experimentation, and concept validation. In the context of digital maturity, Horizon 3 could involve exploring AI-driven autonomous services, quantum computing applications, or completely new digital platform business models that reshape the industry.

4. **Allocate Resources Across Horizons:** Strategically distribute leadership attention, funding, and talent across the three horizons. This step is critical to avoid over-investing in Horizon 1 at the expense of future growth, or conversely, diverting too many resources to speculative Horizon 3 projects without sufficient returns. A balanced portfolio ensures both short-term stability and long-term viability. For digital maturity, this means consciously funding digital innovation at all three levels, ensuring resources are available for both optimizing existing digital assets and building future digital capabilities.

5. **Manage Transitions:** Plan how successful Horizon 3 ideas will be nurtured and scaled into Horizon 2, and how mature Horizon 2 initiatives will eventually transition to become part of the core Horizon 1 business. This involves establishing clear criteria for progression, resource reallocation, and integrating new capabilities into the mainstream operations, or gracefully phasing out unsuccessful ventures. For digital maturity, this means having mechanisms to smoothly integrate new digital innovations into the core business, ensuring they are adopted and scaled effectively, avoiding "innovation theater."

6. **Monitor and Adapt:** Continuously review the portfolio of initiatives across all three horizons. Regularly assess market changes, technological advancements, and competitive shifts. Leaders must be prepared to adjust strategies and reallocate resources as needed, ensuring the organization remains agile and responsive to evolving opportunities and threats, thereby sustaining the transformation journey. This continuous monitoring is vital for digital maturity, allowing leaders to quickly pivot investments in digital initiatives based on market feedback and technological shifts, maintaining a dynamic digital strategy.

4. BALANCED SCORECARD (KAPLAN & NORTON)

The Balanced Scorecard, developed by Robert Kaplan and David Norton, is a strategic performance management framework that helps organizations translate their vision and strategy into actionable objectives and performance measures. It moves beyond purely financial metrics to include perspectives on Customer, Internal Business Processes, and Learning & Growth.

For leadership transformation, especially in the pursuit of digital maturity, the Balanced Scorecard is invaluable because it ensures a holistic view of progress. Leaders can use it to communicate the strategic priorities of digital transformation across the entire organization, aligning individual and team efforts towards common goals. It forces a consideration of how digital initiatives impact not just the bottom line (Financial), but also how they enhance customer engagement (Customer), streamline internal digital processes (Internal Business Processes), and foster new digital skills and a culture of innovation (Learning & Growth). By defining specific Key Performance Indicators (KPIs) and targets for each perspective, leaders can track the effectiveness of their digital investments and strategic shifts. For example, a digital transformation might aim to improve customer satisfaction through a new mobile app (Customer perspective), which relies on efficient backend systems (Internal Business Processes) and requires employees to learn new digital skills (Learning & Growth), all contributing to financial gains. The Balanced Scorecard provides a structured way to monitor these interconnected aspects, ensuring that digital transformation efforts are not isolated technology projects but integrated strategic initiatives that drive comprehensive organizational change and measurable progress towards digital maturity.

BALANCED SCORECARD

1. **Define Strategic Objectives:** Clearly articulate the organization's overarching goals for transformation. These objectives should be derived directly from the strategic vision and cover all critical areas of the business, not just financial ones. Examples include "Enhance customer satisfaction," "Improve operational efficiency," or "Foster a culture of innovation." In the context of digital maturity, this means defining objectives that explicitly leverage digital capabilities, such as "Achieve seamless omnichannel customer experience" or "Automate 50% of core business processes."

2. **Select Measures (KPIs):** For each strategic objective, identify specific, measurable Key Performance Indicators (KPIs) that will track progress. These KPIs must be quantifiable and directly reflect the achievement of the objective. For instance, "Customer Satisfaction Score" for "Enhance customer satisfaction," or "Process Cycle Time" for "Improve operational efficiency." For digital maturity, this means including digital-specific KPIs like "Digital Adoption Rate," "Time to Market for New Digital Features," "Data Quality Index," or "Percentage of Automated Processes."

3. **Set Targets:** Establish ambitious yet achievable targets for each KPI. These targets represent the desired level of performance or the specific outcome that the organization aims to reach within a defined timeframe. Targets provide a clear benchmark against which actual performance can be measured, motivating teams and guiding resource allocation towards specific goals. When pursuing digital maturity, these targets should push the organization beyond incremental improvements towards significant digital leaps, e.g., reducing manual data entry errors by 80% through AI.

4. **Define Initiatives:** Identify the strategic programs, projects, and actions needed to achieve the objectives and hit the established targets. These initiatives are the specific activities that drive the transformation. For example, "Implement new CRM system" or "Launch leadership development program" would be initiatives linked to customer or learning objectives. For digital maturity, these initiatives would heavily involve technology implementation, digital skill development programs, data platform builds, and changes to digital operating models.

5. **Collect Data & Analyze:** Implement robust systems and processes to gather data for all chosen KPIs regularly and accurately. Once collected, analyze this performance data against the set targets to understand progress, identify trends, and pinpoint areas of underperformance or unexpected success. This data-driven approach informs decision-making. Achieving digital maturity means leveraging advanced analytics and automated data collection tools to provide real-time insights into the effectiveness of digital initiatives, moving beyond manual reporting.

6. **Review & Adapt:** Periodically (e.g., quarterly or annually) review the entire Balanced Scorecard with leadership and relevant stakeholders. This review assesses strategic effectiveness, identifies whether the chosen strategies are yielding the desired results, and determines if any objectives, measures, targets, or initiatives need to be adjusted. This iterative process ensures the scorecard remains a dynamic and relevant tool for guiding transformation. In a digitally mature organization, this review process itself is often digitally enabled, allowing for faster analysis and more agile strategic adjustments based on performance data

> "What gets measured gets managed." — **Peter Drucker**, Management Consultant

SWOT Analysis is a foundational strategic planning tool that helps leaders understand the internal and external factors influencing their organization. It involves identifying Strengths (internal positive attributes), Weaknesses (internal negative attributes), Opportunities (external favorable conditions), and Threats (external unfavorable conditions).

While seemingly basic, a thorough and insightful SWOT analysis is critical for any leadership transformation, especially when aiming for digital maturity. It provides leaders with a clear, realistic picture of the organization's current standing in the digital landscape. By systematically categorizing these factors, leaders can identify where the organization has a competitive advantage in digital capabilities (e.g., strong data assets, innovative tech team) and where it lags (e.g., outdated legacy systems, lack of digital skills). It also helps uncover emerging digital opportunities (e.g., AI advancements, new digital markets) and potential digital threats (e.g., new digital-native competitors, cybersecurity risks). This clarity is essential for defining the "burning platform" for digital change and for formulating strategic imperatives. For instance, a strength in data analytics combined with an opportunity in AI could lead to a strategy of developing AI-powered digital products. Conversely, a weakness in legacy IT systems coupled with a threat from digital disruptors highlights an urgent need for cloud migration. The SWOT analysis, particularly when extended into a TOWS matrix (which cross-references the four elements to generate strategies), helps leaders develop targeted action plans that leverage digital strengths, mitigate digital weaknesses, seize digital opportunities, and counter digital threats, thereby building a robust roadmap for achieving comprehensive digital maturity.

SWOT Analysis

1. **Identify Strengths (Internal):** List all internal capabilities, resources, and advantages that give the organization a competitive edge or facilitate transformation. These are positive attributes within the organization's control, such as a strong brand, skilled workforce, efficient processes, proprietary technology, or robust financial health. In a digitally mature context, strengths might include a highly skilled data science team, a modern cloud infrastructure, proprietary AI algorithms, or a strong culture of digital innovation, providing a solid foundation for further digital initiatives.

2. **Identify Weaknesses (Internal):** List all internal limitations, deficiencies, or disadvantages that could hinder the organization's ability to achieve its transformation goals. These are negative attributes within the organization's control, such as outdated technology, skill gaps, inefficient organizational structure, lack of capital, or poor internal communication. For digital maturity, weaknesses often include legacy IT systems that are difficult to integrate, a lack of digital literacy across the workforce, fragmented data silos, or a risk-averse culture that stifles digital experimentation.

3. **Identify Opportunities (External):** List all external factors that the organization could exploit to its advantage to achieve its transformation goals. These are positive external conditions not within the organization's direct control, such as emerging market trends, new technologies, favorable regulatory changes, shifts in customer preferences, or competitor weaknesses. In a digitally mature context, opportunities might include the rise of AI/ML, the increasing adoption of IoT, new digital platforms, or unmet customer needs that can be addressed through digital solutions, opening avenues for digital growth.

4. **Identify Threats (External):** List all external factors that could pose challenges or risks to the organization's transformation efforts. These are negative external conditions not within the organization's direct control, such as new competitors, economic downturns, unfavorable regulatory changes, supply chain disruptions, or evolving customer expectations. For digital maturity, threats specifically include digital disruptors entering the market, cybersecurity risks, rapid technological obsolescence, or a shortage of digital talent, requiring proactive strategic responses from leadership.

5. **Analyze and Strategize (TOWS Matrix):** Use the insights from the SWOT analysis to develop specific strategies. This often involves creating a TOWS matrix:
    - SO (Strengths-Opportunities): How to use digital strengths (e.g., data assets) to capitalize on digital opportunities (e.g., AI).
    - WO (Weaknesses-Opportunities): How to overcome digital weaknesses (e.g., legacy systems) to take advantage of digital opportunities (e.g., cloud adoption).
    - ST (Strengths-Threats): How to use digital strengths (e.g., cybersecurity expertise) to mitigate digital threats (e.g., data breaches).
    - WT (Weaknesses-Threats): How to minimize digital weaknesses (e.g., skill gaps) and avoid digital threats (e.g., competitor disruption).

6. **Develop Action Plans:** Translate the chosen strategies from the TOWS matrix into concrete, actionable plans and initiatives that will drive the transformation journey. These plans should specify responsibilities, timelines, required resources, and expected outcomes, ensuring that the strategic insights are effectively implemented across the organization. For digital maturity, these action plans would prioritize digital investments, talent development, and organizational changes necessary to build and leverage digital capabilities effectively.

Blue Ocean Strategy, developed by W. Chan Kim and Renée Mauborgne, is a powerful framework that challenges organizations to move beyond head-to-head competition in existing, crowded markets ("red oceans") and instead create new, uncontested market space ("blue oceans"). This is achieved by simultaneously pursuing differentiation and low cost, making the competition irrelevant.

For leadership transformation, especially in the context of digital maturity, this framework is profoundly relevant. In a digitally saturated world, many industries become "red oceans" where digital features are commoditized, and competition is fierce on price or incremental improvements. Blue Ocean Strategy provides a methodology for leaders to break free from this cycle. It encourages them to think innovatively about how digital technologies can be leveraged not just to improve existing offerings, but to fundamentally redefine value for customers and create entirely new demand. This might involve using AI to deliver a personalized service that was previously impossible, or leveraging blockchain for a trust-based platform that disrupts traditional intermediaries. The framework's "Four Actions Framework" (Eliminate, Reduce, Raise, Create) is a systematic way to challenge existing industry assumptions and identify new value elements that are uniquely enabled by digital capabilities. By focusing on creating new value rather than just competing on existing digital features, leaders can guide their organizations to achieve true digital differentiation and sustainable growth, a hallmark of advanced digital maturity where technology is used to redefine market boundaries.

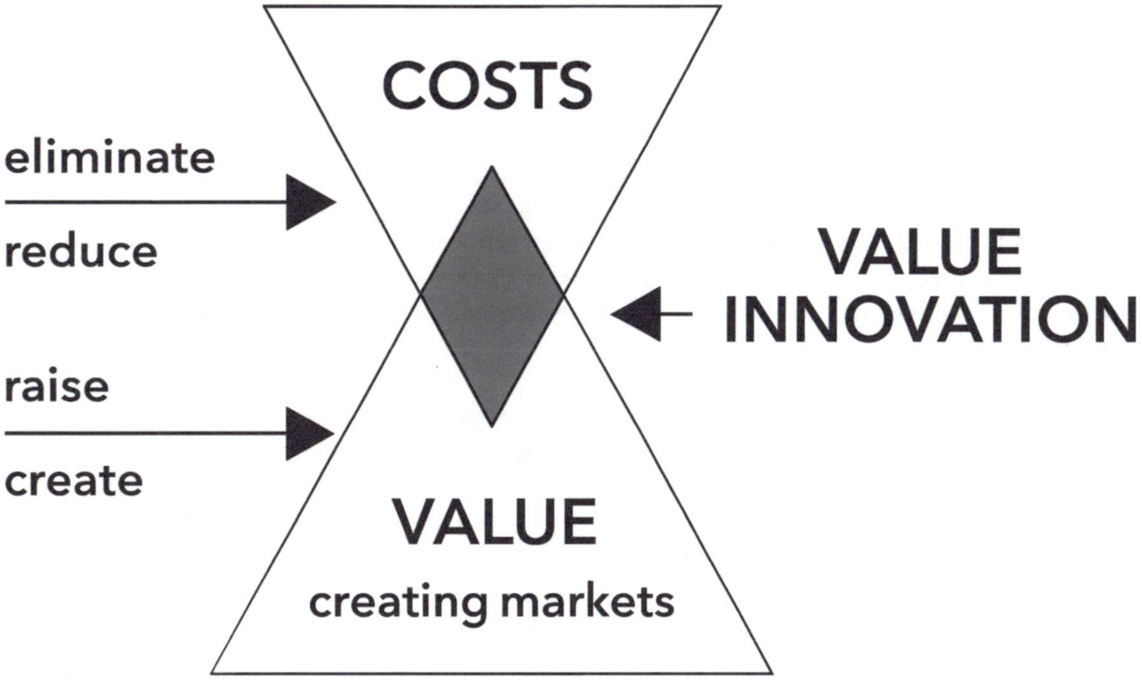

Blue Ocean Strategy Framework

1. Visualize **Current Strategy (As-Is Strategy Canvas):** Begin by mapping the current state of the industry. Plot your organization's offerings and those of your key competitors across the main competing factors that the industry currently invests in and competes on (e.g., price, features, service, digital presence, app functionality, data privacy). This visual representation reveals the "red ocean" – the crowded market space with intense competition and diminishing returns, often characterized by similar digital offerings that struggle to differentiate.

2. **Identify Key Competing Factors:** List all the factors that the industry currently competes on and invests in. These are the attributes or features that customers typically consider when choosing between competing offerings. Understanding these factors is crucial before attempting to redefine the market boundaries. In a digitally mature context, this includes identifying digital features, platform capabilities, and data-driven services that are currently the focus of competition, and which might be ripe for disruption.

3. **Apply Four Actions Framework:** This is the core analytical tool for creating a new value curve. For each identified competing factor, ask four questions:
   - Eliminate: Which factors that the industry has long competed on should be eliminated? (e.g., complex legacy digital features customers don't truly value, or unnecessary digital overhead that adds cost without perceived benefit).
   - Reduce: Which factors should be reduced well below the industry standard? (e.g., over-engineered digital functionalities, or excessive digital marketing spend that yields diminishing returns, allowing for cost savings).
   - Raise: Which factors should be raised well above the industry standard? (e.g., elements that truly differentiate, such as hyper-personalization through AI, seamless cross-platform integration, superior data privacy features, or unparalleled digital customer support).
   - Create: Which factors should be created that the industry has never offered? (e.g., entirely new digital value propositions, such as a platform that connects previously siloed services, or an AI-powered predictive service that anticipates customer needs).

4. **Develop New Value Curve (To-Be Strategy Canvas):** Based on the answers from the Four Actions Framework, plot the new value proposition on a fresh Strategy Canvas. This new curve should look significantly different from the industry's existing value curves, representing a unique and compelling offering that creates a "blue ocean" – uncontested market space. For digital maturity, this new value curve often relies heavily on innovative digital capabilities that redefine customer expectations and create new demand.

5. **Test and Validate:** Conduct experiments, pilot programs, and market research to validate the new value proposition with target customers. This step is crucial to ensure that the innovative offering truly resonates with the market and delivers the intended value before a full-scale launch. Gather feedback and iterate on the concept. For digitally-enabled blue oceans, this involves rapid prototyping and A/B testing of new digital features or services with real users to quickly learn and adapt.

6. **Execute and Integrate:** Implement the new strategy across the entire organization. This involves aligning all internal activities, processes, and people to deliver the blue ocean offering. It requires a significant shift in focus, resource allocation, and often, a cultural transformation to support the new strategic direction and sustain the competitive advantage. For digital maturity, this means building the necessary digital infrastructure, developing new digital skills, and fostering a culture of continuous digital innovation to maintain the unique value proposition and defend the newly created market space.

6x6™ Transformation Blueprint

Decision Velocity Index (DVI)

"In any moment of decision, the best thing you can do is the right thing, the next best thing is the wrong thing, and the worst thing you can do is nothing." --- Theodore Roosevelt, the 26th U.S. President and champion of the "Strenuous Life," Roosevelt proved that in times of radical change, the cost of paralysis is higher than the cost of a corrected mistake. For Leadership, DVI is the antidote to "Trade-off" paralysis; it marks the shift from defensive consensus to a high-velocity culture that treats speed as a survival mechanism for Net Zero.

In the industrial age, leadership was defined by the weight of the "final say." The executive was the ultimate filter, the heroic gatekeeper whose value was measured by their ability to provide the "blessing" or the "veto." This model assumed a world of scarce information and slow-moving markets where the cost of a wrong decision was high, but the cost of a slow decision was manageable. In the AI era, that logic has inverted. Intelligence has become a utility, generated in milliseconds by models that can process quadrillions of data points to identify market shifts, fraud patterns, or supply chain disruptions. Yet, many organizations still force these lightning-fast insights to wait for a Tuesday morning committee meeting. This is the birth of the Identity Chasm: the widening gap between the speed of machine-generated truth and the agonizingly slow pace of human-led execution. To survive this transition, we must redefine the leader's role from a Gatekeeper to a Strategic Orchestrator. The primary metric for this evolution is the Decision Velocity Index (DVI).

The chasm here is the psychological distance between the **Heroic Gatekeeper** and the **Strategic Orchestrator**. For decades, leadership identity was tied to being the "final say"—the bottleneck that provided safety through manual review. The chasm opens when a leader realizes that their personal "gut check" is now slower than the machine's truth. Identifying this is vital because until the leader stops seeing speed as a threat to their importance, they will continue to pay a "Latency Tax" that makes the organization's AI investments useless.

The greatest barrier to high-speed transformation isn't technical; it is the leader's ego. We call this the Heroic Stewardship Bias. Leaders have spent decades being rewarded for their "gut feel" and their ability to personally review every high-stakes move. This creates a psychological "Expert Trap." When an AI system suggests a strategic pivot based on real-time data, the leader feels a visceral need to "check the checker." They believe that their manual oversight adds value, when it often only adds latency.

This hesitation creates a "Leadership Drag" that can be fatal. If the machine identifies a competitive threat in seconds, but the human hierarchy requires six weeks of slide decks and consensus-building to respond, the competitive advantage of the AI is effectively zero. To bridge this chasm, the leader must undergo a psychological shift: realizing that their value is no longer in making the decision, but in designing the framework within which decisions are made autonomously. This is the move from manual control to Strategic Decoupling.

THE RAISON D'ÊTRE:

"TO GUIDE"

By mastering the DVI, the leader fulfills their true purpose: To Guide. When a leader is freed from the burden of micromanaging operational decisions, they can reclaim their status as the Sole Arbiter of Worth. Their energy shifts toward moral authority, long-term stewardship, and ethical guardrails—the things a machine cannot do. We are not automating the leader; we are automating the friction that prevents the leader from leading. In the post-AI era, the best leaders aren't the ones with the most answers; they are the ones who build the most responsive systems.

DEFINING THE DVI: THE LOGIC OF LATENCY

The Decision Velocity Index is a measurement of the time elapsed between an Insight and an Action. Instead of looking at profit margins or headcount, the DVI looks at the "clock speed" of the organization.

The DVI asks one simple question: How long does it take for a proven truth to result in a resource shift? In a traditional company, this "Delta" is measured in months. In a high-velocity, AI-augmented firm, this Delta is measured in minutes. The goal of tracking the DVI is to identify where "Cultural Red Tape" is acting as a tax on the business. If the AI surfaces a 90% confidence signal that a product is failing, and the leadership team waits three months to pull the funding, that three-month gap is your DVI score.

THE OBJECTIVE: TOWARD ORCHESTRATION

The goal of the DVI intervention is to reach a state of Equilibrium we call Orchestration. In this state, the organization operates like a high-performance jet on autopilot. The pilot (the leader) isn't touching the stick for every minor gust of wind; they are setting the destination, monitoring the instruments, and only intervening when the system encounters an anomaly that falls outside its programmed parameters.

Orchestration allows the leader to scale their influence without scaling their workload. By shrinking the DVI, you aren't just moving faster; you are moving with higher precision. You are allowing the "Logic" of the machine to handle the "Function" of the business, so the "Heart" of the leader can handle the "Direction." This is the only way to achieve scale in a world where intelligence moves at the speed of light.

CASE STUDY	THE CHASM REALITY	ACTION & RESULT
1: A GLOBAL RETAIL BANK faced a crisis as fintechs approved loans in minutes, while their process took five days. Senior committees insisted on physical signatures for every loan over $50,000 to "protect integrity."	Leaders were stuck in the **Validation Analyst** trap, believing personal review was the only shield against risk, despite AI having 99% accuracy over years of back-testing.	They moved to **Negative Consent Protocols**: loans meeting all AI safety criteria were auto-funded unless a VP manually "braked" it within 15 minutes. Approval times dropped 94%; volume rose 30%.
2: A HEAVY MANUFACTURING FIRM struggled with "Consensus Constipation." The COO called emergency meetings with six department heads to "align" on supply chain rerouting, taking 48 hours to reach a decision.	This was a **Heroic Gatekeeper** scenario where the leader valued "alignment" over "velocity," ignoring that the cost of delay exceeded the cost of a slightly imperfect decision.	The COO enacted **Threshold Governance**, pre-authorizing the system to auto-reroute shipments under $500,000. By decoupling their signature, they reduced lag 80% and saved $12M annually.
3: A MAJOR HOSPITAL saw discharge times lagging, causing ER bed shortages. The bottleneck was a "Permission-Based" culture where floor managers waited for senior sign-off before cleaning rooms.	The administration charged a **Validation Tax**, requiring manual proof of readiness that digital systems already confirmed. Leaders' identity was tied to being the "Command Center" of all flow.	They shifted to **Strategic Decoupling**, allowing software to trigger intake flows automatically via real-time sensors. This liberated 40 minutes per bed cycle, "creating" 15 new beds without construction.
4: A HIGH-GROWTH E-COMMERCE RETAILER lost margin because the pricing team waited for the CEO's blessing to move prices more than 10%. The CEO feared an algorithm would damage brand perception.	The CEO was an **Intuition Anchor**, valuing "fashion gut" over real-time elasticity data. This created an "11th-Hour Veto" culture that frustrated teams and slowed market response.	The leader ran **Shadow Execution** tests, comparing AI pricing to manual choices. Data proved AI generated 18% more margin; the CEO pivoted to reviewing only swings over 40%.
5: A REGIONAL ENERGY UTILITY faced pressure to reduce downtime. Maintenance leaders personally reviewed every drone inspection photo before dispatching crews, fearing "unnecessary truck rolls."	Leaders were caught in the **Expert Trap**, believing 30 years of experience couldn't be replicated, even though AI vision flagged defects with 95% precision—far faster than human review.	The team adopted **Threshold Governance**, allowing AI to auto-dispatch crews for "Priority 1" safety risks. This cut time-to-repair from 14 hours to 3, improving safety and ending the "Signature Stall."
6: A GLOBAL LAW FIRM struggled with "Status Update" culture. Senior partners spent 60% of their time in meetings with associates for progress reports, despite all data being live in their system.	Partners acted as **Heroic Gatekeepers** of info, using meetings to maintain control rather than using the digital "Source of Truth" to orchestrate higher-level strategy.	The firm set a **Meeting Moratorium** for projects marked "Green" in the data. Partners were forced to use the dashboard, reclaiming 15 hours per week for new business development.

ASSIGNMENT : THE LEADERSHIP CULTURE DIAGNOSTIC AUDIT

The transition from a bottlenecked hierarchy to a high-velocity orchestrator requires a brutal assessment of current behaviors. We often mistake "being busy" with "being effective," but in the AI era, busyness is often just the sound of structural friction. To bridge the Identity Chasm, you must be able to recognize the specific symptoms of Leadership Drag and apply the precise architectural intervention to neutralize them. This audit is designed to move you from theory to action. On the left are ten common Symptoms found in organizations with a low Decision Velocity Index. On the right are ten Strategic Interventions from the Ai6x6 playbook.

The Assignment: Match the symptom observed in your current leadership culture to the intervention required to reach Equilibrium.

THE SYMPTOMS (THE DIAGNOSTIC)	MATCH	THE INTERVENTIONS (THE REMEDY)
1. The 11th Hour Veto: A leader overrides a data-backed recommendation at the last second based on a "bad feeling."		**A. Threshold Governance:** Pre-approving AI-led pivots within specific risk/budget margins to bypass manual sign-offs.
2. Consensus Constipation: High-confidence insights are stalled because every department head must "align" before moving.		**B. Negative Consent Protocols:** Insights are executed automatically unless a human manually "pulls the brake" within a set window.
3. The Validation Tax: Teams spend more time double-checking the AI's work than they do executing the strategy.		**C. Strategic Decoupling:** Separating "Standard Logic" (automated) from "Moral Authority" (human) to prevent operational bottlenecks.
4. Permission-Based Latency: Front-line managers see an opportunity but must wait for three layers of approval to act.		**D. The Evidence Ritual:** Mandating that every meeting begins with a data-driven question that challenges the status quo.
5. The Expert Trap: Senior leaders feel their value is tied to having all the answers, leading to "Assertion Bias."		**E. Algorithmic Guardrails:** Defining the "Safe Zones" where the machine can optimize without needing a human "blessing."
6. Meeting Congestion: The calendar is filled with "Status Updates" that merely recite what the dashboard already shows.		**F. Outcome-Based Agency:** Rewriting roles to reward "Decision Speed" and "Unlearning" rather than "Compliance."
7. Sunk Cost Anchoring: Continuing to fund a failing project because the leader's reputation is publicly tied to its success.		**G. Shadow Execution:** Running AI and human models in parallel to prove machine reliability and build cultural trust.
8. Performance Myopia: Measuring managers on "Activity" (how hard they work) rather than "Velocity" (how fast they pivot).		**H. The Kill-Switch Commission:** An independent body that automatically defunds projects that miss AI-predicted targets.
9. The Clarity Gap: Teams are afraid to act because they aren't sure if the AI's "Truth" overrides the Leader's "Opinion."		**I. Exception-Only Management:** Shifting the leader's focus exclusively to "Black Swan" events that fall outside the model's data.
10. Historical Drift: Using last year's successful playbook to manage today's real-time supply chain disruptions.		**J. Real-Time Feedback Loops:** Replacing quarterly reviews with live performance triggers that adjust strategy instantly.

ANSWER MATCH: 1-A, 2-B, 3-G, 4-C, 5-D, 6-I, 7-H, 8-F, 9-E, 10-J

BLUEPRINT SERIES

SECTION F

Cultural Compass

Ai6x6™ Transformation Blueprint

Story 1 — The Post-War Japanese Industrial Ethos (1945–1960): Japan emerged from the war with limited resources but a collective commitment to national rebuilding. Companies embraced a culture of discipline, apprenticeship, and shared responsibility. Workers identified deeply with their organizations, reinforcing loyalty and craftsmanship. Leaders cultivated humility and long-term thinking. This cultural foundation enabled rapid industrial recovery. It demonstrated how shared values can accelerate capability even before systems mature.

Story 2 — NASA's Mission Culture During Apollo (1961–1972): NASA built a culture defined by precision, accountability, and collective purpose. Engineers and astronauts operated with a shared belief that failure was unacceptable, yet learning was essential. Communication was open, direct, and cross-disciplinary. Rituals, reviews, and norms reinforced psychological safety and technical rigor. This culture enabled teams to navigate uncertainty and complexity. It became a global example of mission-driven cohesion.

Story 3 — The Civil Rights Movement's Organizing Ethos (1954–1968): The movement relied on a culture of discipline, nonviolence, and moral clarity. Communities trained in de-escalation, collective action, and strategic communication. Leaders modeled courage and restraint, shaping norms that guided behavior under extreme pressure. Shared values created unity across diverse groups. This cultural coherence amplified impact far beyond organizational structure. It showed how culture can mobilize change at societal scale.

Story 4 — The Nordic Social Trust Model (1970s–1990s): Nordic countries cultivated cultures rooted in transparency, equality, and civic trust. Institutions operated with high accountability and low hierarchy. Citizens expected fairness and contributed to collective welfare. This cultural fabric enabled strong public services, stable governance, and economic resilience. Trust reduced friction in decision-making and collaboration. It demonstrated how cultural norms can become strategic national assets.

Story 5 — The Rise of Silicon Valley's Innovation Culture (1980s–2000s): Silicon Valley developed a culture that celebrated experimentation, risk-taking, and rapid iteration. Informal networks, open information flow, and meritocratic ideals shaped behavior. Failure was reframed as learning, attracting talent willing to push boundaries. Venture capital reinforced speed and ambition. This culture accelerated technological breakthroughs and entrepreneurial growth. It showed how cultural norms can amplify innovation beyond structural advantages.

Story 6 — The Globalization of Corporate Diversity & Inclusion (2000s–2020s): Organizations worldwide recognized that diverse teams outperform homogeneous ones when supported by inclusive cultures. Companies invested in psychological safety, equitable processes, and representation. Employees gained platforms to voice perspectives and challenge norms. Inclusion became a driver of creativity, retention, and ethical integrity. This cultural shift reshaped leadership expectations and organizational identity. It demonstrated how culture evolves to meet societal and competitive demands.

LESSONS FROM A LEADERSHIP TRANSFORMATION PERSPECTIVE

Leadership maturity in cultural development emerges when leaders recognize that culture is not a slogan but a system of behaviors, norms, and shared meaning. Across decades, leaders learned that culture shapes how people interpret decisions, respond to pressure, and collaborate across boundaries. Effective leaders model the values they expect, reinforce rituals that sustain cohesion, and create environments where trust and accountability coexist. Culture becomes a strategic instrument when leadership aligns purpose, behavior, and identity. Maturity is achieved when culture guides action even without direct oversight—becoming a stabilizing force that strengthens resilience, integrity, and long-term performance.

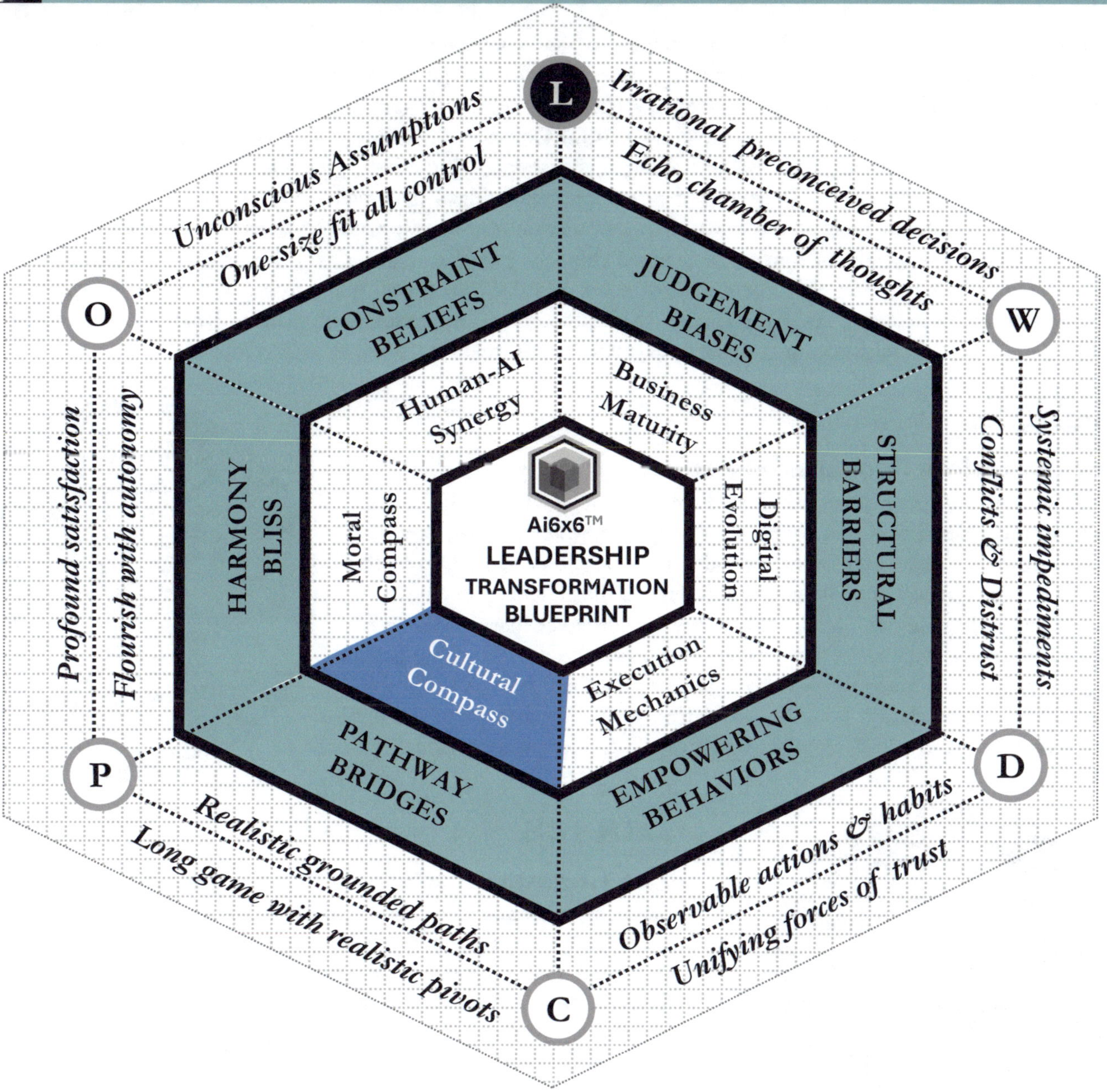

The **Ai6x6 Research Lab** chose this lens because the ultimate legacy of a leader is not found in patents or profits, but in the emotional health of their collective. We reframed this dimension as the stewardship of the organizational "soul," recognizing that culture is the invisible operating system that dictates the success of every algorithm. It is vital because it addresses the invisible shadows—the beliefs, biases, and resistances—that live in the minds of the people. This lens is the bridge between the leader's strategic intent and the workforce's daily reality.

We selected it to ensure leaders build a harmonious and resilient community rather than just an efficient company. By dismantling structural hurdles and building pathways of trust, the leader creates a "greenhouse" for human potential where failure is treated as data. The Lab believes that in an age of automation, the human connection is the only self-sustaining source of energy. This lens forces leaders to look at how their judgment impacts collective behavior. When culture is treated as a discipline, the leader achieves a state of "bliss" where the workforce feels empowered, valued, and fundamentally connected to the mission.

Culture has always been the invisible architecture of an organization, shaping how people think, behave, and make decisions long before any formal strategy takes hold. In the age of AI, this architecture becomes even more consequential. Technology may accelerate capability, but culture determines whether that capability becomes meaningful progress or wasted potential. The **Cultural Compass** lens forces leaders to confront the truth that transformation is not sustained by systems alone; it is sustained by the beliefs people hold, the biases they carry, the barriers they navigate, the behaviors they exhibit, the bridges they build, and the moments of profound satisfaction that anchor their commitment. Without cultural alignment, even the most sophisticated AI strategy collapses under the weight of human resistance. This lens reframes culture as a dynamic system rather than a static identity. It recognizes that culture is not a set of values on a wall but a living pattern of interpretation, interaction, and intention. As AI reshapes workflows, decision cycles, and power dynamics, leaders must understand how cultural forces amplify or undermine transformation.

Beliefs

The Narratives Beneath the Surface

Beliefs and assumptions form the first dimension of the Cultural Compass. They are the quiet forces that shape how people interpret change, how they respond to uncertainty, and how they judge the intentions behind leadership decisions. When beliefs are outdated or misaligned, they distort perception and create friction. Leaders who ignore this dimension discover that people do not resist technology—they resist the meaning they attach to it. Addressing beliefs requires leaders to surface the narratives that operate beneath the surface and reshape them with clarity, honesty, and purpose. In this dimension, cultural truth-telling becomes a leadership act rather than a communication tactic.

Biases

The Cognitive Patterns That Shape Judgment

Biases and mental shortcuts represent the second dimension. These cognitive patterns influence how people evaluate information, assign trust, and make decisions. In an AI-enabled environment, biases can distort both human judgment and the interpretation of machine-generated insights. Addressing them requires intentional design, transparent dialogue, and environments where people feel safe challenging their own assumptions. Leaders must create conditions where cognitive unlearning becomes possible. In this dimension, cultural evolution is driven by awareness, humility, and the willingness to question long-held mental models.

Barriers

The Structural Forces That Restrict Change

Barriers and systemic impediments form the third dimension. These are the structural forces—policies, hierarchies, incentives, and legacy processes—that constrain behavior even when people are willing to change. Leaders often misinterpret resistance as emotional when it is structural. Removing barriers is not about demanding more effort; it is about redesigning the environment so that the desired behaviors become the natural ones. In this dimension, organizational de-friction becomes a strategic priority rather than an operational afterthought, enabling people to move with clarity rather than struggle against constraints.

Behaviors

The Visible Expression of Culture

Behaviors and observable actions represent the fourth dimension. Culture is not what people say; it is what they do consistently. In the AI era, behaviors must evolve to reflect curiosity, adaptability, collaboration, and responsible experimentation. Leaders must model these behaviors visibly and repeatedly. They must demonstrate that learning is valued, that questioning is encouraged, and that accountability is shared. In this dimension, behavioral signaling becomes a powerful tool for shaping norms, reinforcing the idea that culture is enacted through daily choices rather than declared through slogans.

Bridges

The Pathways That Connect People and Purpose

Bridges and cultural pathways form the fifth dimension. These are the connections that allow ideas, trust, and collaboration to flow across the organization. Bridges are built through shared language, shared rituals, and shared purpose. They transform isolated teams into aligned communities. Leaders who invest in cultural pathways create environments where people feel connected to something larger than their role. In this dimension, collective identity-building becomes a catalyst for cohesion, enabling the organization to move as a unified whole rather than a collection of disconnected parts.

Bliss

The Emotional Anchor of Commitment

Bliss and profound satisfaction complete the compass. Bliss is not about comfort; it is about meaning. It is the sense of fulfillment that emerges when people feel their work matters, their contributions are recognized, and their growth is supported. Bliss is the emotional anchor that sustains commitment through uncertainty. Leaders who cultivate it create cultures where people do not merely comply—they care. In this dimension, purposeful fulfillment becomes a strategic asset, strengthening resilience and deepening engagement.

The Cultural Compass lens reminds leaders that transformation is not sustained by technology or process alone. It is sustained by the human experience of change. Culture determines whether people interpret transformation as threat or opportunity, whether they respond with resistance or curiosity, and whether they move in alignment or fragmentation. In a world where AI accelerates both progress and pressure, leaders must treat culture as the primary operating system of the enterprise. When leaders align beliefs, challenge biases, remove barriers, shape behaviors, build bridges, and cultivate meaning, they create cultures capable of evolving continuously. They build environments where people feel psychologically safe, intellectually stimulated, and emotionally invested. Culture becomes the force that amplifies strategy rather than undermining it. It becomes the stabilizing anchor in moments of disruption and the accelerating engine in moments of opportunity.

CULTURAL COMPASS – CULTIVATING JAPANESE GARDEN

The False Climate — BELIEFS

Dismantling "Weather Myths" ensures that unrealistic, universal expectations don't stifle individual growth. Exposing the false belief that all "plants" require the same sun, water, or timing—allowing for specialized care.

The Invasive Weeds — BIASES

Identifying and removing hidden mental shortcuts prevents old habits from choking out innovation. Surgically exposing the leadership judgment and systemic biases that create "shadow" cultures and block progress.

The Hard Clay — BARRIERS

Breaking through structural resistance allows the organizational roots to expand and stabilize. Removing the deeply embedded obstacles like data, automation and AI anxiety that act as physical walls against digital growth.

Calibrated Care — BEHAVIORS

Executing specialized nurturing routines ensures each distinct "species" receives what it needs to thrive. Moving away from "one-size-fits-all" mode to model specific watering, feeding, and light-balancing behaviors required for a diverse ecosystem.

The Trellis System — BRIDGES

Constructing pathways for upward growth provides the support needed for teams to reach their full potential. Building the specific leadership and development paths that allow the workforce to climb through the maturity stages.

The Full Harvest — BLISS

Securing a state of permanent organizational harmony results in a self-sustaining ecosystem of excellence. Reaching the ultimate destination where the "Cultural Soul" operates in a high-fidelity hum, delivering constant value.

6x6™ Transformation Blueprint

Leadership Constraints Beliefs

"The greatest leader is not necessarily the one who does the greatest things. He is the one that gets the people to do the greatest things." - Ronald Reagan

Ronald Reagan (1911–2004) was the 40th U.S. President. He is widely recognized for his charismatic communication style, which earned him the nickname "The Great Communicator." He was known for his optimism and his ability to rally support for his policies.

Leadership constraints beliefs are deeply ingrained, often unconscious assumptions that limit a leader's innovative thinking and actions. These mental models, shaped by past experiences or company culture, are powerful yet difficult to challenge. A common belief is that crucial decisions must be centralized, which creates bottlenecks and stifles innovation. Another is that failure is unacceptable, preventing the risk-taking and experimentation essential for growth. The first step to transformation is self-awareness, where a leader must confront and question these foundational beliefs. As Reagan suggests, an effective leader empowers others to do great things, which is impossible if they believe they must do everything themselves. Releasing these self-imposed constraints allows a leader to delegate, trust, and build a collaborative environment.

This topic explores how Leadership Constraints Beliefs create the strategic direction and foundation for achieving high Business Maturity. It argues that transformation begins not with systems, but with the leaders' non-negotiable convictions. The chapter on Transformational Beliefs explores the courage required to pursue radical, non-linear change, while Foundational and Strategic Beliefs ensure the organizational direction aligns with core tenets and long-term conviction, preventing strategic drift. Furthermore, the book details the impact of Growth Mindsets—seeing employee capabilities and organizational limits as fluid and developable—and the essential need for deep Customer-Centric Beliefs that place the buyer at the center of every decision. Finally, it highlights People-First Beliefs, which treat workforce well-being and development as a strategic asset, acknowledging that mission success depends entirely on human potential. By aligning these core principles, leaders ensure every resource is effectively deployed toward the desired destination.

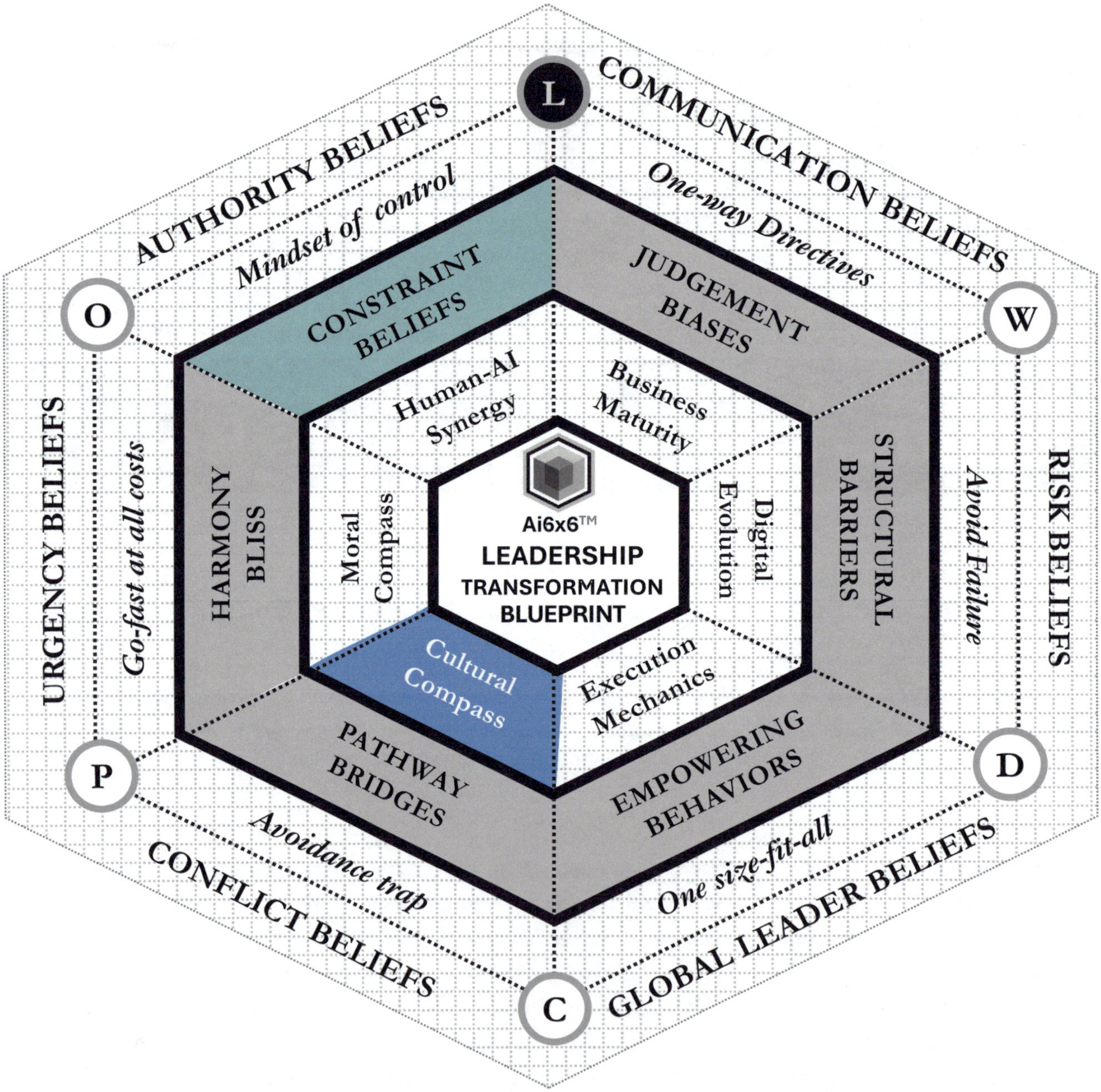

Ai6x6 CULTURAL COMPASS FOR BELIEFS | LEADERSHIP TRANSFORMATION

1. AUTHORITY BELIEFS | The Micromanager's Shadow

When you view authority as a tool for control rather than enablement, you inadvertently create a bottleneck. Your daily life becomes a series of approvals, slowing down high-level execution. This mindset shifts your role from a strategic visionary to a micro-manager, stifling the very talent you hired to scale the organization. Instead of steering the ship, you end up checking the oars. This creates a culture of dependency where no one moves without your nod, ultimately limiting the company's growth to the boundaries of your own personal bandwidth and daily stamina. Over time, your best people leave for environments where they can lead, leaving you with a compliant but uninspired workforce that requires constant, exhausting supervision to function.

DO'S FOR THE LEADER	DON'TS FOR THE LEADER
Empower teams to make their own decisions.	Hoard all the decision-making power.
Give public recognition for good work.	Take all the credit for team success.
Lead by being a good role model.	Dictate without personal involvement.
Actively seek feedback from your team.	Ignore input from others on the team.
Build a culture of mutual trust.	Use fear to enforce compliance and silence.
Clearly define everyone's roles.	Create ambiguity around decision power.

SITUATION	SUGGESTED ACTION
A leader believes they must have all the answers and is reluctant to admit when they're stuck	The leader must model vulnerability by admitting when they're stuck and asking for input. This builds a powerful tool for mutual growth and shows the team that it's okay to not have all the answers.
A team member is reluctant to challenge a senior leader's idea, even if it has flaws	Create a culture where respectful disagreement is expected and valued for successful decision-making. This builds a new norm where people are not afraid to challenge authority with good ideas from anywhere.
A team is waiting for your approval on every minor decision	Consciously step back and empower the team by giving them clear direction without micromanaging the process. Define the **"what,"** but allow them to determine the **"how."**
You are asked for your opinion on a complex problem with many unknowns	Be specific about what you know for sure and what you don't, modeling intellectual humility. This shows the team that it's okay to not have all the answers and that a healthy dose of realism is necessary for success.
A team member is deciding based on the opinion of an authority figure, rather than on data	Coach them to question assumptions and to make decisions based on objective data, not on the opinion of a single person. This helps them to think more critically and to make better decisions.
A team is presenting a unanimous decision and is not considering any alternatives	Challenge them to describe the potential downsides and risks they considered and how they would mitigate them. This encourages critical thinking and ensures that the team has a complete picture.

2. COMMUNICATION BELIEFS | The Information Silo

Assuming information is "need-to-know" creates a vacuum filled by rumors and anxiety. For an executive, this means spending more time on damage control than direction. When you hoard data, you lose the collective intelligence of your team, forcing you to solve every complex problem in isolation. This lack of transparency erodes trust and slows down decision-making at the edges of the organization. In your daily life, you will find yourself constantly repeating instructions because the broader context was never shared, leading to misalignment and a frustrating loss of operational momentum. True power in leadership comes from the speed at which information flows through your organization, not the amount of information you keep locked in your own office or head.

DO'S FOR THE LEADER	DON'TS FOR THE LEADER
Be transparent, share info openly.	Withhold info for personal power.
Listen to understand, not just to reply.	Assume you already know everything.
Use many channels for key messages.	Rely on only one channel for info.
Welcome all questions and concerns.	Make people feel foolish for asking.
Tailor your message to the audience.	Use a generic one-size-fits-all approach.
Follow up on actions for clarity.	Leave conversations without next steps.

SITUATION	SUGGESTED ACTION
A new strategy is being rolled out and needs to be understood by everyone	Lead with a clear, concise, and inspiring vision that is easy for the team to internalize. This helps them to understand the new strategy and to align their efforts toward a single, unified goal.
The team is confused about priorities and is working on too many different things	Establish a clear, single source of truth for all project goals and priorities. This helps the team to focus on what truly matters and to be more productive.
You are asked a difficult question in a public forum	Provide a transparent and honest answer, demonstrating integrity and understanding. This helps you to build trust with the team and to show that you are a leader of integrity.
A project has too many stakeholders, and decisions are constantly being second-guessed	Create a simple, transparent decision-making framework that everyone can understand and follow. This helps to streamline the decision-making process and to ensure that everyone is aligned.
Rumors are spreading within the company	Address the rumors directly and provide factual, clear communication to put them to rest. This helps you to build trust with the team and to show that you are a leader of integrity.
Your team needs to understand the bigger picture and how their work fits into it	Regularly share performance metrics and company-wide updates to provide context and connection. This helps the team to see the meaning in their work and to stay motivated.

A rigid definition of "acceptable risk" often masks a fear of failure. In your daily operations, this manifests as missed first-mover advantages. If you view risk only as a threat to protect against, you'll find yourself managing a stagnating portfolio while more daring competitors redefine your industry's landscape. This cautious stance can feel safe in the short term, but it creates a long-term vulnerability. You risk becoming obsolete as you optimize for a world that no longer exists, failing to realize that in a volatile market, the greatest risk is often taking no risk at all. By reframing risk as a necessary investment in future relevance, you can shift your daily focus from pure protection to the strategic pursuit of growth and innovation.

DO'S FOR THE LEADER	DON'TS FOR THE LEADER
Evaluate risks based on data.	Avoid all risks and miss opportunities.
Let teams take smart, calculated risks.	Punish any and all failures harshly.
Clearly define acceptable risk levels.	Create a completely risk-averse culture.
Learn from failures without blame.	Hunt for a scapegoat when things fail.
Celebrate successful bold decisions.	Criticize attempts that don't pay off.
Plan for worst-case scenarios.	Be overly optimistic and ignore risks.

SITUATION	SUGGESTED ACTION
A team believes that all risk must be eliminated before a project can begin	Remind them that innovation requires taking chances and that learning from failure is a core tenet. This helps the team to overcome their fear of failure and to be more open to new ideas.
A leader believes that a successful outcome is the only measure of a good decision	Create a culture where decisions are judged by the process and information available at the time, not by the outcome. This encourages the team to make bold decisions without fear of future judgment.
The company is hesitant to invest in a new technology, believing it is too risky	Form a small, agile team to quickly prototype and test the new technology with a low-risk pilot. This allows the team to learn about the new opportunity before committing major resources, proving the concept.
The team is afraid to take risks, even if the potential reward is high	Create a space where taking calculated risks is not just tolerated but actively celebrated. This helps to build a culture of innovation and to encourage the team to take more chances.
A project has a tight deadline and no plan B if something goes wrong	Facilitate a pre-mortem to identify all potential risks and create mitigation plans before the project starts. This helps the team to be better prepared for potential problems and to be more agile in your response.
You are asked for your opinion on a complex problem with many unknowns	Be specific about what you know for sure and what you don't, modeling intellectual humility. This shows the team that it's okay to not have all the answers and that a healthy dose of realism is necessary for success.

4. GLOBAL LEADER BELIEFS | Global Blind Spots

Applying a "one-size-fits-all" leadership style to international teams creates cultural friction that halts progress. You may find your directives ignored or misinterpreted. Recognizing that global leadership requires local nuance isn't just a soft skill; it's a logistical necessity to ensure your global strategy lands on the ground. Without this belief, your daily life is plagued by international "misunderstandings" that delay launches and sour partnerships. Success in a global market requires the humility to adapt your style to the cultural context of your teams, ensuring that your vision is translated, not just transmitted. When you embrace local diversity, you unlock a global engine of growth that operates smoothly across borders, time zones, and varied social norms, maximizing your worldwide impact.

DO'S FOR THE LEADER	DON'TS FOR THE LEADER
Respect cultural differences and norms.	Impose your own cultural standards.
Promote diversity in your teams.	Favor people who are like you.
Use empathy to understand perspectives.	Dismiss concerns as just cultural.
Invest in tech for global collaboration.	Underestimate language and time barriers.
Engage directly with global teams.	Lead from a distance and stay disconnected.
Make sure every voice is heard.	Let one person dominate the talk.

SITUATION	SUGGESTED ACTION
Your team spans multiple time zones and cultures, leading to communication challenges	Find satisfaction in leading a diverse team and promote cross-cultural understanding. This helps to build a culture of respect and empathy, which is essential for a global team.
The company is entering a new international market	Articulate a vision that connects the global team to a shared mission, transcending boundaries. This helps the team to see the bigger picture and to work together toward a common goal.
A diverse team is having communication challenges and is struggling to connect	Establish shared principles of communication that are respectful of all cultures. This helps the team to communicate more effectively and to build a stronger sense of unity.
You are leading a project with a global impact	Find joy in uniting people from different backgrounds toward one shared, inspiring goal. This helps you to stay motivated and to lead a diverse team effectively.
You notice cultural biases affecting team decisions and creating friction	Facilitate open dialogue to help the team understand and overcome biases. This helps to build a culture of respect and empathy, which is essential for a global team.
The team struggles to see beyond their local market	Share success stories from other global regions to inspire them and broaden their perspective. This helps the team to see the bigger picture and to work together toward a common goal.

If you believe harmony is the goal, you'll avoid the "necessary friction" that drives innovation. Your meetings will feel productive but result in bland, safe decisions. This avoidance creates a culture of artificial harmony where real issues rot beneath the surface, eventually exploding into crises that demand your immediate attention. Daily, this means you are surrounded by "yes-people" who won't tell you when a strategy is failing. By reframing conflict as a source of intellectual diversity, you move from managing comfortable silences to navigating the productive tensions that lead to market-disrupting breakthroughs. High-performing teams don't avoid conflict; they master it, using disagreement as a sharpening stone for their best ideas, ultimately saving you from making expensive, unchallenged strategic errors.

DO'S FOR THE LEADER	DON'TS FOR THE LEADER
Address conflicts quickly and directly.	Avoid all difficult conversations.
Facilitate respectful, open dialogue.	Take sides and create a winner/loser.
Focus on the issue, not the person.	Make conflicts personal or emotional.
Use conflict to improve and innovate.	See all conflict as a negative sign.
Set ground rules for handling disputes.	Let emotions spiral out of control.
Mediate to help find resolutions.	Leave teams to solve conflicts alone.

SITUATION	SUGGESTED ACTION
Two team members are in a heated, personal conflict, hurting morale and creating a toxic work environment.	Mediate with a neutral stance, helping them find a respectful compromise and a positive way forward. This helps to de-escalate the situation and to build a stronger sense of unity.
A disagreement is causing friction on a project and is slowing down the project's progress.	Facilitate a structured conversation focused on mutual goals and respect, not personal attacks. This helps the team to find a solution that works for everyone and to move forward with the project.
There is unspoken tension in the team, they are communicating, and there is a lot of passive-aggressive behavior.	Address the conflict directly and create an open space for a healthy resolution. This helps you to build trust with the team and to show you are a leader who is willing to handle difficult conversations.
Employees are lacking a sense of psychological safety and reluctant to speak up in meetings or challenge ideas	Actively and consistently solicit input from all team members, not just the most vocal ones. Validate their input by thanking them and exploring their ideas further.
A team member is avoiding a difficult conversation with a colleague, creating a lot of tension and holding them up	Coach them on how to have a difficult conversation in a respectful and productive way. Help them to see that a difficult conversation is an opportunity for growth and stronger relationship.
A project is stalled due to a conflict with team not communicating effectively and preventing them from moving forward.	Guide the team to a solution that respects everyone's contributions. This helps the team to find a solution that works for everyone and to move forward with the project.

6. URGENCY BELIEFS | The Burnout Cycle

Equating speed with productivity often leads to "false urgency," where everything is a priority. Your daily life becomes a blur of reactive fire-fighting. This pace prevents deep strategic thinking and forces your team into a state of permanent "sprint," which eventually erodes the quality of your most critical outputs. When urgency is the only gear, innovation suffers because there is no time for reflection or experimentation. You end up running very fast toward the wrong goals, mistaking activity for progress and burning out your best talent before they can reach the finish line of the fiscal year. True leadership involves the wisdom to distinguish between "fast" and "hurried," ensuring that your organization maintains a sustainable, high-velocity pace that doesn't sacrifice long-term health for short-term optics.

DO'S FOR THE LEADER	DON'TS FOR THE LEADER
Prioritize tasks effectively and wisely.	Create a false sense of urgency.
Promote a healthy, sustainable pace.	Push for constant, unrealistic speed.
Set clear, realistic deadlines.	Set arbitrary, unachievable deadlines.
Protect your team from scope creep.	Add new tasks to projects endlessly.
Reward quality of work, not just speed.	Reward speed at the cost of quality.
Build a culture of focused priorities.	Allow constant distractions and new tasks.

SITUATION	SUGGESTED ACTION
A team believes that faster is always better, leading to rushed, low-quality work	Prioritize a healthy, sustainable work pace over an aggressive, burnout-inducing timeline. Adjust the scope or deadline to be more realistic.
Employees believe in constant availability is leading to burnout and a lack of a healthy work-life balance	Set a clear expectation of "no-work" after hours and communicate that resting is a priority. This sends a clear message that a healthy work-life balance is a core part of the company's culture.
A team is struggling to keep up with an ever-increasing workload, leading to a feeling of being overwhelmed	Remove non-essential tasks from the team's plate, allowing them to focus on what truly matters. This helps the team prioritize their work and ensures that they have a sustainable workload
A team is deciding too quickly under a false sense of urgency, without considering all the information	Remind the team that a well-informed decision is more valuable than a fast one. Encourage them to take the time they need to gather all the necessary information before deciding.
A leader is deciding based on a gut feeling, rather than on data, which can lead to a major mistake.	Encourage the leader to make decisions based on objective data and a thorough risk analysis. This helps to ensure that decisions made with a balanced perspective and understanding of risks and rewards.
A project is behind schedule due to a lack of resources, its overwhelmed and is not sure how to get back on track.	Prioritize the most important tasks and work with the team to identify what can be removed from the project. This helps the team to focus on what truly matters and to get back on track.

6x6™ Transformation Blueprint

Leadership Judgment Biases

"When faced with a decision, the best course of action is often to suspend judgment and seek out the opinions of smart people who disagree with you." - Ray Dalio

Ray Dalio (b. 1949) is an American billionaire investor and founder of Bridgewater Associates. He is famous for his book Principles and his philosophy of "radical transparency" and "radical open-mindedness," which he credits for his success.

Leadership judgment biases are mental shortcuts that lead to flawed, irrational decisions. These are a natural part of human cognition that can have devastating effects if unchecked. A classic example is Confirmation Bias, where a leader seeks information that validates their existing beliefs while dismissing conflicting evidence. This can lead to a leader stubbornly pursuing a failing strategy, blind to signs of its collapse. Another pitfall is the Sunk Cost Fallacy, where a leader continues to invest in a project because of resources already spent, rather than its future potential. To overcome them, leaders must implement structured processes that force them to pause, reflect, and actively seek diverse perspectives. As Dalio advises, this creates a culture of intellectual honesty where decisions are made through a rigorous, objective process.

This foundational chapter establishes the framework's strategic basis by exploring how Leadership Judgement Biases create cognitive shortcuts that derail strategic direction and impede organizational Business Maturity. It argues that flawed decisions stem less from bad intent and more from systemic thinking errors. The analysis of Confirmation Bias shows how leaders resist evidence that challenges their assumptions, preventing radical change and hindering competitive positioning. It addresses the paralyzing effect of Sunk Cost Bias, which causes organizations to continue funding failing investments simply due to prior commitment, and the dangers of Overconfidence Bias, which leads to an exaggerated belief in success and insufficient risk planning. The chapter also details how Hindsight Bias distorts learning from past events, while Optimism Bias creates unwarranted positive outlooks that mask threats. Finally, it examines Loss Aversion Bias, explaining why the fear of small, immediate losses can prevent large, necessary strategic gains, ultimately locking the enterprise into suboptimal directions.

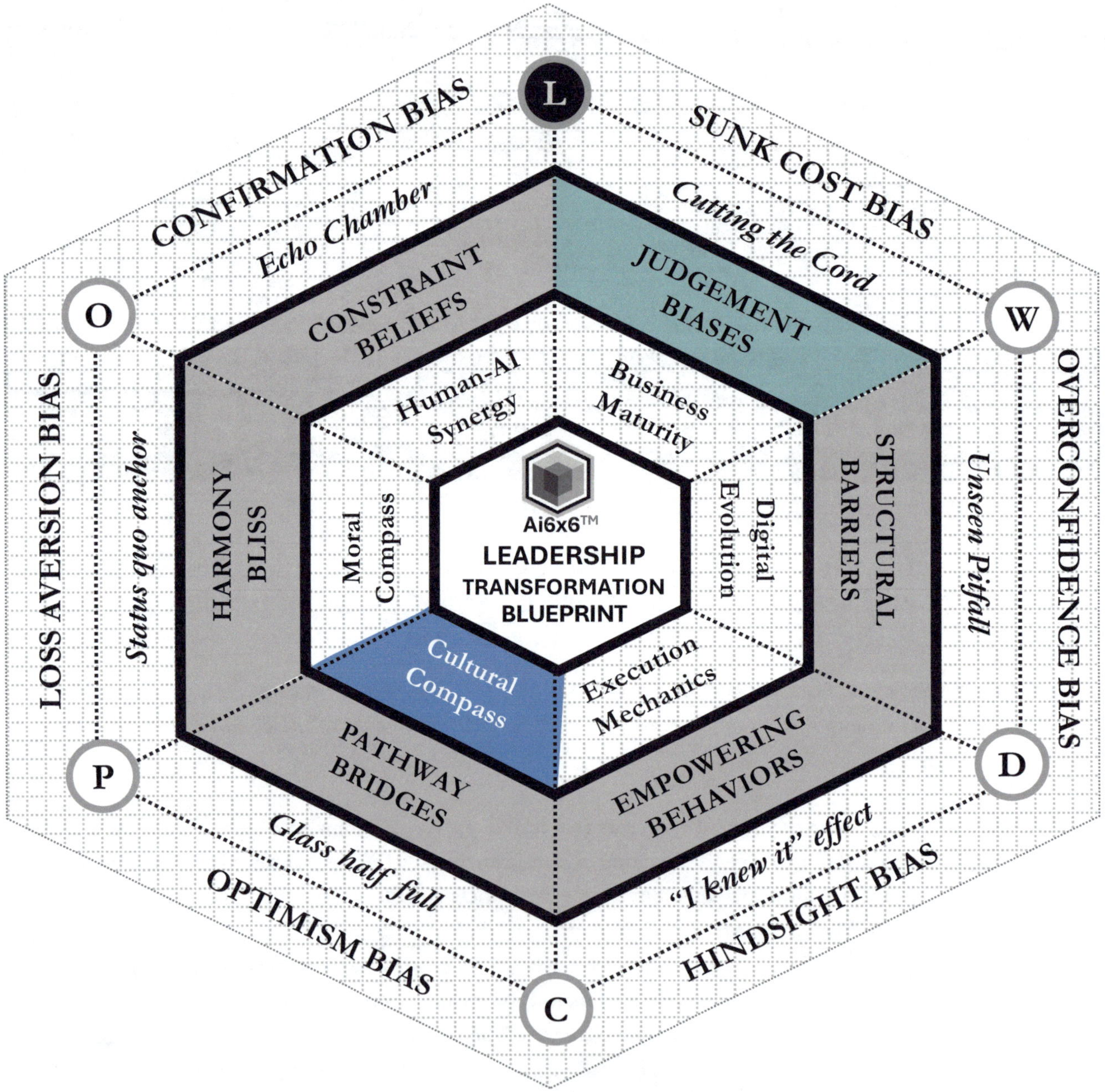

Ai6x6 CULTURAL COMPASS FOR BIASES | LEADERSHIP TRANSFORMATION

This bias acts as a filter that only lets in news you want to hear. In your daily decision-making, it creates a dangerous echo chamber. You risk investing millions into strategies based on flawed data because you ignored the "red flags" that contradicted your initial gut feeling or stated vision. As an executive, this bias makes you vulnerable to blind spots that competitors will eventually exploit. To counter this, you must intentionally surround yourself with "devil's advocates" who challenge your assumptions, ensuring that your daily path is guided by objective reality rather than a self-fulfilling prophecy of success. By actively seeking out information that proves you wrong, you become a more resilient and intellectually honest leader, capable of pivoting your strategy before reality forces your hand.

DO'S FOR THE LEADER	DON'TS FOR THE LEADER
Actively seek information that challenges.	Only seek info that confirms beliefs.
Ask for diverse opinions on every topic.	Surround yourself with "yes-men."
Encourage debate and respectful dissent.	Punish anyone who disagrees with you.
Look for flaws in your own ideas.	Assume your initial idea is the best.
Use data to disprove your assumptions.	Cherry-pick data that supports your view.
Consider all perspectives, not just one.	Listen only to a select few sources.

SITUATION	SUGGESTED ACTION
A team is presenting data supporting their favored project, ignoring contradictory evidence, confirming belief	Actively seek out information that could disconfirm the project's success. Consult a "devil's advocate" or an expert whose job it is to find flaws in your plan to ensure a balanced perspective.
An expert's opinion differs from a core belief you hold about the market, and you dismiss it as it contradicts your own.	Ask for their strongest evidence and rationale, approaching their perspective with a genuinely open mind. This allows you to challenge your own assumptions and consider a new perspective.
The market seems predictable, and you're planning based on expected outcomes, : not considering disruptions or changes.	Explore what specific data points would signal an impending disruption or a change in the market. This helps you to prepare for a range of scenarios and to be more agile in your response.
A new hire has a different approach to a problem, but the team is quick to dismiss it, which is a barrier to innovation.	Encourage the team to test the new approach in a low-stakes environment. By allowing them to see the benefits of a new approach, they can overcome their bias and be more open.
A product is receiving negative customer feedback, but no changes are being made, which is frustrating for customers.	Trust the team to make real-time adjustments to the product based on customer feedback and data. This shows that the company values customer input and is agile to respond quickly.
You have a strong gut feeling about a new hire, but skills and past performance data do not fully support the decision.	Set your personal feelings aside and focus solely on their skills and past performance data. This helps you to make a more objective decision and to avoid making mistake based on bias.

Refusing to kill a failing project because of the time and capital already spent is a trap. It keeps you anchored to the past. Daily, this drains resources—both financial and emotional—that should be pivoted toward future growth. True leadership is having the courage to write off yesterday's mistakes to fund tomorrow's successes. When you fall for this bias, you send a message that persistence is more important than performance. Breaking free requires a cold, rational look at future potential rather than past investment, ensuring your daily resource allocation is always optimized for the highest possible future return. An executive's job is not to justify the costs of the past, but to maximize the opportunities of the future, even if that means admitting a prior plan failed.

DO'S FOR THE LEADER	DON'TS FOR THE LEADER
Evaluate projects based on future value.	Continue a project because of past cost.
Be willing to abandon failing projects.	Refuse to quit a project you invested in.
Set clear criteria for project success.	Use a vague definition of success.
Focus on future opportunities.	Dwell on past investments and effort.
Conduct rational exit reviews.	Ignore all signs that a project is failing.
Reallocate resources to new ideas.	Keep pouring money into a losing project.

SITUATION	SUGGESTED ACTION
A long-running project has consumed vast resources with little progress with a fear of admitting failure	Evaluate its current merits and future potential as if it were a brand-new idea, free from past investment. This allows you to make an objective decision based on its current value, not resources spent.
A core business strategy is failing to deliver the results & this fear of admitting failure is a major barrier	Call for a rapid review of the strategy and be open to a major course correction. Instead of clinging to a failing plan, encourage the team to stop & try new approach showing agility and growth mindset.
A legacy product is losing money and market share, but the company is hesitant to retire it because of pride.	Determine if the resources tied to this product could be better used to fund a more promising initiative. This helps you to make a rational decision based on the company's potential, not on its past.
You personally championed a failing initiative, consuming lot of resources and are afraid of admitting failure.	Publicly announce the decision to stop and explain the rationale, showing a willingness to admit failure. This models a growth mindset and allows the company to cut its losses and move on.
An investment is not yielding the expected returns and continues to drain resources:	Cut losses quickly based on objective performance data rather than on the amount of past spending. This allows the company to free up resources and invest in more promising opportunities.
A team is emotionally attached to a non-viable project, preventing it from working on more promising projects.	Use a clear, objective checklist to decide whether to continue or quit, removing emotional influence. This helps the team to make a rational decision based on data, not on emotions.

Overestimating your "hit rate" leads to aggressive targets without realistic safeguards. You may find yourself promising results to the board that your infrastructure cannot support. This gap between your confidence and reality creates high-stress environments and eventually damages your professional credibility when the inevitable misses occur. While confidence is a leadership requirement, overconfidence blinds you to the complexities of execution. In your daily life, this bias can lead to a dismissive attitude toward valid concerns from your team, causing you to overlook the very operational details that determine whether a visionary strategy succeeds or fails. Humility is the antidote to this bias, allowing you to pair your bold vision with a grounded, realistic assessment of what it takes to win in the real world.

DO'S FOR THE LEADER	DON'TS FOR THE LEADER
Use humble realism to gauge success.	Overestimate your own abilities.
Seek input from others with expertise.	Act as if you are the only expert.
Plan for potential setbacks.	Believe nothing can go wrong.
Recognize and admit your own limits.	Never admit when you're wrong.
Value evidence over personal opinion.	Trust your gut feeling more than data.
Ask for critical feedback.	Avoid any and all criticism.

SITUATION	SUGGESTED ACTION
An employee is overconfident in their abilities, overestimating their own ability, which can lead to a major mistake.	Assign them a project with clear, measurable outcomes and provide them with regular, specific feedback, helping them to see the gap between their perceived ability and actual performance
A team presents a unanimous decision and is not considering any alternatives as they are confident in their decision	Challenge them to describe the potential downsides and risks they considered and how they would mitigate them. This encourages critical thinking and ensures team success.
A project's success is celebrated, but the team believes they are "perfect" and have nothing more to learn	Challenge the team to identify areas where they could improve, even after a big success. This helps them to stay humble and to not take future successes for granted.
A junior employee is presenting a solution that is technically flawed, but they are confident in their answer	Instead of dismissing their idea, ask probing questions to guide them toward finding their own solution. This encourages critical thinking and helps them build the skills they need in the future.
A leader is overconfident and is not willing to listen to their team, and this is a barrier to a healthy and productive team.	Coach the leader on the value of humility and the importance of listening to their team. Help them to see that their team is their greatest asset and that a humble leader is a more effective one.
A new hire has a different approach to a problem, but the team is quick to dismiss it which is a major barrier to innovation.	Encourage the team to test the new approach in a low-stakes environment. By allowing them to see the benefits of a new approach, you can help them be more open to new ideas.

4. HINDSIGHT BIAS | The "I Knew It" Effect

Claiming "I knew it all along" after a failure prevents genuine post-mortem learning. If you believe outcomes were predictable, you'll blame individuals rather than fixing systemic flaws. This bias stunts your growth as a strategist because it convinces you that you already have the answers, making you less observant of the nuanced variables that drive modern markets. Daily, this manifests as a "blame game" that discourages your team from reporting early warning signs. By admitting that the future is inherently uncertain, you foster a culture of vigilance and curiosity rather than one of defensive, retroactive justifications. This shift allows your organization to build a more sophisticated understanding of probability and risk, turning every unexpected outcome into a valuable lesson for future decision-making.

DO'S FOR THE LEADER	DON'TS FOR THE LEADER
Practice authentic learning from events.	Falsely believe an outcome was predictable.
Focus on what you learned from a failure.	Say, "I knew that was going to happen."
Objectively review past decisions.	Rationalize past events after they happen.
Document decisions and their reasoning.	Rewriting history to make you look good.
Create a blameless post-mortem process.	Blame a person for a bad outcome.
Look forward, not backward.	Dwell on how a bad event could be avoided.

SITUATION	SUGGESTED ACTION
A project fails, and the team is quick to say they "knew it would happen all along"	Redirect the conversation to focus on what was known *before* the event and what new information was gained. This encourages an honest post-mortem and genuine learning.
A successful product launch is attributed solely to the team's "brilliant" strategy, ignoring luck.	Actively acknowledge the role of luck and other external factors in the success to keep a grounded perspective. This helps the team to be humbler and to not take future successes for granted.
A competitor's perceived "obvious" success prevents team from genuine competitive analysis and learning	Facilitate a genuine, non-defensive analysis of the competitor's strengths and what they did right. This helps the team to learn from their rivals and to improve their own strategy.
After a failed investment, a leader says, "We should have seen the signs", creating a culture of blame	Reframe the discussion to focus on the information available and the decision-making process. This helps the team to learn from the event without creating a toxic environment, and open to taking risks.
The team is overly critical of past decisions, creating culture of second-guessing believing in better choices.	Remind the team that decisions are made with the information available at the time. Encourage them to focus on how to improve the decision-making process for the future.
A project is stalled due to a fear of making a decision that might be judged negatively later	Create a culture where decisions are judged by the process and information available at the time, not by the outcome, encouraging . bold decisions without fear of future judgment.

While vision requires hope, blind optimism ignores the "pre-mortem" planning needed for resilience. You might overlook supply chain vulnerabilities or market shifts, assuming things will just "work out." This leaves you unprepared for volatility, turning a manageable setback into a catastrophic disruption of your daily operations and long-term goals. For an executive, the cost of excessive optimism is often a lack of contingency planning. True strategic health involves balancing your inspiring "sunny day" vision with a rigorous assessment of "stormy day" scenarios, ensuring that the organization can survive the reality of a messy, unpredictable business environment. By tempering your optimism with realism, you provide your team with a roadmap that is both inspiring and navigable, regardless of the external conditions you face.

DO'S FOR THE LEADER	DON'TS FOR THE LEADER
Engage in prepared, realistic planning.	Be overly positive, leading to poor prep.
Acknowledge potential risks and problems.	Believe bad things will only happen to others.
Create contingency plans.	Assume everything will work out fine.
Be a realistic, not a hopeful, leader.	Ignore negative data in your reports.
Listen to your team's concerns.	Dismiss concerns as being negative.
Seek input from pessimists.	Only listen to optimistic people.

SITUATION	SUGGESTED ACTION
A team is confident that they can meet deadline, not considering all risks and roadblocks	Lead a critical discussion about potential roadblocks and the realistic resources required. This helps the team to create a more realistic timeline and to be better prepared for potential problems.
A leader is not being honest about the risks and challenges of the new initiative, leading to a lack of trust	Present a balanced view that includes both the opportunities and the realistic challenges. This helps to build trust with the team and to ensure that everyone is aligned on the risks and rewards.
An employee underestimating time and resources required for a project leading to major delay or failure	Coach the employee on how to create a more realistic project plan. Help them to see the value of a realistic timeline and to be better prepared for potential problems.
A business plan is based on the most +ve case scenario, without a backup plan for a downturn	Create a formal "what-if" plan for a range of market scenarios to prepare for any eventuality. This helps you to be better prepared for a market downturn and to be more agile in your response.
The team is not preparing for a worst-case scenario and is not considering any risks or roadblocks	Facilitate a pre-mortem to identify all potential risks and create mitigation plans before the project starts. This helps the team to be better prepared for potential problems and to be agile in responses.
A leader is deciding based on a gut feeling, rather than on data	Encourage the leader to make decisions based on objective data and a thorough risk analysis. This helps to ensure that decisions are made with a balanced perspective and an understanding of risks and rewards.

The fear of losing what you have often outweighs the potential of what you could gain. This bias keeps you playing defense. In your daily life, it means rejecting innovative ideas because of a small potential downside, effectively handing the future of the market over to your bolder competitors. You might find yourself protecting declining legacy products while ignoring the disruptive technologies that will eventually replace them. Overcoming loss aversion requires a shift in perspective: seeing the failure to innovate as the greatest potential loss of all. By prioritizing the "cost of inaction" over the "cost of failure," you ensure that your organization remains on the offensive, constantly seeking out new territory rather than just trying to hold onto a shrinking past.

DO'S FOR THE LEADER	DON'TS FOR THE LEADER
Take calculated risks for gains.	Avoid all risks to prevent losses.
Weigh potential gains against losses.	Focus only on the possibility of a loss.
Frame opportunities in terms of gain.	Frame all choices in terms of a potential loss.
Encourage bold, well-reasoned decisions.	Punish people for making a risky choice.
Create a safety net for failed risks.	Make it feel like one mistake is the end.
Focus on long-term growth.	Be paralyzed by short-term losses.

SITUATION	SUGGESTED ACTION
The company is hesitant to retire a legacy product that is losing money & market share, even if its net –ve	Determine if the resources tied to this product could be better used to fund a more promising initiative. This helps you to make a rational decision based on the company's future potential, not on its past.
A project is behind schedule, team morale dropping and focusing on blame game exercise.	Shift the team focus from assigning blame to providing support. Instead of asking "Why are we behind?" you should ask, "What support or resources are needed to get back on track?"
An employee struggling in a role they've had for years is hesitant to leave, impacting professional growth	Honestly assess if a new role or career path would be better for their and the company's future. This helps you to make a rational decision based on the employee's potential, not on their past performance.
An investment is not yielding the expected returns and continues to drain resources	Cut losses quickly based on objective performance data rather than on the amount of past spending. This allows the company to free up resources and invest in more promising opportunities.
The team is hesitant to change inefficient processes with a "we've always done it this way" mindset	Encourage a mindset of "test and learn" by running small experiments with new processes before adopting them. This allows the team to see the benefits of change firsthand and overcome resistance to new ways.
A competitor is gaining market share, you want to be defensive about it, dismissing their success	Conduct a genuine, non-defensive analysis of their strengths and what they are doing right. This helps you to learn from your competitors and to create a better product or service.

6x6™ Transformation Blueprint

Leadership Structural Barriers

"The distance between number one and number two is always a constant. If you want to improve the organization, you have to improve yourself and the organization gets pulled up with you." - Indra Nooyi

Indra Nooyi (b. 1955) is an Indian-American business executive who served as the CEO and Chairwoman of PepsiCo. She is known for her "Performance with a Purpose" strategy, which focused on environmental sustainability and corporate social responsibility.

Leadership structural barriers are systemic, institutional, or cultural impediments that prevent a leader from being effective. These are organizational dysfunctions that can neutralize even inspiring leadership. A Blame Culture Barrier is an environment where mistakes are punished rather than seen as learning opportunities, which stifles risk-taking and honest communication. A Bureaucratic Barrier is an overly rigid set of rules and processes that create friction and slow down decision-making. A Siloed Barrier is where departments operate in isolation, hoarding information and preventing the cross-functional collaboration essential for problem-solving. Overcoming these barriers requires more than a change in personal style; it demands a concerted effort to redesign the organization's systems and culture. As Indra Nooyi suggests, to improve the organization, you must improve yourself and, by doing so, pull the organization up with you.

This foundational component explores Leadership Structural Barriers—ingrained organizational obstacles that prevent strategic change and compromise Business Maturity. The most formidable barrier is Strategic Inertia, representing the deep-seated inability to pivot or change direction quickly, often rooted in past successes. Success also hinges on sustained Sponsorship: transformation fails rapidly without high-level executive support to champion the cause and absorb political risk. Crucially, this component analyzes Misaligned Incentives, detailing how rewarding the status quo or short-term gains directly conflicts with long-term goals, creating internal sabotage. This is compounded by significant Resource Constraints, where insufficient budget or talent starves the change initiative. The framework addresses Inadequate Communication, the failure to articulate a compelling, consistent vision across organizational tiers, leading to confusion and resistance. Finally, it tackles Political Resistance, where power struggles prioritize internal status over the strategic mission, fundamentally derailing directional clarity.

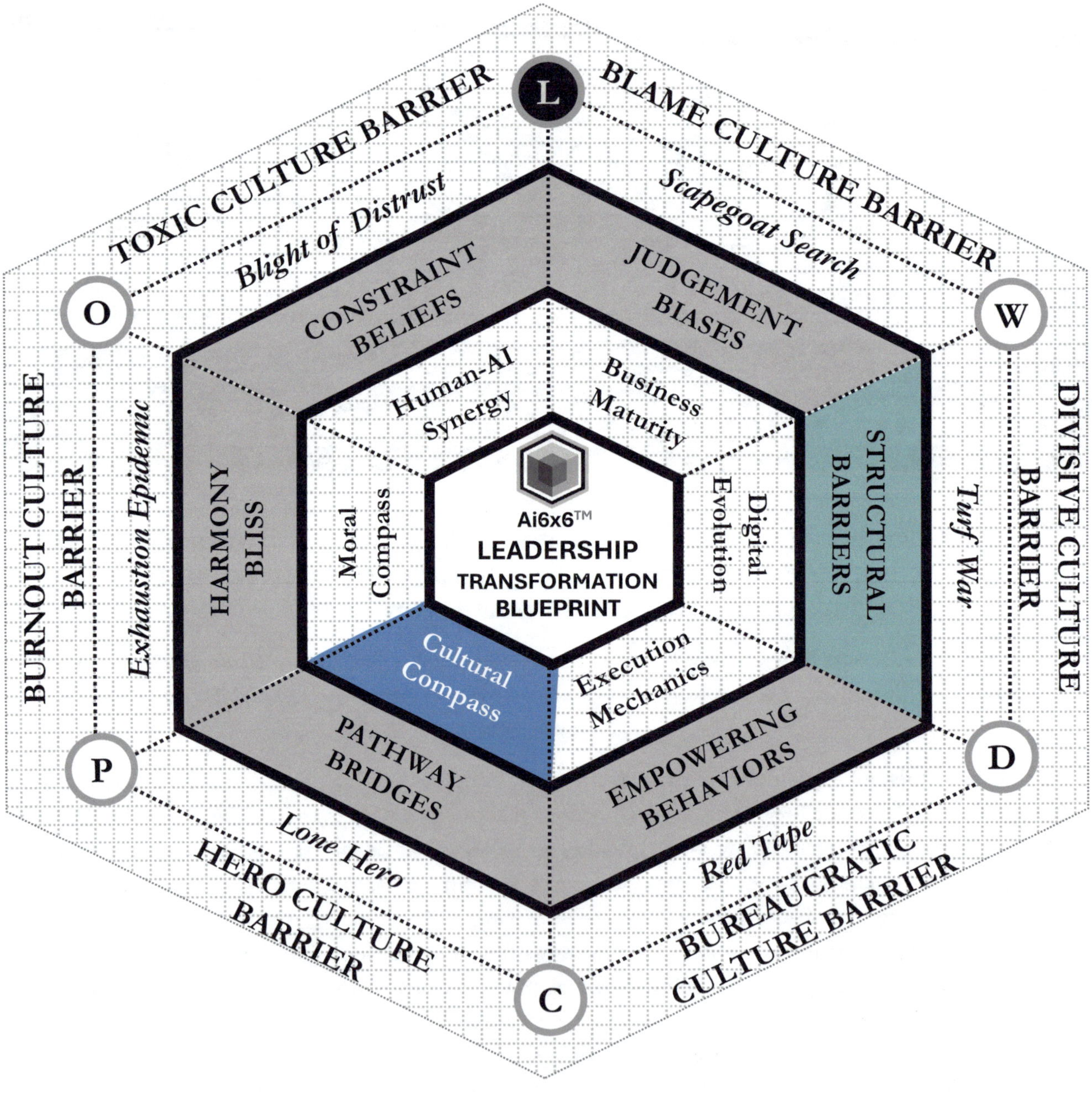

Ai6x6 CULTURAL COMPASS FOR BARRIERS | LEADERSHIP TRANSFORMATION

1. TOXIC CULTURE BARRIER | The Blight of Distrust

A toxic environment acts as a tax on every transaction. You'll find that talent acquisition is harder and turnover is constant. Your daily energy is spent managing interpersonal drama and HR escalations rather than driving growth. It turns your organization into a place people flee rather than a place they build. Over time, a toxic culture repels the high-performers who have other options, leaving you with a workforce that is disengaged or actively undermining your goals. Fixing this requires more than a memo; it requires a fundamental audit of how people are treated at every level of the hierarchy. As an executive, you must realize that culture is not a "soft" issue—it is the operating system that determines the speed and quality of everything your company does.

DO'S FOR THE LEADER	DON'TS FOR THE LEADER
Promote a positive, safe environment.	Allow a negative environment to persist.
Address negative behaviors directly.	Ignore bullying or harassment.
Foster psychological safety.	Let people feel unsafe to speak up.
Lead by example with kindness.	Engage in gossip or back-stabbing.
Listen to employee concerns.	Dismiss employee unhappiness.
Create clear policies on behavior.	Have vague or no rules on conduct.

SITUATION	SUGGESTED ACTION
Employees are engaging in gossip, passive-aggressive behavior, and back-stabbing, creating toxic environment	Publicly address the behavior, clearly stating that it is unacceptable and will not be tolerated, sending clear message that a toxic work environment is bad, and healthy team is priority.
A bully on the team is hurting morale and causing good employees to leave	Address the person directly, try coaching and mentoring route, and, if needed, remove them from the team to protect the culture, sending a clear message.
Management is showing favoritism, leading to resentment, that employees feel like they are not being treated fairly	Establish clear, transparent, and objective criteria for all promotions and raises. This ensures that all employees are treated fairly and that the company is a meritocracy.
You notice a lack of respect and empathy in team communication and there is a lot of tension.	Mediate a conversation focused on empathy and respect, creating a new norm for communication. This helps to build a culture of respect and empathy, which is essential for healthy work-life.
A team member is constantly complaining and has a negative attitude, bringing down the morale of the entire team	Address the negativity directly and work with the employee to find a positive path forward. If they are unwilling to change, it may be time to remove them from the team to protect culture.
The company is experiencing a high turnover rate; employees are leaving the company due to a toxic environment.	Conduct an honest, transparent audit of the company's culture and work to address the root causes of the toxicity. This helps you to build a healthy and productive team that employees want.

When mistakes are punished, innovation dies. Your team will stop taking risks and start hiding errors. You'll be the last to know when a project is off the rails because people are too afraid to speak up. This barrier forces you to lead in the dark, without honest feedback. Daily, you are managing "surprises" that could have been mitigated if people felt safe enough to flag them early. A blame-free culture doesn't mean a lack of accountability; it means focusing on solving the problem rather than finding a scapegoat, allowing for faster recovery and more honest communication. By shifting the focus from "who" to "how," you create a resilient organization where employees are empowered to fix systems rather than fear for their jobs.

DO'S FOR THE LEADER	DON'TS FOR THE LEADER
Value learning over assigning blame.	Punish mistakes, which stifles innovation.
Focus on the process, not the person.	Hunt for a scapegoat after a failure.
Celebrate experiments, even if they fail.	Make people feel afraid to try new things.
Create a blameless post-mortem.	Engage in public shaming of employees.
Provide feedback as a growth opportunity.	Use feedback to criticize and blame.
Take responsibility for team failures.	Point fingers at your employees.

SITUATION	SUGGESTED ACTION
A project fails, a fear of blame spreads with employees pointing fingers with no one taking responsibility.	You must intervene immediately and shift the focus from "who" is at fault to "what" went wrong. This action models a growth mindset and creates a safe space for the team to be vulnerable.
An employee is afraid to try anything new after a project failure, preventing them from taking on new challenges.	Frame the failure as a vital data point for future success and celebrate the learning that came from it. This helps them to see failure as a necessary part of growth & be open to new challenges.
A manager makes a public error and tries to quickly move past it signaling to the team that mistakes are shameful.	The manager must model vulnerability by openly and transparently addressing the mistake. By admitting to an error, they normalize it and show the team that it's an acceptable and part of growth.
An employee receives negative feedback and takes it as a personal attack becoming defensive	Teach them to reframe feedback from a judgment of their character to a data point for growth. This helps them to see feedback as a tool for improvement, not as a personal attack.
Collaboration is stifled by a sense of competition and fear of judgment making employees working in silos	Empower the team to establish their own social contracts and collaborative norms. Facilitate a workshop where they collectively decide how to communicate and give feedback.
The team is afraid to give their leader bad news in a culture where bad news is punished and reprimanded.	Create a safe space for the team to share bad news without fear of punishment. This helps you to get a more accurate picture of what is happening and to address problems before they escalate.

Internal silos turn departments into warring factions. You'll spend your days mediating "turf wars" between Sales and Product, or Finance and Operations. This friction slows down every cross-functional initiative, making simple tasks feel like Herculean efforts and diluting the impact of your overarching corporate strategy and vision. When departments view each other as competitors rather than partners, the organization's collective energy is wasted on internal friction rather than external competition. Breaking these barriers requires you to align incentives across the entire organization, ensuring that everyone is rewarded for shared outcomes rather than isolated departmental wins. Your goal is to turn "us vs. them" into a unified "us vs. the market," ensuring that every part of the company is pulling in the same direction.

DO'S FOR THE LEADER	DON'TS FOR THE LEADER
Foster a sense of shared purpose.	Create internal conflicts and silos.
Encourage cross-functional teamwork.	Pit teams against each other.
Promote open communication between teams.	Allow departments to operate alone.
Celebrate company-wide achievements.	Only celebrate one team's success.
Break down departmental barriers.	Reinforce "us vs. them" mentalities.
Create common goals for all to work toward.	Let departments have conflicting goals.

SITUATION	SUGGESTED ACTION
Departments are operating in silos, leading to inefficiency and promising features engineering cannot deliver.	Create cross-functional teams to tackle a shared problem that requires diverse skills. By giving them a single, shared objective, they are forced to collaborate and communicate in a way they wouldn't normally.
Key information is not being shared between teams, leading to redundant work and a lack of innovation	Implement a centralized knowledge management system where all teams can easily share and access information. This ensures that all teams have access to the same information and can work together.
A team is unwilling to share its tools or resources with other teams, showing "hoarding" behavior	Create a culture of transparency and collaboration by celebrating and rewarding teams that share their resources and knowledge with others. This helps to break down silos and to build a stronger sense of unity.
A disagreement is causing friction on a project, where team is not on the same page, slowing down progress.	Facilitate a structured conversation focused on mutual goals and respect, not personal attacks. This helps the team to find a solution that works for everyone and to move forward with the project.
Employees feel a sense of loyalty to their department than the company, creating "us versus them" mentality	Encourage cross-departmental mentorship programs and social events to help employees build relationships and empathy for their colleagues in other departments, helping build a sense of unity and purpose.
Two team members are in a heated, personal conflict, leading to a drop in productivity.	Mediate with a neutral stance, helping them find a respectful compromise and a positive way forward. This helps to de-escalate the situation and to build a stronger sense of unity.

A "rules-first" mindset creates a paralysis that prevents agility. You'll see your best ideas die in committee. For an executive, this means watching opportunities pass you by while your team navigates endless red tape. It replaces meaningful outcomes with mindless compliance, turning your innovators into administrators. In your daily life, this barrier adds layers of unnecessary complexity to every decision, making the organization slow and brittle. To lead effectively, you must constantly prune the "organizational debt" of outdated rules and processes, ensuring that your structure serves your strategy rather than the other way around. Agility requires a culture of "freedom within a framework," where people are trusted to use their judgment to achieve goals rather than blindly following a handbook of procedures.

DO'S FOR THE LEADER	DON'TS FOR THE LEADER
Streamline processes for efficiency.	Create a rigid, rule-bound culture.
Empower teams to make their own choices.	Require approval for every small decision.
Use technology to automate processes.	Rely on manual, paper-based systems.
Simplify your workflows.	Make things more complex than they need.
Promote a culture of action.	Be paralyzed by endless red tape.
Listen to ideas for process improvements.	Ignore feedback on slow processes.

SITUATION	SUGGESTED ACTION
A rigid project plan is unable to adapt to new information, slowing down progress & preventing quick response.	Empower the team to adapt the plan in real-time, giving them flexibility to respond to new data. Trust their judgment to make changes as needed without being slowed down by bureaucracy.
A competitor launches a new feature that threatens your market position causing significant disadvantage.	Empower your team to quickly pivot and respond to the change without being slowed down by bureaucracy. This allows them to make decisive moves and stay competitive in the fast-paced market.
The team is hesitant to change inefficient processes, causing delays and frustration.	Encourage a mindset of "test and learn" by running small experiments with new processes before adopting them. This allows the team to see the benefits of change firsthand.
A product is receiving negative customer feedback that is being ignored by leadership.	Trust the team to make real-time adjustments to the product based on customer feedback and data. This shows that the company values customer input and is agile enough to respond quickly.
An innovative idea is dismissed out of hand due to a lack of budget without capitalizing on innovative ideas.	Form a small, agile team to quickly prototype and test the new idea with a low-risk pilot. This allows the team to learn about the new opportunity before committing major resources.
A leader is micromanaging their team, but they don't realize it but their team feels stifled and disempowered.	Provide the leader with a clear and honest assessment of their behavior and its impact on the team. This helps them to see the problem and to adjust their behavior accordingly.

Depending on a single "superstar" (even yourself) creates a single point of failure. If that person leaves or burns out, the system collapses. Your daily life becomes a series of dependencies where you cannot delegate because the "hero" holds all the knowledge, preventing the organization from scaling independently. This culture discourages teamwork and prevents the development of a robust middle-management layer. To move beyond this, you must transition from a "hero" model to a "system" model, where success is built into the processes and the collective capability of the team rather than the individual heroics of a few. True leadership isn't about being the hero; it's about creating an environment where everyone has the tools and the support to be a hero when their moment arrives.

DO'S FOR THE LEADER	DON'TS FOR THE LEADER
Empower all team members.	Over-rely on a single leader for success.
Share authority and responsibility.	Make one person the sole hero.
Recognize collective effort and success.	Only give credit to the "star" performer.
Develop future leaders.	Have only one strong person on the team.
Build a resilient, self-sufficient team.	Create a team that cannot function without one person.
Delegate important tasks.	Do all the important tasks yourself.

SITUATION	SUGGESTED ACTION
A leader is not in touch with the day-to-day reality of the team, and this is a major barrier to addressing the problem.	Implement a regular "skip-level" meeting where you meet with direct reports of your direct reports to helps you to stay in touch with the team and to address problems before they escalate.
A leader is overconfident and is not open to feedback from their team and the leader is not willing to listen to their team	Coach the leader on the value of humility and the importance of listening to their team. Help them to see that their team is their greatest asset and that a humble leader is a more effective one.
A team is stuck on a difficult problem and is expecting you to provide a solution; giving an opportunity to learn	Instead of providing a solution, ask probing questions to guide them toward finding their own solution. This encourages critical thinking and teach them to handle future issues autonomously.
A key employee on a high-stakes project seems nervous, doubting their ability to succeed & constantly seeking reassurance	Have a one-on-one conversation to express your unwavering confidence in their capabilities. Be specific and genuine by referencing past successes and the unique skills they possess.
An employee has a great idea but lacks the authority or resources to act on it hence feeling disengaged and insulted.	Give them the support and budget to lead a small pilot project to test the viability of their idea. This empowers the employee to take ownership of their idea and value creativity and initiative.
A manager is afraid to delegate & acting as a bottleneck, with everything having to go through them to control everything	Coach the manager on how to delegate decisions, empower and help the team to see that their role is to enable the team, not to control every detail, freeing them to focus on strategic tasks.

Rewarding exhaustion as a badge of honor eventually leads to a catastrophic loss of institutional memory as people quit. You'll see a decline in the quality of decision-making across the board. Leading a burned-out team is like driving a car with an empty tank—you won't reach your destination. Daily, this manifests as low morale and a lack of creative energy. As an executive, you must model sustainable habits, recognizing that peak performance requires recovery. A culture that ignores the limits of human endurance will eventually face a talent crisis that no amount of hiring can solve. By prioritizing the mental and physical well-being of your team, you ensure that the organization has the sustained energy and clarity needed to win the marathon of business, not just a sprint.

DO'S FOR THE LEADER	DON'TS FOR THE LEADER
Promote a sustainable work pace.	Create an environment that leads to burnout.
Encourage taking vacation and breaks.	Discourage taking time off.
Ensure workloads are reasonable.	Overload employees with too many tasks.
Protect your team's personal time.	Expect constant availability from staff.
Celebrate rest and recovery.	Reward long hours and working on weekends.
Model healthy work habits.	Work 80-hour weeks and expect others to.

SITUATION	SUGGESTED ACTION
A team is consistently working long hours, leading to exhaustion and low morale on the verge of burnout.	Review project scope and deadlines and actively protect the team's time by enforcing healthy boundaries. This ensures that the team has a sustainable workload and the time they need to rest and recharge.
Employees are showing signs of stress, not taking their vacation time & feeling guilty for taking time off	Encourage and role-model taking full time off and disconnecting completely to recharge. By taking your own vacation and talking about its benefits, you send a clear message that resting is a priority.
A project team is overwhelmed by the expectations, and the pressure is dropping their quality of work.	Prioritize a healthy, sustainable work pace over an aggressive, burnout-inducing timeline by adjusting deadline to be more realistic. This shows that you value the team's well-being over timelines
A high-performing employee is consistently overworking and showing signs of burnout	Coach them on how to delegate tasks and create a better work-life balance, seeing that their long-term health is more important than short-term gains and empower them to set healthy boundaries.
An employee is sabotaging their own success by procrastinating or making small mistakes	Coach them on how to manage their workload and to set healthy boundaries. Help them to see that success is a positive thing and that they can handle the increased responsibility.
You see late-night emails from your team, always signaling a culture of 24x7 overworking and show-off.	Set a clear expectation of "no-work" after hours (except crisis) and communicate that resting is a priority. This sends a clear message that a healthy work-life balance is a core part of the company's culture.

6x6™ Transformation Blueprint

Leadership Empowering Behaviors

"If your actions inspire others to dream more, learn more, do more and become more, you are a leader." - John Quincy Adams

John Quincy Adams (1767–1848) was the sixth U.S. President. He is known for his remarkable career as a diplomat, senator, and member of the House of Representatives, and for his dedication to public service.

Leadership empowering behaviors are active, observable actions that directly counter negative leadership beliefs, biases, and barriers. These practices build trust, foster collaboration, and inspire every employee to contribute their best. Trust Behavior is a foundational practice where a leader shows confidence in their team by delegating responsibility and avoiding micromanagement, acting as a direct antidote to a toxic culture. Agile Behavior demonstrates a commitment to flexibility, allowing the organization to respond quickly to a dynamic market. Empowerment Behavior is the most transformative, deliberately shifting authority to teams. As John Quincy Adams suggests, a true leader's actions inspire others to "dream more, learn more, do more and become more." These behaviors are the tangible practices that build a healthier reality for the organization.

This foundational framework explores Leadership Behaviors, detailing the specific actions required to establish a clear strategy and achieve Business Maturity. Central to this is Visionary Behavior, where leaders articulate a compelling and inspiring vision for the future, unifying efforts across the enterprise and providing directional clarity. It mandates Risk-Taking Behavior, embracing calculated risk and experimentation to push boundaries and find new growth vectors, rather than being paralyzed by fear of failure. Leaders must exhibit Transparent Behavior, communicating openly and honestly about both challenges and successes to build fundamental trust and organizational stability. Furthermore, the framework emphasizes Empowering Behavior, delegating authority and responsibility to the front lines to foster ownership and encourage high-performing, decentralized teams. This is balanced by Accountable Behavior, requiring leaders to take ownership of outcomes, both good and bad, setting a non-negotiable standard for the entire organization. Finally, Resilient Behavior is crucial, demonstrating the capacity to bounce back quickly from market setbacks and adapt swiftly to strategic changes.

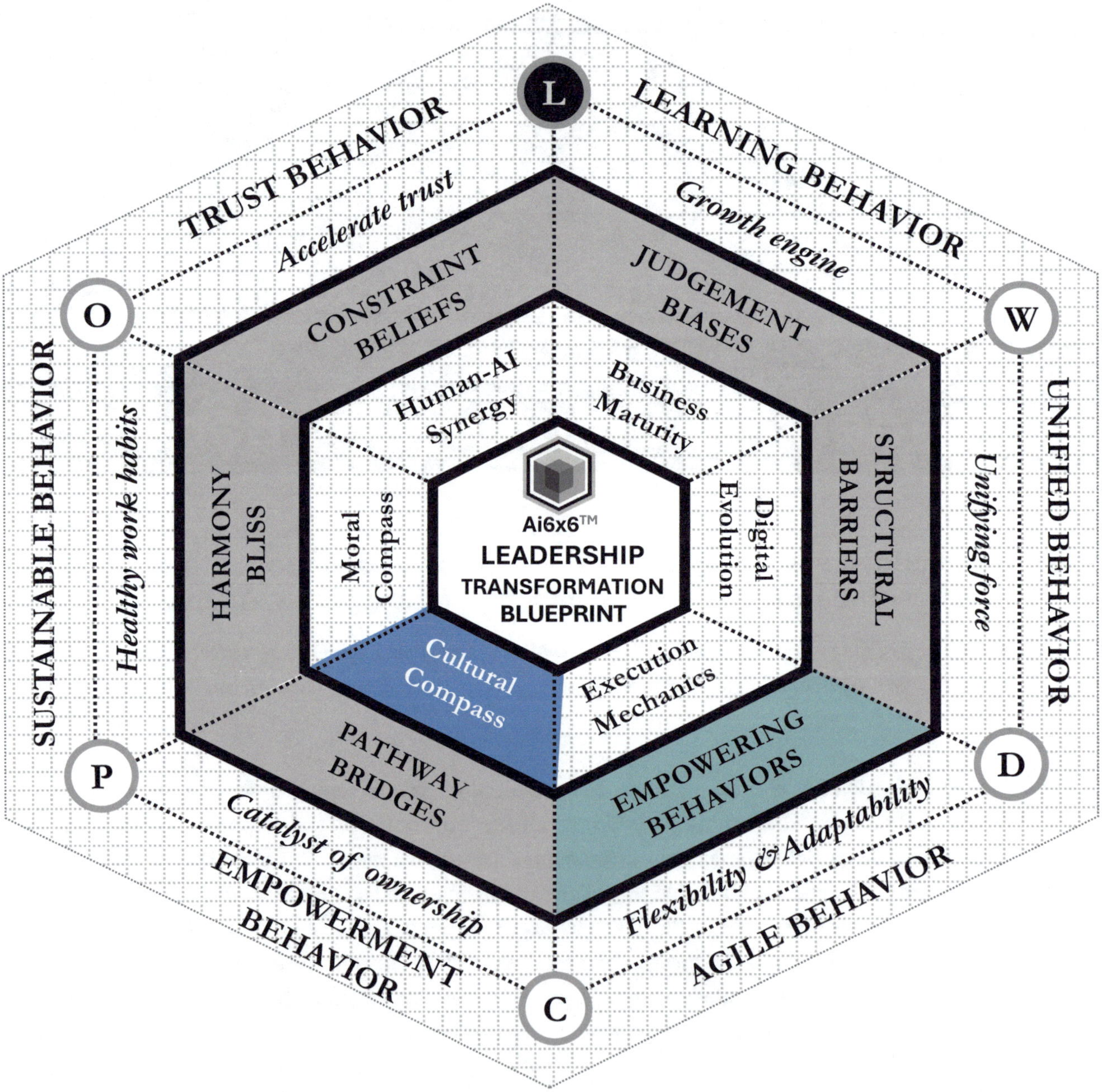

Ai6x6 CULTURAL COMPASS FOR BEHAVIORS | LEADERSHIP TRANSFORMATION

1. TRUST BEHAVIOR | The Trust Accelerator

By fostering psychological safety, you unlock the full potential of your team. In your daily life, this means you don't have to "check in" constantly. High-trust environments move faster and cost less to operate, allowing you to focus on the future while your team handles the present with excellence. Trust is the lubricant of a high-performance organization; without it, every interaction requires verification, which slows down the entire system. When you lead with trust, you empower your employees to take ownership of their work, leading to higher engagement and more innovative solutions to the complex problems your business faces. This behavior creates a virtuous cycle where trust builds competence, and competence further strengthens the bond of trust between you and your leadership team.

DO'S FOR THE LEADER	DON'TS FOR THE LEADER
Create an uplifting, safe work environment.	Foster a culture of fear and blame.
Be consistent and reliable.	Be unpredictable and unreliable.
Be transparent with all information.	Withhold information from your team.
Delegate tasks and show confidence.	Micromanage every single decision.
Honor your commitments and promises.	Make promises you don't keep.
Admit mistakes and take responsibility.	Never admit to being wrong.

SITUATION	SUGGESTED ACTION
A project failure triggers blame, hesitation, and defensiveness across the team	Shift the team from "who" is at fault to "what" happened, reinforcing a learning mindset and restoring psychological safety so people feel comfortable speaking openly again
Employees hesitate to speak up or challenge assumptions, leading to quiet meetings and early signs of groupthink	Invite input from all voices and validate contributions to normalize healthy debate, reduce groupthink, and show that constructive dissent is a valued part of decision-making
Team members defer to you for routine decisions, signaling dependency and reduced ownership	Set direction but let them determine the "how" to build autonomy, accelerate execution, and signal that you trust their judgment and capability
A key employee on a critical project shows self-doubt and repeatedly seeks reassurance	Reinforce their strengths with specific examples to rebuild confidence, helping them shift from anxiety to ownership and perform at the level the project requires
Collaboration weakens as competition and subtle criticism push people into silos	Facilitate team-defined norms and agreements to replace silos with shared expectations, creating a more supportive environment where people feel safe contributing
Lack of peer-to-peer support appears as individuals focus narrowly on their own tasks	Highlight and praise supportive behaviors to signal that helping colleagues is expected, strengthening team cohesion and encouraging a culture of mutual accountability

When you treat failures as data points rather than crimes, you accelerate the pace of improvement. This behavior changes your daily conversations from "Who messed up?" to "What did we find?" It creates a resilient organization that evolves in real-time, keeping you ahead of market shifts and internal stagnancy. By prioritizing learning, you foster a culture of curiosity and experimentation. This means your team is constantly upgrading their skills and processes, ensuring the organization remains competitive. In your daily life as an executive, this leads to a more adaptable and intelligent workforce that solves problems before they reach your desk. You become a leader of a "learning machine" that treats every market challenge as an opportunity to gain a deeper, more actionable understanding of your industry.

DO'S FOR THE LEADER	DON'TS FOR THE LEADER
Value growth and learning over blame.	Punish mistakes and criticize failures.
Encourage curiosity and asking questions.	Discourage people from asking questions.
Provide training and development.	Have a strict, unforgiving work process.
Celebrate learning from setbacks.	Only celebrate success and perfection.
Invest in education and new skills.	Cut training budgets and resources.
Model a growth mindset yourself.	Have a fixed mindset about people.

SITUATION	SUGGESTED ACTION
A product launch fails and blame spreads, preventing open discussion and blocking meaningful learning	Lead a no-blame debrief that focuses on shared learning, root causes, and constructive forward movement rather than punishment or defensiveness
A manager hides a public mistake, signaling errors should be concealed and weakening team trust	Encourage transparent acknowledgment of the mistake to model vulnerability, normalize learning, and strengthen psychological safety across the team
A project falls behind schedule and morale drops as people focus on blame instead of solutions	Reframe conversations toward needed support and resources, shifting attention from fault-finding to collaborative problem-solving and renewed momentum
An employee fears proposing a bold idea, worrying it will be dismissed as too radical	Support a small pilot to safely test the idea, encouraging creativity while minimizing risk and building confidence in experimentation
Team members avoid asking for help, believing it signals weakness and reducing collective efficiency	Model curiosity by seeking input yourself, reinforcing that asking for help is strength and building a culture of shared learning
The organization struggles to adapt to market shifts, delaying decisions and missing emerging opportunities	Invest dedicated time and budget in continuous learning to strengthen adaptability, encourage growth, and embed improvement into organizational culture

3. UNIFIED BEHAVIOR | The Unifying Force

Rallying everyone around a "North Star" eliminates the need for constant micro-alignment. When the purpose is clear, your daily role shifts from a driver to a navigator. Decisions are made faster at every level because everyone knows exactly what the organization is trying to achieve and why it matters. This unified behavior reduces internal friction and ensures that all efforts are directed toward the same strategic goals. For you, it means fewer meetings about "what we are doing" and more high-level discussions about "where we are going next," maximizing the impact of your leadership on the company's long-term trajectory. A unified team acts as a single organism, responding to challenges with a collective intelligence that is far greater than the sum of its individual parts.

DO'S FOR THE LEADER	DON'TS FOR THE LEADER
Foster a shared sense of purpose.	Create internal conflicts and silos.
Create common goals for all teams.	Let teams have conflicting objectives.
Encourage cross-functional collaboration.	Pit departments against one another.
Share information across the company.	Keep information in isolated departments.
Promote a common mission and vision.	Focus on personal or team glory.
Celebrate achievements together.	Only celebrate one team's success.

SITUATION	SUGGESTED ACTION
Departments operate in silos, creating conflict and inefficiency as teams act independently without shared alignment	Form cross-functional groups with a single objective to strengthen collaboration, build empathy, and create shared purpose across departments through coordinated problem-solving
Employees feel disconnected from the mission and cannot see how their work contributes to the broader vision	Reinforce the "why" behind the work, linking individual tasks to the mission so employees regain meaning, clarity, and renewed engagement
Teams pursue conflicting goals, causing internal friction and preventing unified progress toward organizational priorities	Align all departmental goals beneath one organizational objective, eliminating competition and ensuring every team contributes to a unified direction
Employees distrust stated values, seeing them as empty buzzwords disconnected from daily behavior and real culture	Involve employees in defining authentic values through workshops, creating ownership and transforming values into a meaningful cultural guide
A new strategy creates confusion because it was communicated through dry data without a compelling narrative	Share the strategy through a clear story explaining the need for change, the future vision, and each employee's essential role
Teamwork is weak as employees focus only on individual tasks, causing mistakes and fragmented project execution	Use shared-goal team activities that require collaboration, helping employees appreciate interdependence and strengthening unity in daily work

Embracing flexibility allows you to pivot before the market forces you to. Daily, this looks like shorter feedback loops and iterative planning. It reduces the "heavy lifting" of large-scale changes because your team is already accustomed to constant, incremental adjustments, making your organization much more responsive to external shocks. In an executive context, agility means being comfortable with ambiguity and willing to change course when the data suggests a better path. This behavior ensures that your organization remains relevant in a fast-paced economy, allowing you to seize new opportunities as they arise rather than being tied to outdated plans. Agile leaders recognize that a perfect plan is less valuable than a flexible system that can learn and adapt as new information becomes available in the market.

DO'S FOR THE LEADER	DON'TS FOR THE LEADER
Embrace flexibility and adaptability.	Stick to rigid, inflexible plans.
Respond quickly to changes in the market.	Ignore market changes and competitors.
Empower teams to make decisions.	Require approval for every small change.
Use iterative development cycles.	Use a long, waterfall development.
Encourage learning and experimentation.	Discourage trying new things.
Adapt processes to new information.	Refuse to change old processes.

SITUATION	SUGGESTED ACTION
A rigid project plan prevents adaptation as new information emerges, slowing progress and limiting effective response	Enable real-time plan adjustments by trusting team judgment, reducing bureaucracy, and keeping the project responsive to changing conditions
A competitor launches a feature, and slow internal approvals delay your ability to respond effectively	Empower rapid pivots by removing approval bottlenecks, enabling decisive action that maintains competitiveness in a fast-moving market
A core strategy fails but the company continues investing due to fear of admitting failure	Initiate a rapid strategic review, embrace course correction, and encourage learning to replace stagnation with forward momentum
New technology presents opportunity, but hesitation prevents exploration and slows innovation across the organization	Form a small team to prototype quickly, using a low-risk pilot to learn before committing major resources
The team resists changing inefficient processes, relying on outdated habits that slow progress significantly	Promote small experiments to demonstrate benefits, helping the team overcome resistance and adopt improved ways of working
A product receives negative feedback, but no changes occur due to perceived effort required	Trust the team to adjust quickly based on customer input, showing responsiveness and reinforcing commitment to user needs

Pushing decision-making authority down to the front lines frees up your calendar. You stop being the bottleneck and start being a mentor. This behavior develops the next generation of leaders while ensuring that decisions are made by those closest to the information, resulting in more accurate and timely outcomes. When you empower your team, you increase their sense of autonomy and accountability. This not only improves operational efficiency but also builds a more motivated and capable workforce. For your daily life, it means you can focus on high-stakes strategic decisions, knowing the day-to-day operations are in competent, empowered hands. Empowerment is the ultimate form of leverage, allowing your influence to extend far beyond the limits of your own personal time and direct oversight.

DO'S FOR THE LEADER	DON'TS FOR THE LEADER
Shift authority to teams.	Make all decisions yourself.
Trust employees to do their job.	Micromanage and check in constantly.
Provide resources and support.	Give responsibility without support.
Encourage team autonomy.	Force people to follow strict rules.
Delegate decisions and projects.	Do everything yourself to ensure quality.
Recognize initiative and ownership.	Take credit for your team's ideas.

SITUATION	SUGGESTED ACTION
A team waits for your approval on every decision, creating dependency, slowing progress, and limiting initiative	Define clear boundaries and grant full authority within them, enabling faster decisions, greater confidence, and stronger ownership across the team
A manager becomes a bottleneck by avoiding delegation, preventing their team from acting autonomously and growing effectively	Coach the manager to delegate decisions, empowering their team and freeing themselves to focus on higher-value strategic responsibilities
An employee has a strong idea but lacks authority or resources, causing disengagement and lost innovation potential	Provide support and budget for a small pilot, empowering ownership and demonstrating that creativity and initiative are valued
A team faces a difficult problem and looks to you for answers instead of developing their own solutions	Ask probing questions to guide their thinking, strengthening critical-thinking skills and building confidence in solving future challenges independently
Team members avoid taking initiative or proposing ideas, showing low motivation and limiting innovation across the organization	Publicly recognize employees who show initiative, reinforcing desired behavior and encouraging others to contribute more proactively
High turnover occurs because employees feel excluded from decisions, reducing ownership and weakening commitment to the team	Involve the team in meaningful decisions, increasing investment, strengthening morale, and building loyalty through shared ownership

Prioritizing health and balance ensures your team—and you—can play the long game. Daily, this means setting boundaries that protect mental clarity. Sustainable leadership prevents the "boom and bust" cycles of productivity, leading to a steady, high-performing culture that doesn't collapse under the weight of its own ambition. By modeling sustainable behavior, you show your team that long-term success is built on a foundation of well-being. This reduces turnover, minimizes errors caused by fatigue, and ensures that the organization has the sustained energy needed to pursue its most ambitious goals over many years, not just a few quarters. In an executive role, being "always on" is a liability; true effectiveness requires the clarity of mind that only comes from intentional recovery and a balanced approach to work.

DO'S FOR THE LEADER	DON'TS FOR THE LEADER
Prioritize healthy work habits.	Promote a culture of burnout.
Encourage regular breaks and rest.	Expect constant work and availability.
Ensure workloads are reasonable.	Overload people with too many tasks.
Respect personal time and weekends.	Make employees work late and on weekends.
Recognize and prevent burnout.	Ignore signs of stress and exhaustion.
Model a healthy work-life balance.	Work excessive hours yourself.

SITUATION	SUGGESTED ACTION
A team rushes through work without pausing to reflect, losing insight and repeating avoidable mistakes	Build structured reflection moments that capture learning, strengthen judgment, and create a rhythm supporting continuous improvement and sustainable performance
Employees treat every task as urgent, creating constant pressure and eroding thoughtful, strategic decision-making	Establish clear urgency criteria, helping the team differentiate true priorities and operate with steadier focus and healthier pacing
The organization celebrates speed over quality, encouraging shortcuts that undermine long-term credibility and trust	Redefine success metrics to include craftsmanship, reliability, and impact, reinforcing sustainable excellence instead of reactive speed
Leaders make rapid commitments without considering capacity, overwhelming teams and creating chronic strain	Introduce capacity checks before accepting new work, ensuring commitments align with available energy, resources, and strategic priorities
Teams avoid raising early risks, fearing they will be seen as blockers or negative thinkers	Normalize early risk-surfacing as responsible leadership, creating psychological space for proactive problem-solving and healthier project trajectories
Employees hesitate to question outdated practices, allowing inefficiencies to persist and drain collective energy	Encourage constructive challenge by rewarding thoughtful improvements, signaling that evolution is expected and essential for sustainable progress

6x6™ Transformation Blueprint

Leadership Pathway Bridges

"The function of leadership is to produce more leaders, not more followers." - Ralph Nader

Ralph Nader (b. 1934) is an American political activist and attorney. He is a leading figure in consumer protection and is known for his book Unsafe at Any Speed, which was highly influential in creating auto safety laws.

Leadership pathway bridges are specific processes and frameworks that provide a clear pathway for a leader's personal and professional transformation. They are the practical tools that turn good intentions into real, measurable outcomes. The Objective Inquiry Bridge is a formal mechanism for actively seeking evidence that challenges a leader's beliefs, a direct counter to confirmation bias. The Rational Exit Bridge provides an objective evaluation framework to help leaders decide when to abandon a failing project, overcoming the sunk cost bias. A Growth Mindset Bridge fosters a belief in learning and developing skills throughout the organization, which counters the stagnation caused by fixed aptitude beliefs. As Ralph Nader's quote suggests, the function of leadership is to "produce more leaders, not more followers." These bridges make this a reality by creating a structured approach to self-correction and continuous improvement.

This foundational framework explores Leadership Pathway Bridges, detailing the crucial connections leaders must build to establish strategic Directional clarity and secure Business Maturity. The Communication Bridges are paramount, establishing formal, reliable channels for sharing the vision and ensuring the strategic message is received, understood, and internalized across all organizational tiers. The Stakeholder Bridges focus on proactive relationship management with key internal and external decision-makers, ensuring continuous buy-in, resource allocation, and political risk mitigation. Trust Bridges are the essential moral foundation, building confidence through consistency, integrity, and shared vision, without which change is impossible. Furthermore, the framework emphasizes Mentorship Bridges, which establish formal pathways for guiding and developing talent specifically tailored for the transformed state, cultivating future leaders. Vision Bridges are crucial mechanisms for connecting the abstract, long-term future state to tangible, near-term milestones that teams can easily contribute to. Finally, Feedback Bridges mandate the creation of formal, psychologically safe mechanisms that allow unfiltered data and crucial operational insights to flow back up to leadership, enabling responsive course correction.

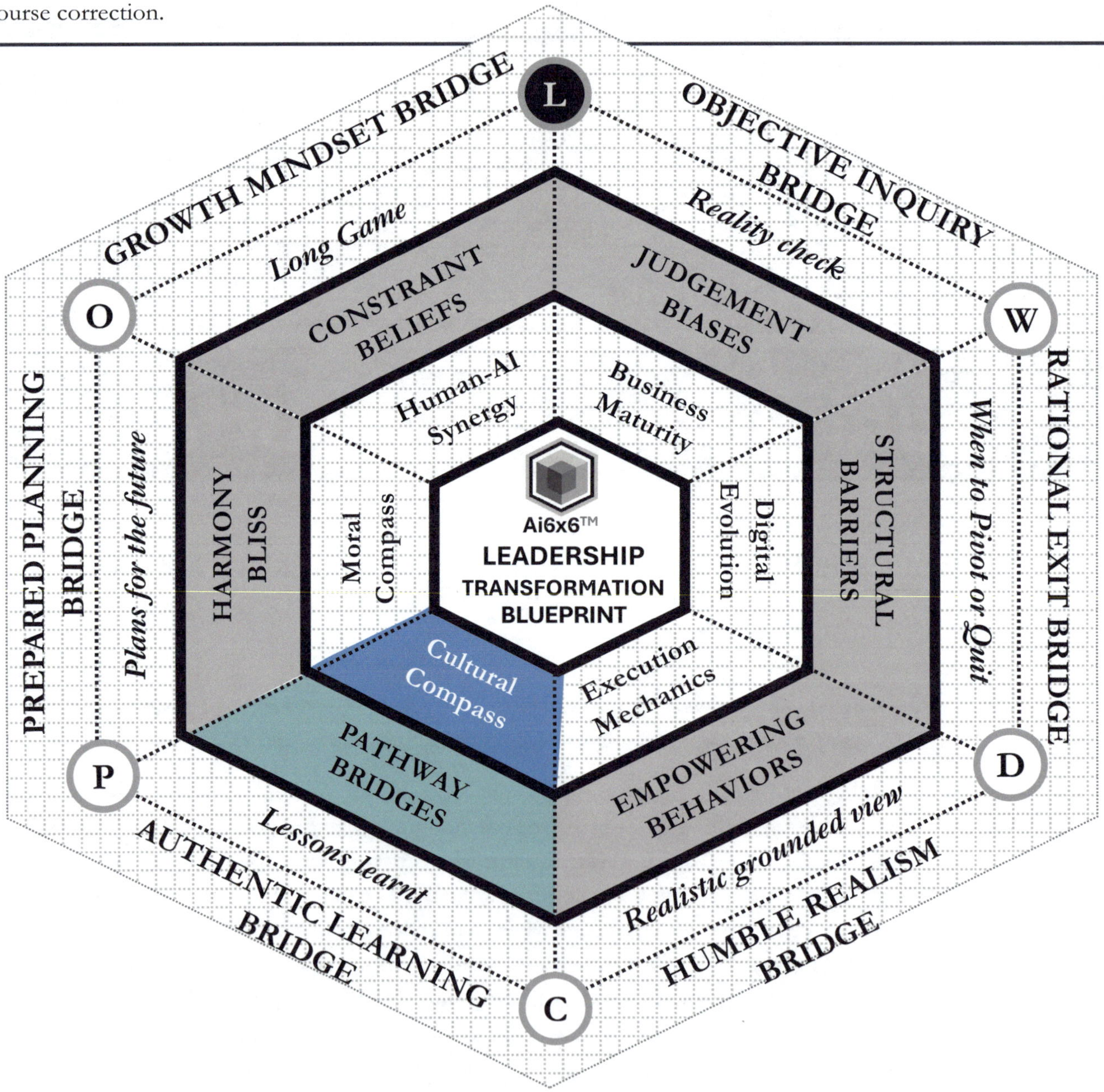

Ai6x6 CULTURAL COMPASS FOR BRIDGES | LEADERSHIP TRANSFORMATION

Believing that skills can be developed changes how you view your talent pipeline. You stop looking for "perfect" hires and start building a "learning machine." This daily focus on development increases employee engagement and ensures your leadership team is always evolving to meet the next set of global challenges. This bridge connects where your team is now to where they need to be in the future. It fosters a culture of resilience and ambition, where challenges are viewed as opportunities to grow rather than threats to avoid. For you, it means leading a team that is constantly expanding its own capabilities, ensuring that the organization's most asset—its people—never stops appreciating in value as the business landscape becomes increasingly complex.

DO'S FOR THE LEADER	DON'TS FOR THE LEADER
Foster a belief in learning and skills.	Believe that skills are innate and fixed.
Encourage taking on new challenges.	Have a fear of failure and new tasks.
Provide training and development.	Only hire people with a specific skill.
Celebrate effort over talent alone.	Only praise people for their innate ability.
Create a safe space for mistakes.	Punish people for making mistakes.
Model a growth mindset yourself.	Have a fixed mindset about people.

SITUATION	SUGGESTED ACTION
An employee avoids proposing ideas because they believe only perfect suggestions deserve attention	Create a forum where early, unfinished thinking is expected, helping them practice contribution without pressure or fear of judgment
A team treats expertise as fixed, assuming senior members should always know more than junior colleagues	Rotate teaching roles so everyone occasionally leads a short lesson, reinforcing that knowledge is shared, evolving, and collective
A manager resists new tools because learning something unfamiliar feels uncomfortable and threatening	Pair them with a tech-curious peer for a brief learning sprint, showing that discomfort is a natural part of expanding capability
A team avoids ambitious goals because they fear falling short and appearing unprepared or incapable	Introduce aspirational targets with no penalty for misses, emphasizing exploration, creativity, and learning over flawless execution
An employee repeats the same approach despite stagnating results, believing consistency alone signals competence	Run a controlled experiment comparing their current method with a new one, demonstrating that improvement comes from iteration, not repetition
A team interprets slow progress as failure, losing motivation before meaningful results can emerge	Highlight early indicators of improvement and teach them to track leading signals, reinforcing that growth often appears gradually before accelerating

By actively seeking "disconfirming evidence," you protect yourself from expensive mistakes. In your daily routine, this means asking, "What would have to be true for this to fail?" It turns you into a more rigorous thinker, ensuring your strategies are battle-tested before you commit significant resources or political capital. This bridge helps you cross from the island of bias to the mainland of reality. It forces you to look at the data that makes you uncomfortable, which is often exactly where the most important insights are hidden. Ultimately, this practice makes your leadership more resilient and your decisions significantly more robust, as they are based on a comprehensive view of the evidence rather than a curated selection that supports a pre-conceived notion.

DO'S FOR THE LEADER	DON'TS FOR THE LEADER
Actively seek evidence that challenges.	Only seek info that confirms your views.
Create a process for critical review.	Reject feedback that opposes your opinion.
Encourage debate and intellectual honesty.	Shut down all dissent and disagreement.
Ask "what if we are wrong?" often.	Assume your own ideas are perfect.
Use data to disprove your own theories.	Cherry-pick data that supports your view.
Empower teams to question assumptions.	Punish people for challenging your views.

SITUATION	SUGGESTED ACTION
You strongly prefer one vendor because you've worked with them before, even though competing proposals offer stronger value	Run a blind comparison of all proposals without vendor names. This forces you to evaluate purely on merit and reduces familiarity bias
A team insists their solution is the "industry standard," but they cannot articulate alternatives they ruled out	Require them to present at least two contrasting options with pros, cons, and assumptions. This widens the decision frame and exposes hidden trade-offs
You believe a product feature is essential because customers "always ask for it," but you haven't reviewed recent data	Examine fresh usage patterns and customer feedback to validate whether the demand still exists. This prevents outdated assumptions from driving decisions
A senior leader's opinion is dominating the discussion, and others are agreeing without offering independent analysis	Ask each person to write their recommendation privately before group discussion. This surfaces original thinking and reduces conformity pressure
You feel unusually confident about a forecast because it aligns with your long-held view of the market	Identify which data points would prove your forecast wrong. This builds intellectual discipline and protects against confirmation bias
A team claims a project is low-risk because "similar projects succeeded before," ignoring contextual differences	Ask them to list what is different this time — environment, constraints, stakeholders, timing. This reveals hidden variables and sharpens risk assessment

3. RATIONAL EXIT BRIDGE | The Project Pivot

This bridge gives you a "pre-planned" way out of failing ventures. It removes the emotion from quitting. Daily, this allows you to reallocate resources from dead-end projects to high-growth opportunities. It is the hallmark of an executive who values future returns over the preservation of past pride or ego. By establishing clear "exit criteria" at the start of a project, you avoid the trap of throwing good money after bad. This bridge ensures that your organization remains lean and focused, always moving toward the most promising horizons rather than being weighed down by the ghosts of projects that didn't work. Knowing when to stop is just as important as knowing when to start, and this bridge provides the intellectual framework to do both with total clarity.

DO'S FOR THE LEADER	DON'TS FOR THE LEADER
Objectively evaluate projects.	Continue a project due to past investment.
Set clear criteria for success.	Use vague metrics that are easy to meet.
Be willing to abandon failing projects.	Refuse to quit a project you invested in.
Focus on future opportunities.	Dwell on past costs and effort.
Conduct blameless project reviews.	Engage in a witch hunt after a project fails.
Reallocate resources to new ideas.	Keep pouring money into a losing cause.

SITUATION	SUGGESTED ACTION
A project continues out of habit even though recent data shows diminishing returns and declining strategic relevance	Pause execution and run a rapid value reassessment, comparing current investment against updated strategic priorities to determine whether continuation is justified
A team is emotionally attached to a legacy initiative and resists acknowledging that market conditions have fundamentally shifted	Facilitate an evidence-based review that separates sentiment from strategy, helping the team see the objective case for redirecting effort
Early indicators show the project is drifting off-track, but the team keeps pushing forward hoping momentum will fix underlying issues	Conduct a structured checkpoint focused on root causes, not effort, and decide whether to pivot scope, reset assumptions, or exit entirely
A major dependency outside your control becomes unstable, threatening timelines and outcomes, yet the team continues planning as if nothing changed	Reevaluate feasibility with updated constraints and explore alternative paths, ensuring decisions reflect present realities rather than outdated assumptions
Leadership pressure encourages continuation because of sunk costs, even though the project no longer aligns with the company's direction	Reframe the conversation around opportunity cost, highlighting what resources could achieve if redirected toward higher-value initiatives
A promising project hits repeated roadblocks, and the team interprets pivoting as failure rather than strategic adaptation	Normalize pivoting as intelligent course correction, not retreat, and redesign the project around validated learning rather than original assumptions

4. HUMBLE REALISM BRIDGE | The Grounding Lens

Seeing the world as it is, not as you wish it were, leads to better forecasting. This bridge keeps you grounded. Your daily operations benefit from realistic timelines and budgets. It builds trust with stakeholders because they know your assessments are based on data and humility rather than hype. When you cross this bridge, you move away from performative optimism and toward actual performance. It allows you to address the "brutal facts" of your business environment while maintaining the faith that you will prevail in the end. This balance is critical for any executive navigating a complex market landscape, as it allows you to lead with a steady hand that is neither blinded by hope nor paralyzed by fear but always guided by the truth.

DO'S FOR THE LEADER	DON'TS FOR THE LEADER
Promote a realistic view of a situation.	Maintain an overly optimistic view.
Acknowledge potential risks and problems.	Ignore all negative data and info.
Plan for potential setbacks.	Believe nothing will go wrong.
Admit when you don't have all the answers.	Act as if you know everything.
Listen to your team's concerns.	Dismiss concerns as negative thinking.
Seek input from different perspectives.	Only listen to people who agree with you.

SITUATION	SUGGESTED ACTION
A leader assumes a new initiative is universally supported, but quiet resistance surfaces in informal conversations	Invite private, one-on-one check-ins to uncover unspoken concerns, grounding decisions in the team's actual sentiment rather than perceived alignment
A team reports strong progress, yet key milestones keep slipping without clear explanation	Ask for a transparent breakdown of blockers and assumptions, shifting the conversation from optimistic summaries to concrete operational reality
A leader believes a partnership is thriving, but the partner's engagement has noticeably decreased	Reach out directly to understand their evolving priorities, ensuring decisions reflect the true state of the relationship
A team claims workload is manageable, but subtle signs of strain appear in quality and communication	Conduct a candid workload review to surface hidden pressure points and recalibrate expectations before issues escalate
A leader assumes customer satisfaction is high based on selective positive feedback	Request a full distribution of customer sentiment, including neutral and negative signals, to anchor decisions in the complete picture
A team insists a process is efficient because "it has always worked," despite growing delays and friction	Map the workflow end-to-end to reveal bottlenecks, grounding improvement efforts in observable reality rather than legacy assumptions

Genuinely deconstructing past events ensures mistakes aren't repeated. This isn't about blaming; it's about systemic improvement. In your daily life, it turns every setback into a competitive advantage. You become a leader who builds "compounding knowledge," where the organization gets smarter and more efficient with every project completed. This bridge connects past experiences to future strategies in a meaningful way. It requires an environment where people feel safe being honest about what went wrong. By building this bridge, you ensure that your company is a "learning organization" that never makes the same expensive mistake twice. You transform the "tuition" paid in failures into the "dividends" of future success, ensuring that every experience contributes to the organization's long-term intelligence and market dominance.

DO'S FOR THE LEADER	DON'TS FOR THE LEADER
Genuinely learn from past events.	Rationalize past mistakes away.
Focus on what was learned from failure.	Say, "I knew that would happen."
Objectively review past decisions.	Rewrite history to make you look better.
Document decisions and their reasoning.	Let history fade away and be forgotten.
Learn from both successes and failures.	Only analyze things that went well.
Create a safe space for honest feedback.	Punish people for being honest.

SITUATION	SUGGESTED ACTION
A team member rushes to solutions without understanding the underlying problem, missing key learning opportunities	Slow the pace and require a brief problem-framing exercise, helping them uncover root causes and build deeper analytical capability
A leader avoids admitting they don't know something, preventing the team from exploring the topic openly	Model intellectual honesty by acknowledging uncertainty and inviting shared inquiry, creating a culture where learning is collective rather than performative
A team repeats a familiar process even though results have plateaued, assuming consistency equals mastery	Introduce a structured review of assumptions and encourage experimentation with alternative approaches to reignite learning and improvement
An employee seeks quick answers from others instead of attempting their own reasoning first	Ask them to propose an initial hypothesis before receiving guidance, reinforcing active learning rather than passive information consumption
A team celebrates completion but cannot articulate what they learned or would change next time	Conduct a short retrospective focused on insights, not blame, ensuring each project strengthens future judgment and capability
A leader pushes for speed and unintentionally discourages thoughtful exploration, causing shallow understanding	Create protected time for deeper inquiry and signal that quality thinking is valued as much as rapid execution

Accounting for "Plan B" and "Plan C" reduces your daily anxiety. When the unexpected happens, you don't panic—you execute the contingency. This bridge allows you to lead with a sense of calm that permeates the organization, providing the stability necessary for others to perform at their best. It bridges the gap between your ideal scenario and the unpredictability of the real world. For an executive, this means you are always three steps ahead of a potential crisis. This level of preparation doesn't slow you down; it makes you faster because you have already mentally rehearsed the response to various obstacles. You move from a state of "reaction" to a state of "response," which is the hallmark of a seasoned, strategic leader.

DO'S FOR THE LEADER	DON'TS FOR THE LEADER
Create plans that account for risks.	Operate with an overly optimistic view.
Conduct a thorough risk analysis.	Ignore all potential risks.
Create contingency plans.	Have no plan for when things go wrong.
Think through worst-case scenarios.	Believe that the best case is the only case.
Involve your team in the planning.	Plan in a vacuum without team input.
Be adaptable when plans need to change.	Stick to the original plan at all costs.

SITUATION	SUGGESTED ACTION
A team begins a project with enthusiasm but no shared understanding of constraints, dependencies, or sequencing	Establish a structured kickoff that clarifies assumptions, maps dependencies, and aligns expectations so execution begins with a grounded, unified plan
A leader reacts to issues only after they escalate, creating avoidable fire-drills and rushed decisions	Build a forward-looking review rhythm that identifies emerging risks early, enabling calm adjustments instead of reactive scrambling
A team repeatedly underestimates timelines because they plan for ideal conditions rather than realistic scenarios	Require scenario-based planning that includes buffers, constraints, and contingencies, reinforcing disciplined forecasting over optimistic guessing
A project moves forward without clarity on decision rights, causing confusion, delays, and duplicated effort	Define explicit ownership, escalation paths, and decision boundaries before execution begins, ensuring the team operates with confidence and speed
A leader assumes cross-functional partners are aligned, but later discovers misinterpretations that derail progress	Conduct alignment checkpoints with each partner to validate shared understanding, preventing hidden misalignment from surfacing at critical moments
A team focuses heavily on launch activities but neglects post-launch support, leading to operational strain	Build a full lifecycle plan that includes stabilization, monitoring, and iteration, ensuring long-term success rather than short-term completion

6x6™ Transformation Blueprint

Leadership Harmony Bliss

"A great person attracts great people and knows how to hold them together." - Johann Wolfgang Von Goethe

Johann Wolfgang von Goethe (1749–1832) was a German polymath and writer. He is widely regarded as one of the greatest figures in German literature, best known for his dramatic play Faust and his significant contributions to philosophy and science.

Leadership harmony bliss is the positive state where leadership is a source of profound satisfaction and success. It is the culmination of a successful transformation journey, where a leader has transcended negative beliefs and biases and successfully dismantled organizational barriers. This state is characterized by several key elements. Blissful Empowerment is the deep joy a leader feels from building a culture of trust and empowering their team to succeed. Blissful Clarity represents a state of complete transparency and understanding, where decisions are based on clear data and open communication. As Goethe's quote suggests, a great person attracts great people and knows how to hold them together. This state is not about personal glory but about the collective achievement of a high-performing, ethical, and inspiring organization.

This foundational framework explores Leadership Harmony Bliss, detailing the strategic outcomes that result from effective directional guidance and secure Business Maturity. The highest achievement is Strategic Alignment, resulting in a clear, unified vision across the entire organization, eliminating conflicting goals and ensuring all resources pull in the same direction for faster execution. Success is measured by Sustainable Growth, defined as consistent, repeatable, and profitable expansion that is resilient to market volatility and market shifts, providing long-term stakeholder confidence. The framework targets Market Leadership, securing a dominant and recognized position within the industry based on superior value delivery and differentiation. Leadership Bliss is rooted in Brand Equity, cultivating a strong, trusted, and highly valued organizational reputation among all stakeholders, which acts as a strategic buffer against competition. A critical internal outcome is a Purpose-Driven Culture, where the workforce is united and motivated by a shared, inspiring mission that goes beyond profit. Finally, the ultimate capability is an Innovation Engine, ensuring a constant, reliable stream of new ideas, features, and products that sustain competitive advantage and future-proof the entire organization against disruption.

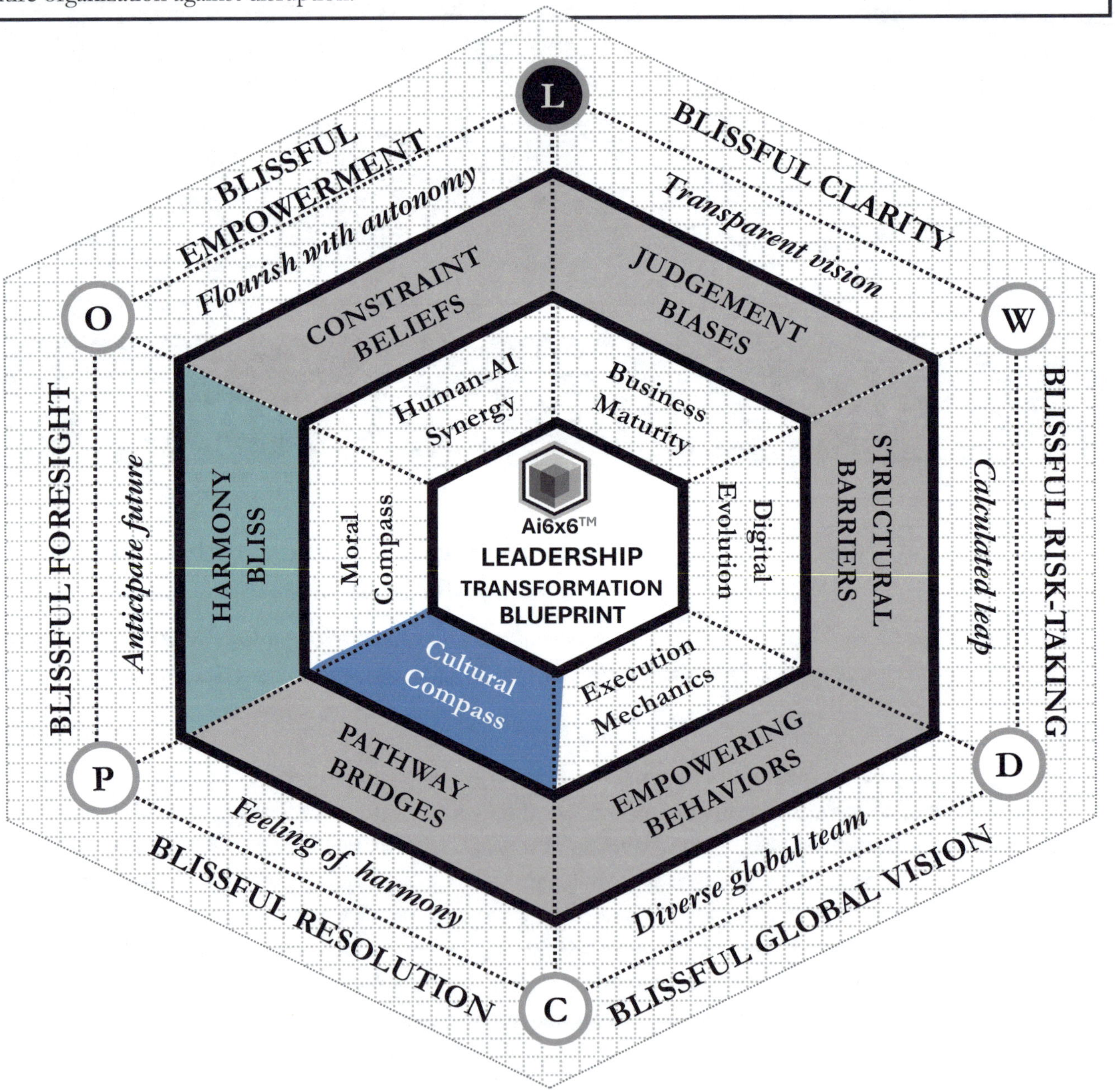

Ai6x6 CULTURAL COMPASS FOR BLISS | LEADERSHIP TRANSFORMATION

There is a profound sense of relief when you realize you don't have to be the smartest person in the room. This bliss comes from watching your team succeed independently. It transforms your daily work from a burden of responsibility into a rewarding experience of coaching, mentorship, and collective triumph. This state of harmony is reached when you have successfully transferred both authority and capability to those you lead. You are no longer the "hero" saving the day; you are the architect of a system that generates its own success, allowing you to lead with more ease and much higher impact. This bliss is the ultimate reward for a leader who has invested in the growth and autonomy of others.

DO'S FOR THE LEADER	DON'TS FOR THE LEADER
Give teams the freedom to decide.	Micromanage and take control back.
Trust employees completely.	Distrust employees and their work.
Celebrate team ownership and autonomy.	Make all decisions on behalf of others.
Provide resources for teams to succeed.	Give responsibility without support.
Recognize and reward team initiative.	Take credit for the team's ideas.
Encourage risk-taking for innovation.	Punish people for taking a risk.

SITUATION	SUGGESTED ACTION
A leader hesitates to delegate meaningful responsibilities, limiting team growth and confidence	Assign a strategic task with clear boundaries and supportive guidance, enabling capability growth while reinforcing trust through meaningful ownership
A team member waits for direction because they're unsure independent initiative is genuinely encouraged	Clarify decision rights and invite their proposed approach, reinforcing autonomy and strengthening confidence through validated independent action
A leader intervenes too quickly during challenges, unintentionally undermining team confidence and development	Provide coaching without taking control, allowing space for problem-solving and reinforcing capability through supportive, confidence-building feedback
A team avoids experimentation because they fear disappointing leadership with imperfect outcomes	Celebrate thoughtful attempts and highlight learning value, normalizing experimentation as essential for growth and strengthening confidence through exploration
A leader centralizes decisions to maintain control, slowing progress and limiting shared ownership	Shift key decisions to the team and review outcomes collaboratively, reinforcing accountability and accelerating execution through distributed leadership
A team member underestimates their potential and avoids stretch responsibilities due to fear	Assign a manageable stretch role with structured support, reinforcing progress and strengthening empowerment through meaningful, confidence-building achievement

When information flows freely, the "fog of war" lifts. You experience the peace of knowing exactly where the organization stands. This clarity reduces the mental load of second-guessing and allows you to make bold, strategic moves with total confidence, knowing your foundation is transparent and your data is solid. This is the bliss of alignment—where every part of the organization is "singing from the same songbook." In your daily life, this means you spend less time clarifying and more time executing. You can see the path ahead with such vividness that the obstacles become manageable and the destination certain, leading to a profound sense of strategic calm and operational excellence.

DO'S FOR THE LEADER	DON'TS FOR THE LEADER
Create a state of transparency.	Withhold information from your team.
Communicate the mission and vision.	Keep your goals vague and unclear.
Ensure everyone understands the plan.	Let people guess what to do next.
Foster profound understanding.	Create confusing or complex systems.
Provide clear, objective feedback.	Give vague and unhelpful feedback.
Link tasks to a larger purpose.	Make work seem meaningless.

SITUATION	SUGGESTED ACTION
A team is confused about priorities because multiple initiatives appear equally urgent	Rank priorities transparently and explain trade-offs clearly, giving the team a grounded sequence that strengthens focus, alignment, and coordinated execution
A leader communicates strategy abstractly, leaving the team unsure how it connects to daily responsibilities	Translate strategy into concrete actions and relatable examples, helping the team understand expectations clearly and connect broader vision to practical, everyday execution
A team avoids asking clarifying questions because they fear appearing uninformed or unprepared	Invite clarifying questions openly and acknowledge complexity, reinforcing that understanding drives excellence and creating a safer environment for transparent, confident dialogue
A leader assumes alignment because no concerns are raised, yet subtle misunderstandings persist	Conduct a brief alignment check to surface confusion early, correct misinterpretations quickly, and ensure the entire team moves forward with shared clarity and confidence
A team receives inconsistent updates from different leaders, creating ambiguity and unnecessary rework	Consolidate communication into a unified message that provides consistent direction, reduces confusion, and strengthens alignment across all stakeholders involved in execution
A leader shares goals without explaining constraints, leaving the team guessing feasibility boundaries	Explain aspirations and limitations together, enabling realistic planning, grounded decision-making, and a clearer understanding of what success requires within current operational constraints

Making a high-stakes decision that is rooted in rigorous logic—rather than desperation—is exhilarating. This bliss comes from the mastery of your craft. It turns your daily challenges into a high-level game of strategy where the stakes are high, but your preparation and intuition are even higher. This isn't the stress of gambling; it's the joy of a calculated move that pays off. It's the feeling of being "in the zone" as an executive, where your experience and your analysis meet to create a decisive competitive advantage. This bliss validates your skills and provides a deep sense of professional fulfillment, as you successfully navigate the complex uncertainties of the market with a steady, confident hand.

DO'S FOR THE LEADER	DON'TS FOR THE LEADER
Celebrate making well-calculated decisions.	Criticize any decision that has risk.
Encourage bold decisions for growth.	Have a fear of failure and risk.
Create a safe space for failed risks.	Punish people for mistakes.
Weigh potential gains against losses.	Focus only on the possibility of a loss.
Provide resources to manage risks.	Offer no support for risky ventures.
Model a healthy attitude towards risk.	Be a completely risk-averse leader.

SITUATION	SUGGESTED ACTION
A team avoids bold ideas because previous attempts were dismissed quickly and without discussion	Create a protected forum for ambitious proposals, evaluate potential thoughtfully, and reinforce that courageous thinking is valued and essential for meaningful progress
A leader hesitates to pursue innovation due to fear of stakeholder criticism or perceived instability	Launch a small, low-risk pilot to validate assumptions, gather evidence, and build confidence before expanding into a larger, more visible initiative
A team relies on familiar methods even though results have plateaued and momentum has slowed	Encourage controlled experiments comparing new approaches with existing ones, reinforcing that improvement emerges through exploration rather than habitual repetition
A leader delays decisions while waiting for perfect information, slowing momentum and reducing opportunity	Define a threshold for sufficient data, commit once reached, and reinforce decisive action as a disciplined leadership behavior
A team fears accountability if a bold initiative underperforms or produces mixed results	Frame risk-taking as shared learning, emphasize collective responsibility, and highlight insights gained rather than focusing solely on outcomes
A leader avoids challenging industry norms despite clear opportunities for differentiation and competitive advantage	Conduct a structured competitive scan, identify gaps, and pursue a bold move grounded in evidence and strategic conviction

Seeing a diverse, international team move in unison toward a grand goal is the pinnacle of leadership. It's the "symphony" effect. This bliss provides a deep sense of purpose, reminding you that your work transcends cultural barriers and contributes to a larger, globally connected story of success. There is a unique harmony in witnessing different perspectives and backgrounds coalesce into a single, powerful direction. It validates your role as a global citizen and a leader, showing that the right vision can unite anyone, anywhere, in the pursuit of something truly meaningful. This bliss is the realization that your leadership has the power to bridge divides and create a shared future for people all over the world.

DO'S FOR THE LEADER	DON'TS FOR THE LEADER
Lead diverse teams toward a shared goal.	Let teams have conflicting objectives.
Create a shared, inspiring mission.	Have a vague, uninspiring vision.
Connect teams across the globe.	Create silos between international teams.
Promote a culture of inclusion.	Be insensitive to other cultures.
Listen to all perspectives and ideas.	Only listen to people similar to you.
Travel and meet with international teams.	Remain disconnected from the global team.

SITUATION	SUGGESTED ACTION
A global team struggles to connect because cultural nuances are misunderstood or overlooked during collaboration	Create structured cultural exchange sessions that build mutual understanding, strengthen trust, and improve cross-regional collaboration through shared learning and deeper appreciation of diverse perspectives
A leader sets a global goal but fails to show how each region contributes meaningfully	Map regional roles clearly, explain interdependencies, and reinforce how each team advances the shared mission with purpose, clarity, and aligned expectations
A team prioritizes local wins over global alignment, creating fragmented progress and inconsistent execution	Reinforce shared metrics, highlight cross-regional achievements, and strengthen unity by celebrating progress that advances the collective mission and long-term organizational direction
A leader communicates in one cultural frame, limiting resonance across diverse global audiences and contexts	Adapt messaging to reflect regional perspectives, ensuring inclusivity, clarity, and shared meaning across all cultural environments and communication styles
A global initiative stalls due to misaligned time zones and inconsistent communication rhythms across regions	Establish predictable collaboration windows and asynchronous workflows that respect regional differences, maintain momentum, and strengthen global execution discipline
A team feels disconnected from the broader mission due to intense local pressures and competing demands	Reconnect daily work to the global purpose, reinforcing meaning, alignment, and shared ambition across all regions through consistent narrative clarity

Turning a heated conflict into a breakthrough is incredibly satisfying. It replaces the "weight" of disagreement with the "energy" of a new solution. This bliss improves your daily relationships, transforming potential enemies into allies and ensuring that the organization's social fabric is strengthened rather than torn by debate. This is the harmony of "constructive tension." You learn to appreciate the sparks that fly when smart people disagree, because you know how to guide those sparks into a fire that illuminates the way forward. It makes your daily interactions more authentic, deeper, and ultimately more productive for everyone involved, as you foster a culture where the truth is more important than being right.

DO'S FOR THE LEADER	DON'TS FOR THE LEADER
Resolve conflicts with respect.	Let conflicts fester and become toxic.
Facilitate open and honest dialogue.	Avoid all difficult conversations.
Find solutions that benefit everyone.	Create a winner and a loser.
Use conflict as a catalyst for growth.	View all conflict as negative and bad.
Create a clear path for resolution.	Leave teams to their own devices.
Mediate with a neutral and fair stance.	Take sides or show favoritism.

SITUATION	SUGGESTED ACTION
Two team members avoid addressing tension, creating subtle friction that undermines collaboration and trust	Facilitate a structured conversation that surfaces concerns respectfully, rebuilds understanding, and restores productive working relationships through clarity, empathy, and shared commitment
A leader dismisses early conflict signals, hoping issues resolve themselves without intervention or structured dialogue	Intervene early with curiosity, explore underlying concerns, and model healthy conflict engagement that prevents escalation and strengthens team cohesion and trust
A team debates aggressively without listening, creating defensiveness and slowing meaningful progress across discussions	Introduce listening rounds requiring participants to summarize others' views before responding, strengthening empathy, improving dialogue quality, and enabling more constructive outcomes
A leader prioritizes speed over resolution, leaving issues unresolved beneath the surface and affecting morale	Allocate time for closure, clarify agreements, and ensure conflicts end with shared understanding, renewed commitment, and strengthened working relationships
A team member dominates discussions, silencing quieter voices and creating imbalance in decision-making processes	Establish equal-voice protocols that ensure all perspectives are heard, valued, and incorporated into decisions, strengthening fairness and team cohesion
A leader avoids difficult conversations to maintain harmony, allowing issues to compound over time	Approach the conversation with empathy and directness, reinforcing that respect includes honesty, timely resolution, and constructive accountability

The feeling of seeing a trend months before it hits and positioning your company to catch the wave is pure professional joy. This foresight validates your hard work and strategic thinking. It gives you a sense of control over the future, providing a calm, steady confidence in your leadership journey. This is the bliss of the "long view." While others are reacting to today's news, you are already living in next year's reality. It reduces the stress of volatility and replaces it with the satisfaction of being prepared. Leading with foresight is the ultimate expression of executive harmony and strategic excellence, allowing you to shape the future rather than just survive it.

DO'S FOR THE LEADER	DON'TS FOR THE LEADER
Lead with a clear, strategic vision.	Have a vague, short-term plan.
Make decisions with the future in mind.	Make decisions based on short-term gains.
Use data to forecast future trends.	Ignore data and operate on intuition alone.
Anticipate future challenges and needs.	Be constantly reactive to new problems.
Communicate the long-term strategy.	Keep the big picture a secret.
Create a culture of proactive planning.	Have a habit of "winging it" on projects.

SITUATION	SUGGESTED ACTION
A leader focuses heavily on immediate tasks and loses sight of long-term direction and priorities	Set aside weekly strategic reflection time to reconnect with future priorities, emerging opportunities, and long-term organizational positioning
A team executes well but lacks understanding of where the organization is ultimately heading	Share long-term scenarios and explain how current work contributes to future positioning, strategic advantage, and sustained organizational relevance
A leader reacts to market shifts only after competitors move first, reducing agility and strategic influence	Build an early-signal dashboard to anticipate trends, enabling proactive decisions, stronger positioning, and more confident long-term planning
A team struggles to prioritize because they cannot see which initiatives matter most strategically	Clarify strategic anchors and align work around the few initiatives that drive the greatest long-term impact and organizational value
A leader underestimates emerging risks because they seem distant or abstract within current operations	Conduct periodic risk-mapping sessions to visualize long-term threats and opportunities, strengthening preparedness, resilience, and strategic adaptability
A team becomes comfortable with current success and stops scanning for future possibilities or disruptions	Introduce horizon-scanning rituals that keep the organization alert, curious, and consistently prepared for emerging opportunities and challenges

6x6™ Transformation Blueprint

Leadership Diversity

"A diverse mix of voices leads to better discussions, decisions, and outcomes for everyone."

— *Sundar Pichai, CEO of Alphabet and Google. It emphasizes that diversity is a direct driver of superior strategic outcomes, which is critical for the complex decisions required during a digital transformation.*

For **Leadership Transformation**, diversity is a strategic imperative that inoculates the enterprise against failure in the face of rapid digital change. A leadership team diverse in thought, background, and experience (spanning gender, culture, and professional domain) possesses the wide-angle perspective necessary to challenge long-held assumptions and identify threats that a homogenous group would overlook. This cognitive diversity is the best defense against crippling **groupthink**, which can lead to catastrophic digital investment decisions. Diverse leadership is better equipped to interpret ambiguous market signals and scenario-plan for disruption, ensuring the transformation strategy is robust and ethically sound. Furthermore, when executives model inclusive behavior and show that divergent opinions are valued, it establishes the **psychological safety** needed throughout the entire organization. This culture of inclusion empowers all employees to embrace new digital processes, contribute unique insights without fear, and actively champion the transformation, making the entire company more agile and resilient.

In today's interconnected world, diversity is a fundamental pillar of strength and innovation. Often simplified to visible differences, true diversity is a rich tapestry of attributes shaping how individuals perceive and contribute. Understanding this multifaceted nature is crucial for fostering environments where every voice thrives. This essay explores diversity through six critical dimensions: **demographic, cognitive, experiential, geographic, functional, and perspective diversity**. By examining these distinct facets, we can better appreciate how a holistic approach to diversity enriches teams and communities, driving robust solutions.

First, **Demographic Diversity** encompasses varied inherent and acquired characteristics. This includes age, bringing diverse generational insights, and gender, race, ethnicity, sexual orientation, ensuring a wide range of lived experiences. Socioeconomic status, religion, disability, and nationality further contribute to a richer collective understanding. Valuing demographic diversity ensures equitable representation and leverages unique insights from varied personal journeys.

Secondly, **Cognitive Diversity** highlights how people think, process information, and approach problems. This includes varied thinking styles like analytical vs. intuitive approaches, and different problem-solving or learning methods. When a group embraces cognitive diversity, it tackles problems from multiple angles, leading to more innovative solutions and a thorough understanding of complex issues, free from a single dominant thought process.

Next, **Experiential Diversity** focuses on the breadth of life and work experiences individuals bring. This covers varied work histories across industries and roles, providing a rich pool of professional knowledge. Diverse educational backgrounds, from different disciplines or institutions, also offer distinct frameworks. Beyond formal credentials, personal life experiences like travel and cultural immersion contribute significantly, equipping teams with a wide array of skills and adaptable approaches.

Geographic Diversity emphasizes the impact of place on an individual's perspective. This includes differences from regional or local backgrounds, shaping values and priorities. National backgrounds introduce distinct cultural norms, while urban versus rural upbringings offer varying insights into community dynamics and societal needs. These varied geographical lenses provide a broader understanding of global and local markets and behaviors.

Furthermore, **Functional Diversity** pertains to the variety of professional roles, specialized areas, and departmental expertise within a team. This dimension recognizes contributions from different functions like engineering, marketing, or finance, each bringing specific domain knowledge. Variations in hierarchical roles and industry experience also inform cross-pollination of best practices. Leveraging functional diversity ensures all necessary technical skills are present, fostering comprehensive problem-solving.

Finally, **Perspective Diversity** encompasses the fundamental differences in viewpoints and mental models individuals hold, often influenced by other diversity forms. This includes varied ideological stances, cultural interpretations, and distinct value systems. Understanding diverse stakeholder perspectives—from customers to employees—is crucial for holistic decision-making. Ultimately, perspective diversity allows for problems to be framed in multiple ways, challenging assumptions and encouraging empathetic, effective outcomes.

By intentionally cultivating and integrating these six dimensions—demographic, cognitive, experiential, geographic, functional, and perspective—organizations and societies can harness the profound power of true diversity. This comprehensive approach not only fosters equity but also enhances creativity, resilience, and innovation, demonstrating that a multitude of viewpoints is the clearest path to collective success.

1. DEMOGRAPHIC DIVERSITY

Embracing **demographic diversity** in leadership means intentionally cultivating a top-tier group that mirrors the rich tapestry of society and the organization's broader workforce and customer base. This goes beyond mere representation; it's about leveraging the inherent value that comes from leaders with varied ages, genders, racial and ethnic backgrounds, physical abilities, socio-economic upbringings, and sexual orientations. These visible and non-visible differences aren't just checkboxes; they are sources of unique experiences, perspectives, and problem-solving approaches. Leaders from different demographic groups often bring distinct insights into market needs, cultural sensitivities, and employee experiences, which are crucial for effective transformation. When a leadership team lacks this breadth, it risks developing strategies that are narrow in scope, fail to resonate with diverse markets, or overlook the needs of significant employee populations. True demographic diversity in leadership fosters a more empathetic, relatable, and ultimately, more effective organization, better equipped to navigate complex global challenges and connect authentically with diverse stakeholders. It enhances the legitimacy and credibility of transformation initiatives by demonstrating a genuine commitment to inclusivity from the very top.

Topic	Tangible steps	Challenges you may face
AGE DIVERSITY IN SUCCESSION PLANNING		
Varied age groups in leadership impact future role readiness & transformation continuity. Diverse generational views boost leadership capabilities & organizational resilience.	Implement mentorship for junior/senior leaders. Set skill-based succession criteria, minimizing age bias. Conduct cross-generational knowledge transfer workshops. Promote inclusive leadership training.	Resistance from entrenched leaders. Difficulty convincing senior talent to mentor. Overcoming biases favoring specific age groups, hindering equitable succession.
GENDER DIVERSITY IN DECISION-MAKING		
Balanced gender representation influences leadership decisions. Diverse gender perspectives lead to comprehensive problem-solving & innovative outcomes.	Set targets for gender representation in leadership committees. Mandate inclusive meeting practices. Implement unconscious bias training for hiring/promotion panels.	Overcoming ingrained stereotypes. Lack of female candidates in senior pipelines. Resistance to changing established norms & power structures.
RACIAL & ETHNIC DIVERSITY IN GLOBAL TEAMS		
Diverse racial & ethnic backgrounds in multinational leadership foster cultural intelligence & market empathy, reflecting global customers & employees.	Develop cultural immersion programs for global leaders. Establish affinity groups for mentorship/support. Prioritize diverse representation in international project teams.	Systemic racism. Difficulty attracting/retaining diverse talent. Lack of cultural understanding causing unintentional exclusion.

Cognitive diversity within leadership is paramount for navigating the inherent complexities of transformation. It acknowledges that true innovation and robust problem-solving stem not just from varied experiences or demographics, but fundamentally from differences in how people think, process information, and approach challenges. A leadership team composed solely of analytical thinkers might excel at optimizing existing processes but could struggle with disruptive innovation. Conversely, a team too reliant on intuitive thinking might miss critical data points or overlook practical constraints. By fostering a mix of analytical, creative, systemic, and neurodiverse cognitive styles, leaders are better equipped to challenge assumptions, identify unseen opportunities, predict unforeseen risks, and design more comprehensive and resilient strategies. It's about building a leadership brain trust where different cognitive strengths complement each other, leading to more thorough evaluations, more innovative solutions, and a greater capacity for adaptive learning in the face of rapid change. This richness in thought ensures that strategic decisions are thoroughly vetted from multiple angles, preventing blind spots and fostering intellectual agility crucial for sustained transformation success.

Topic	Tangible steps	Challenges you may face
ANALYTICAL VS. INTUITIVE LEADERSHIP		
Balancing data-driven & gut-feeling approaches shapes strategy & risk. Leads to nuanced decision-making, combining factual rigor with insightful foresight.	Design frameworks requiring data/qualitative input. Encourage leaders to articulate reasoning. Form diverse teams combining cognitive styles.	Friction between styles. Difficulty quantifying intuition. Culture valuing one style over another hinders collaboration.
SYSTEMIC THINKERS IN CHANGE		
Leaders seeing interconnectedness & long-term implications drive comprehensive transformation. Ensures holistic, sustainable changes.	Provide systems thinking training. Assign systemic thinkers to cross-departmental initiatives. Use "pre-mortem" exercises to identify potential downstream impacts.	Difficulty identifying/developing systemic thinking. Resistance to long-term views. Overwhelm from too much interconnected info.
CREATIVE PROBLEM-SOLVERS IN INNOVATION		
Leaders with imaginative thinking inspire novel solutions to complex challenges. Crucial for breakthrough innovations & disruptive strategies.	Establish innovation labs led by creative thinkers. Implement brainstorming for unconventional ideas. Create time for blue-sky thinking.	Risk aversion stifling new ideas. Difficulty translating creative ideas into actionable plans. Perceived lack of "seriousness."

Experiential diversity among leaders is a powerful catalyst for successful transformation, providing a rich tapestry of practical insights drawn from varied career paths, industries, educational backgrounds, and personal life stages. Leaders who have navigated different organizational cultures, worked across diverse sectors (e.g., startup to corporate, public to private), or gained international exposure bring a broader repertoire of solutions and a more nuanced understanding of complex challenges. This breadth of experience prevents insular thinking, encourages the adoption of "best practices" from seemingly unrelated fields, and fosters resilience by drawing on lessons learned from past successes and failures in varied contexts. For instance, a leader with crisis management experience can bring invaluable composure and clear-headedness during periods of intense organizational upheaval. By intentionally cultivating a leadership team with diverse professional and personal journeys, an organization gains a formidable collective wisdom, enhancing its ability to foresee opportunities, mitigate risks, and execute transformation initiatives with greater adaptability and effectiveness. It promotes a learning culture where varied practical wisdom is valued and applied to new strategic challenges.

Topic	Tangible steps	Challenges you may face
CROSS-INDUSTRY LEADERSHIP INSIGHTS		
Leaders from different industries bring fresh perspectives & transferable solutions to transformation hurdles. Broadens solution set.	Recruit leaders with diverse industry backgrounds. Organize peer-learning forums for cross-industry insights. Encourage participation in external industry associations.	Difficulty translating insights to new contexts. Resistance to "outsider" approaches. Lack of understanding of new industry nuances.
STARTUP VS. CORPORATE BACKGROUNDS		
Impact of leaders from agile startup vs. structured corporate environments on transformation. Blending these fosters innovation & stability.	Integrate startup leaders into corporate innovation units. Create joint ventures pairing corporate/startup leadership. Foster internal "intrapreneurship" programs.	Culture clash between agile & traditional structures. Difficulty integrating fast decisions. High expectations for immediate results.
GLOBAL VS. LOCAL EXPERIENCE IN STRATEGY		
Leaders with diverse international & local market exposure balance global standardization & regional adaptation. Optimizes strategy.	Implement global leadership rotation. Establish regional advisory boards with local market knowledge. Encourage dual reporting lines.	Complex balance of global vs. local. Resistance from local leaders to global mandates. Misinterpretation of cultural nuances.

Geographic diversity in leadership is indispensable for organizations operating in a globalized world, especially during periods of significant transformation. Leaders deeply rooted in varied regions, countries, or even urban versus rural settings bring critical local market understanding, cultural nuances, and insights into unique regulatory environments and logistical challenges. Without this localized perspective at the top, global strategies risk being homogenous, culturally insensitive, or simply impractical in execution. For example, a leader experienced in emerging markets might bring crucial insights into scalable, low-cost solutions, while someone from a highly regulated economy understands the complexities of compliance. This diversity ensures that transformation initiatives are not just conceptualized centrally but are designed with built-in adaptability for regional variations. It empowers the organization to navigate geopolitical risks, understand localized customer behaviors, and foster stronger relationships with diverse regional stakeholders. Leaders with varied geographic exposure contribute to strategies that are globally coherent yet locally relevant, maximizing impact and minimizing missteps in a fragmented world.

Topic	Tangible steps	Challenges you may face
REGIONAL MARKET UNDERSTANDING		
Leaders from different regions offer unique insights into local market dynamics, consumer behaviors, & competition. Vital for global strategies.	Establish regional leadership councils. Mandate global strategies be reviewed/adapted by local leaders. Support leader travel/immersion in diverse regions.	Overlooking nuances for standardization. Difficulty adapting to rapid changes. Resistance from central leadership to delegate.
NAVIGATING GLOBAL REGULATORY LANDSCAPES		
Leaders familiar with diverse international legal/compliance frameworks are crucial for multi-national transformations. Ensures adherence & mitigates risks.	Create global compliance leadership committees. Invest in continuous education on international regulations. Consult local legal experts.	Keeping up with rapidly changing laws. Reconciling conflicting requirements. Risk of non-compliance due to inadequate local understanding.
LOCAL INFRASTRUCTURE AND RESOURCE KNOWLEDGE		
Leaders understanding local infrastructure (digital, logistics) & resources impact operational transformation. Optimizes local implementation.	Empower local leaders for infrastructure investments. Develop centralized knowledge of regional capabilities. Foster local service provider partnerships.	Underestimating varied local infrastructure impact. Inaccurate resource assumptions. Difficulty scaling global solutions.

Functional diversity at the leadership level is critical for executing comprehensive and interconnected business transformations. It ensures that strategic decisions are informed by the deep expertise and unique perspectives of various professional disciplines, rather than being confined to a single departmental silo. A leadership team comprising experts from finance, technology, human resources, operations, marketing, and legal functions can collectively identify opportunities, anticipate challenges, and design solutions that are holistic, practical, and sustainable. For instance, a finance leader ensures fiscal viability, while a technology leader champions digital adoption, and an HR leader focuses on people-centric change management. Without this blend, transformation efforts might optimize one area at the expense of another, leading to fragmented implementation or unintended negative consequences. By fostering cross-functional leadership, an organization can break down traditional barriers, enhance collaboration, and create truly integrated strategies that leverage the strengths of every part of the business, ensuring that the transformation is not just conceptually sound but also executable across all organizational facets.

Topic	Tangible steps	Challenges you may face
FINANCE-LED TRANSFORMATION GOALS		
Leaders with strong financial backgrounds prioritize cost optimization & ROI as primary transformation drivers. Ensures fiscal responsibility.	Appoint finance leaders to co-lead initiatives. Establish clear financial metrics & ROI targets. Implement rigorous cost-benefit analyses for investments.	Over-emphasis on short-term cuts. Resistance if financial goals overshadow innovation. Difficulty projecting ROI for intangible benefits.
TECHNOLOGY-DRIVEN INNOVATION LEADERSHIP		
Leaders with deep tech expertise champion digital adoption & disruptive innovation. Essential for driving tech-enabled change.	Elevate technologists to strategic roles. Foster "tech literacy" for leaders. Create cross-functional innovation teams combining tech/business experts.	Resistance to tech change. Rapid evolution challenges skills. Difficulty integrating new tech with legacy systems.
HR'S ROLE IN PEOPLE-CENTRIC CHANGE		
HR leaders focus on cultural shifts, talent management, & employee well-being during transformation. Ensures human capital is central to success.	Position HR leaders as strategic partners in planning. Develop robust change management/communication. Implement well-being/reskilling programs.	Business leaders resist prioritizing people. Difficulty measuring HR ROI. Challenges addressing employee anxiety/resistance.

Perspective diversity within leadership is arguably one of the most subtle yet profound drivers of successful transformation. It goes beyond demographics, experience, or function, delving into the varied worldviews, core values, ethical frameworks, and risk tolerances that shape how leaders interpret information and prioritize objectives. A team with diverse perspectives challenges inherent biases, avoids groupthink, and considers a wider spectrum of possibilities and potential consequences. For example, balancing leaders who prioritize short-term gains with those focused on long-term sustainability, or those who are risk-seeking with those who are risk-averse, leads to more robust and balanced strategic decisions. This diversity ensures that transformation is not just about efficiency or growth, but also about ethical responsibility, employee well-being, and genuine stakeholder value. It encourages leaders to approach problems from fundamentally different angles, leading to more resilient strategies, better stakeholder engagement, and a more adaptive and morally grounded organization capable of navigating complex and ambiguous future landscapes. It cultivates a culture of constructive challenge and holistic decision-making.

Topic	Tangible steps	Challenges you may face
CUSTOMER-CENTRIC VS. PRODUCT-CENTRIC VIEWS		
Leadership balances internal product focus with external customer needs. Ensures products solve customer problems & resonate with market.	Implement "Voice of Customer" programs. Conduct leadership "customer immersion" days. Establish KPIs balancing product features with satisfaction.	Internal resistance to shifting from product-centric. Difficulty capturing diverse customer needs. Leaders struggle balancing desires vs. feasibility.
SHORT-TERM VS. LONG-TERM VISION		
Leaders' differing time horizons influence transformation pace, scope, & success. Balancing these ensures both immediate gains & future viability.	Develop multi-year roadmap with short-term milestones. Assign leaders with complementary focuses. Use scenario planning to visualize impacts.	Pressure for immediate results. Leaders may prioritize short-term gains. Difficulty articulating long-term value to boards.
RISK-SEEKING VS. RISK-AVERSE APPROACHES		
Varied risk tolerance impacts pursuit of disruptive vs. incremental improvements. A healthy balance drives calculated growth.	Foster culture where calculated risks are encouraged. Establish "fail fast, learn faster." Create portfolio balancing high-risk/reward with safer bets.	Decision paralysis when tolerances conflict. Dominance of extreme risk views. Difficulty setting clear, unified risk parameters.

BLUEPRINT SERIES

SECTION G

Moral Compass

Ai6x6™ Transformation Blueprint

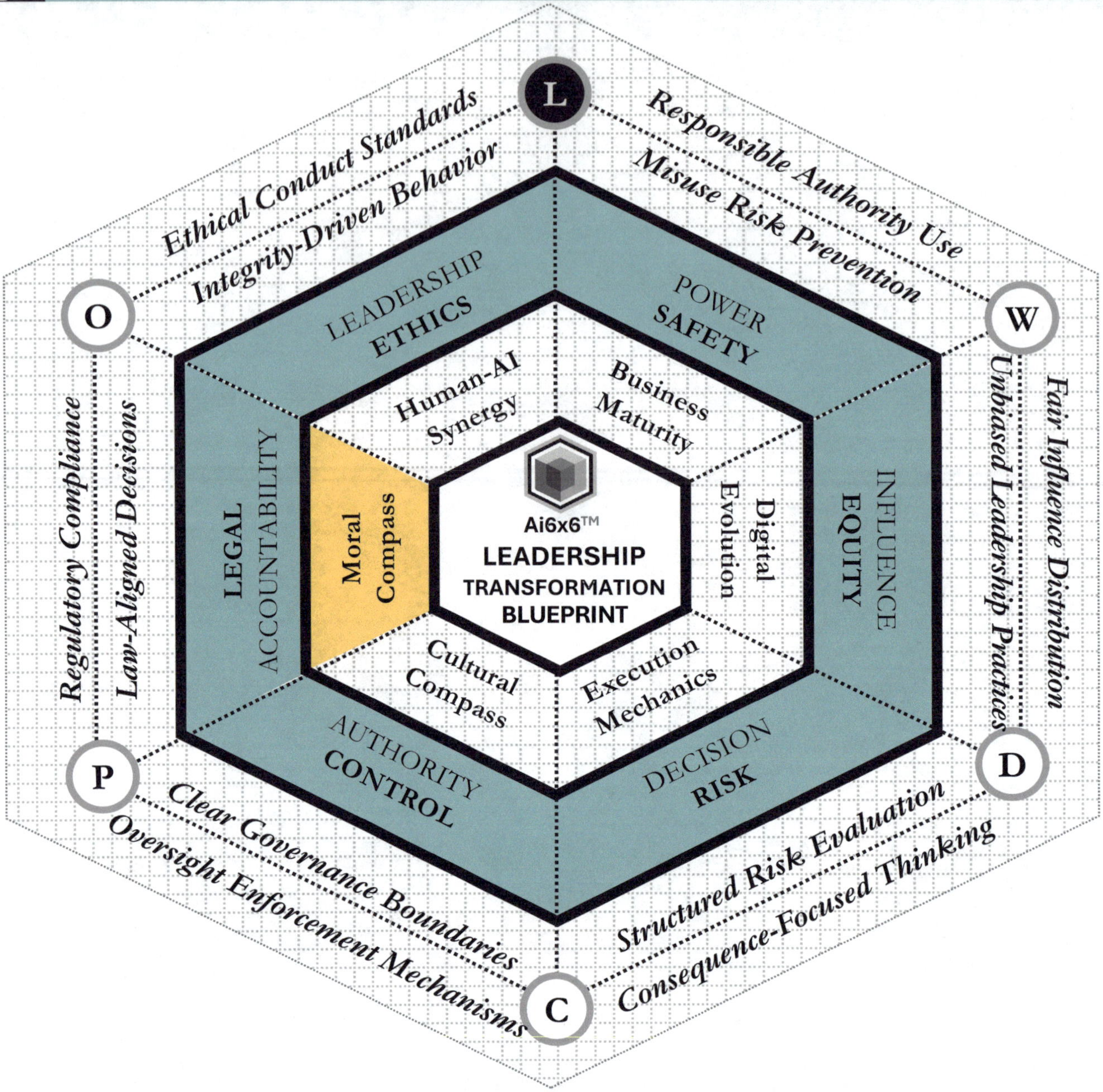

The **Ai6x6 Research Lab** identified the Moral Compass as the most vital lens because it serves as the final guardian of the leader's honor. This is the ultimate filter for every decision, ensuring the power of accelerated intelligence never outpaces the bedrock of human principle. We chose this because a leader's legacy is built on the integrity of the path they took, not just the height of the peaks they climbed. It is vital because it provides the moral authority required to lead through the ethical "fog" of autonomous systems.

This lens is the heart's final check—a rigorous audit of power safety, equity, and accountability. It ensures the transformed enterprise is not only a titan of industry but a beacon of public trust. The Lab believes that in the 2026 landscape, a leader's greatest risk is not a market failure, but a moral one. By focusing on decision risk and regulatory accountability, we provide the tools to navigate the heavy responsibility of influence. This lens is about the courage to be honorable when the temptation for speed is highest. Ultimately, it allows a leader to look back with pride, knowing they built an organization that was as ethical as it was intelligent.

Story 1 — The Nuremberg Trials and Ethical Accountability (1945–1946): After World War II, the world confronted atrocities committed under the guise of legal authority. The Nuremberg Trials established that individuals and institutions are accountable for actions that violate fundamental human rights. Legal frameworks were strengthened by moral reasoning, not replaced by it. Judges balanced evidence with ethical principles to define new global norms. The trials demonstrated that morality must guide power. They became a foundation for modern ethical governance.

Story 2 — The Tuskegee Syphilis Study Exposure (1932–1972): For decades, researchers withheld treatment from Black men to observe disease progression, violating consent and dignity. Public exposure triggered national outrage and forced reforms in medical ethics. Oversight boards, informed consent, and patient rights became mandatory. Machines and data had played roles in tracking outcomes, but human ethics had been absent. The scandal showed how systems fail when morality is excluded. It reshaped global standards for ethical research.

Story 3 — The Environmental Movement and Earth Day (1970): Industrial growth created pollution that threatened ecosystems and public health. Scientists provided data, but moral arguments mobilized action. Citizens demanded accountability, and governments introduced regulations to protect air, water, and wildlife. The movement reframed environmental protection as a shared ethical duty. It demonstrated how moral clarity can transform policy, industry, and public behavior. The cultural shift established sustainability as a long-term societal obligation.

Story 4 — The Anti-Apartheid Global Solidarity Movement (1950s–1994: Apartheid enforced racial segregation and oppression in South Africa. Activists, governments, and institutions worldwide united to challenge the regime through sanctions, protests, and moral pressure. Human stories amplified the injustice, while global networks coordinated action. The movement showed how collective moral conviction can influence political systems. It proved that ethical solidarity can reshape international norms. The end of apartheid became a landmark in moral mobilization.

Story 5 — The Rise of Corporate Social Responsibility (1990s–2000s): Globalization exposed labor abuses, environmental harm, and opaque supply chains. Consumers demanded transparency, and companies adopted codes of conduct, sustainability reporting, and ethical sourcing. Machines tracked compliance, but humans defined the standards. CSR shifted from public relations to strategic responsibility. Organizations learned that long-term value depends on ethical behavior. It marked a turning point where morality became integral to corporate identity.

Story 6 — The AI Ethics Movement (2015–2025): As AI systems influenced decisions in hiring, policing, healthcare, and finance, concerns grew about bias, transparency, and accountability. Researchers, policymakers, and civil society pushed for ethical frameworks to guide development. Machines provided capability, but humans defined boundaries. The movement emphasized fairness, explainability, and human oversight. It demonstrated that emerging technologies require moral governance as much as technical innovation. AI ethics became a global priority shaping the future of intelligent systems.

LESSONS FROM A LEADERSHIP TRANSFORMATION PERSPECTIVE

Leadership maturity in moral domains emerges when leaders recognize that authority must be guided by ethical responsibility, not just strategic intent. Across decades, leaders learned that moral clarity strengthens legitimacy, protects institutions, and prevents harm. Effective leaders create environments where ethical reasoning is embedded in decision-making, not treated as an afterthought. They balance power with accountability, ensuring that systems serve people rather than exploit them. Moral leadership requires courage to confront wrongdoing, humility to learn from failure, and discipline to uphold principles under pressure. Maturity is achieved when leaders institutionalize ethics as a non-negotiable foundation for action.

Morality has always shaped how organizations behave, but in the age of AI, it becomes a structural requirement rather than an aspirational ideal. The **Moral Compass** lens forces organizations to confront a truth that is often ignored: intelligence—whether human or machine—does not create ethical behavior. It amplifies whatever moral foundation already exists. When an organization is grounded in integrity, clarity, and principled action, AI strengthens it. When the foundation is inconsistent, opaque, or misaligned, AI exposes those weaknesses with unforgiving speed. Morality is no longer a philosophical preference; it is the architecture that determines whether accelerated decisions, automated systems, and scaled consequences lead to progress or harm. This lens reframes morality as an enterprise-wide discipline, not a leadership trait. It applies equally to workforce practices, data governance, customer interactions, product design, and operational systems.

Ethics

Establishing the Moral Center

Ethics forms the first dimension of the Moral Compass. Ethics is not simply about avoiding misconduct; it is about establishing a moral center that guides every decision, every interaction, and every use of technology. Across all six pillars, ethics determines how data is collected, how employees are treated, how customers are served, how products are built, and how operations are executed. Organizations that prioritize ethics create cultures where integrity is expected, not encouraged. Ethical clarity becomes the anchor that ensures AI is deployed with intention, restraint, and respect. In this dimension, principled behavior becomes the foundation of moral authority.

Safety

Protecting People, Systems, and Society

Safety represents the second dimension. As AI accelerates decision-making and expands operational reach, the potential for harm increases. Safety is not limited to physical environments; it includes psychological safety for the workforce, data safety for customers, model safety for AI systems, and operational safety for mission-critical processes. Safety ensures that no individual, team, or automated system can cause harm through unchecked influence or unintended consequences. It is not about limiting innovation; it is about ensuring that innovation does not outpace responsibility. In this dimension, responsible protection becomes essential to moral operation.

Equity

Ensuring Fairness Across Stakeholders

Equity forms the third dimension. Influence in modern organizations is no longer tied solely to hierarchy; it flows through data pipelines, algorithmic recommendations, and automated workflows. Equity ensures that these systems do not privilege certain groups, perspectives, or outcomes at the expense of others. It applies to workforce advancement, customer treatment, product accessibility, and data representation. Equity is not about equalizing results; it is about ensuring that the system itself is not biased. In this dimension, fairness becomes a structural requirement rather than a social aspiration.

Risk

Navigating Uncertainty With Discipline

Risk represents the fourth dimension. AI-enabled decisions carry greater speed, greater reach, and greater uncertainty. Risk is not about avoiding uncertainty; it is about navigating it with discipline. Organizations must anticipate consequences, stress-test assumptions, and understand the risks embedded in both human and machine judgment. This applies to operational risks, product risks, customer risks, workforce risks, and data risks. Organizations that embrace this dimension act decisively without acting blindly. In this dimension, risk-aware decision-making becomes a moral responsibility.

Control

Defining Governance and Accountability

Control forms the fifth dimension. As organizations adopt AI, authority becomes more complex: who decides, who approves, who monitors, and who intervenes when systems drift? Control ensures that governance structures are clear, decision rights are defined, and accountability is transparent. It applies to data governance, product governance, operational governance, and workforce governance. Control is not about bureaucracy; it is about ensuring that power—human or machine—is exercised with clarity and accountability. In this dimension, structured oversight becomes the backbone of moral operation.

Legal

Upholding Lawful and Responsible Conduct

Legal completes the Moral Compass. In a world where data, automation, and intelligence shape every action, legal obligations become inseparable from organizational obligations. Compliance is not a constraint; it is a safeguard. Legal integrity ensures that decisions align with regulatory expectations, societal norms, and ethical standards across all six pillars. It protects the organization from missteps and reinforces public trust. In this dimension, lawful stewardship becomes the final test of moral credibility.

The Moral Compass lens reminds organizations that moral authority is not a byproduct of good intentions; it is the result of deliberate, disciplined choices. Morality must be engineered into systems, not assumed in people. When organizations prioritize ethics, enforce safety, ensure equity, manage risk, establish control, and uphold legal accountability, they create environments capable of navigating complexity with confidence and integrity. They build systems where people feel protected, customers feel respected, and society feels considered. They create the moral conditions for transformation to take root and endure. In this maturity, morality becomes an enterprise capability—one that strengthens trust, accelerates adoption, and stabilizes transformation. It becomes the differentiator that determines whether AI amplifies value or amplifies harm. Organizations that master the Moral Compass do not merely deploy intelligent systems; they deploy them responsibly. They create futures where intelligence and integrity advance together, ensuring that progress is not only possible but principled.

Family Values | **ETHICS**

The foundational moral code—religious, cultural, and social—that dictates the ground rules for "Right vs. Wrong." Establishing the core integrity of the home so that all who belong understand the non-negotiable standards of conduct.

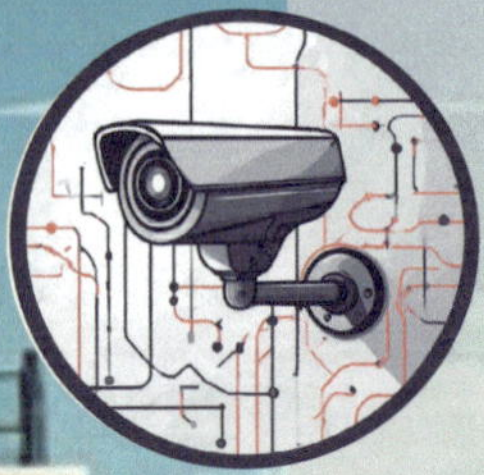

Home Security | **SAFETY**

Integrated protection—cameras, sensors, and smoke detectors—that provide total awareness of the premises. Deploying proactive warning systems and delegation guardrails to protect the inhabitants from "hidden fires."

Balanced Contribution | **EQUITY**

Ensuring everyone helps with the cooking and upkeep, with expenses shared equitably by each earning member. Maintaining participation and representation so that no member feels excluded from the collective success.

The Insurance Shield | **RISK**

Comprehensive fire, flood, and auto coverage to provide long-term peace of mind for the entire household. Using stress testing and "what-if" scenarios to protect the family assets from unforeseen market storms.

Access & Keys | **CONTROL**

Ensuring all adults have keys and codes, with clear emergency rules and a defined line of command for disasters. Establishing rigid RACI controls to ensure that a human always has the final word on the most critical locks.

The Estate Will | **LEGAL**

Recording Wills, Trusts, and Medical Mandates with a legal custodian to safeguard the valuables for the future. Securing the organization's legacy through regulatory accountability, ensuring the estate is legally unassailable for generations.

6x6™ Transformation Blueprint

Leadership Ethics

"Ethics is no longer a guardrail for technology — it is the steering wheel."

Tim Cook, CEO of Apple, known for championing privacy and responsible tech.

Leadership Ethics requires a disciplined moral foundation, and Ethical Conduct Standards define the behavioral expectations leaders must uphold in every decision, interaction, and directive. They establish the nonnegotiable norms that govern honesty, transparency, and principled action. This matters because leaders set the ethical tone for the entire organization. Ethical Conduct Standards prevent moral drift by anchoring leadership behavior to clear expectations. They also reinforce fairness by ensuring all leaders operate under the same ethical obligations. When ignored, leadership behavior becomes inconsistent and unpredictable. Teams experience confusion about what is acceptable. Over time, Leadership Ethics deteriorates as standards lose authority and ethical boundaries blur. Integrity Driven Behavior ensures leaders act in alignment with the organization's stated values.

Leadership is the practice of auditing your own shadow before it falls over the organization. By sitting with Ethics Case Logs in a distraction-free print environment, a leader ensures their private reflection matches their public authority. This topic moves beyond passive compliance into active integrity, identifying "Gifts and Influence Restrictions" that could cloud judgment. It is about building a foundation of trust so robust that the organization's intent is never questioned. In the AI era, where logic is increasingly automated, the leader's primary duty is to remain the unassailable moral anchor that defines the "Why" behind every automated "How." By using a print-based tool, you remove the noise of digital notifications, allowing for the deep, slow thinking required to navigate complex moral trade-offs. This ensures that your leadership legacy is built on a bedrock of unshakeable character, which is the only scarcity in a world of machine logic.

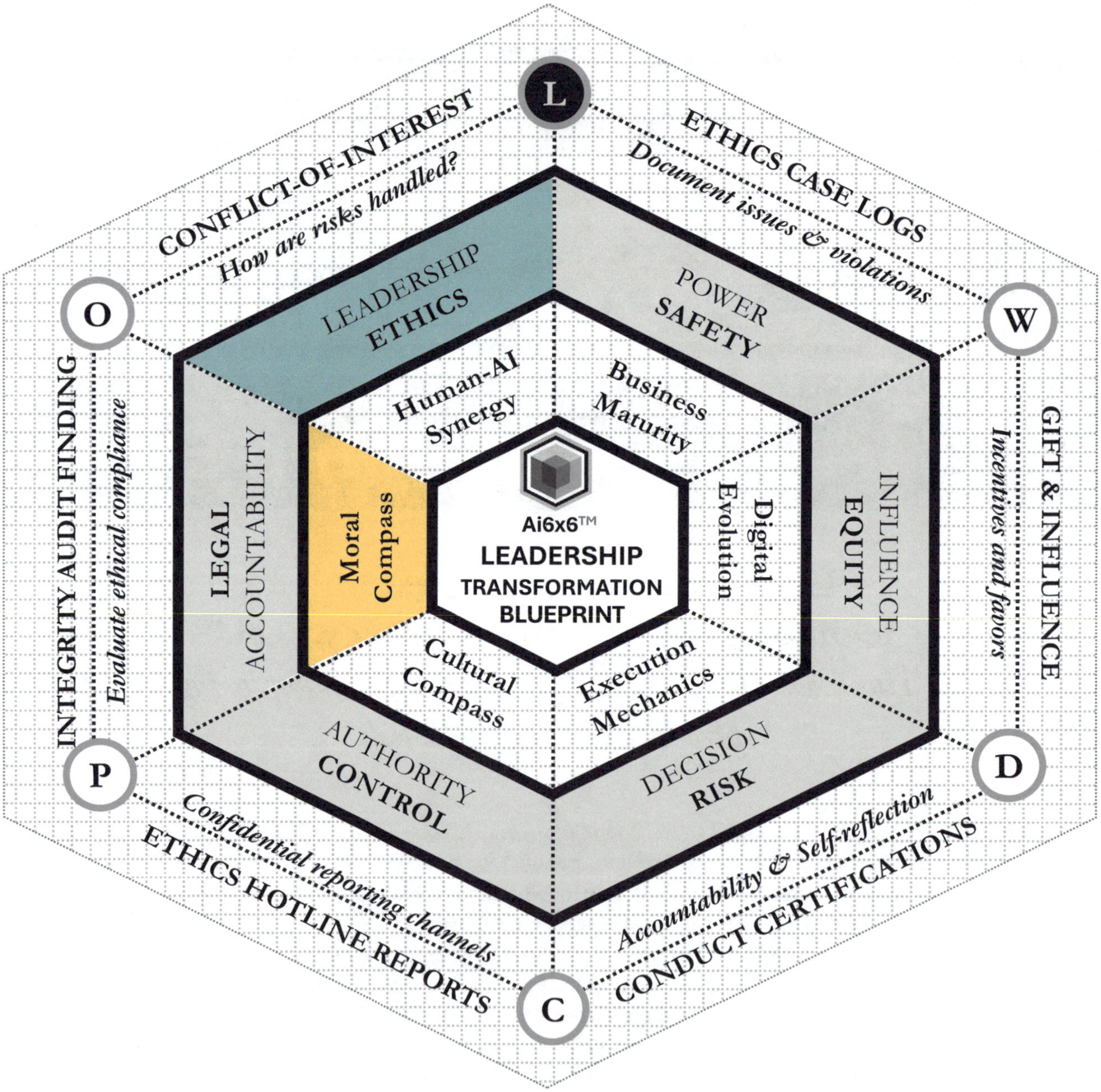

Ai6x6 LEADERSHIP ETHICS FOR MORAL COMPASS | LEADERSHIP TRANSFORMATION

1. CONFLICT-OF-INTEREST REVIEWS

Conflict of Interest Reviews identify situations where personal, financial, or relational interests could compromise leadership judgment. They define how risks are disclosed, evaluated, and mitigated to preserve impartiality. This process ensures that decisions remain aligned with organizational purpose rather than individual advantage. It matters because conflicts of interest undermine fairness, distort decision quality, and weaken moral authority. Clear reviews prevent hidden incentives from influencing outcomes. They also reinforce the expectation that leadership power must be exercised without private benefit. When ignored, conflicts create suspicion, favoritism, and ethical breaches that damage trust. Teams begin to question whether decisions are made for the right reasons. Over time, the organization experiences cultural erosion as compromised behavior becomes normalized.

When done well, conflict of interest reviews create transparency, fairness, and confidence in leadership decisions. People trust that choices are made objectively and without personal bias. This strengthens cohesion and reduces friction arising from perceived inequities. These reviews reinforce the principle that leadership requires impartiality and disciplined self regulation. They remind leaders that integrity is demonstrated not only in what they do but in what they refuse to exploit. Ultimately, they protect the legitimacy of leadership and the ethical foundation of the organization.

MORAL COMPASS SCENARIO	LEADERSHIP ACTION
Maria is asked to evaluate a vendor proposal submitted by a company owned by her former colleague. She realizes the relationship could influence her judgment even if unintentionally.	The leader reassigns the evaluation to an independent reviewer to eliminate bias. They also document Maria's disclosure in the conflict-of-interest log to maintain transparency.
David is asked to approve a budget increase for a project led by his brother. He feels pressured because the team expects a quick decision.	The leader instructs David to formally recuse himself from the approval process. They escalate the decision to a neutral executive and record the recusal for audit purposes.
Priya recommends a consulting firm where she previously worked but does not initially disclose the connection. Her peers notice the recommendation seems unusually enthusiastic.	The leader pauses the procurement process and requests a written disclosure from Priya. They then rerun the vendor comparison using an unbiased selection panel.
Jordan is serving on a hiring panel where one of the candidates is a close friend. He believes he can remain objective but understands the optics are problematic.	The leader removes Jordan from the panel and assigns an alternate reviewer to preserve fairness. They also update the conflict-of-interest log to prevent similar oversights.
Elena receives an expensive gift from a supplier shortly before contract renewal discussions. She feels uncomfortable but unsure how to handle the situation.	The leader documents the gift and instructs Elena to return it in accordance with policy. They then transfer negotiation responsibilities to a leader with no exposure to the supplier.
Marcus is asked to approve performance bonuses for employees who also support his private side business. He recognizes that approving the bonuses could appear self-serving.	The leader shifts bonus decisions to HR or an independent executive to avoid bias. They also initiate a conflict-of-interest review to determine whether policy violations occurred.

2. ETHICS CASE LOGS

Ethics Case Logs document ethical issues, dilemmas, and violations that arise across the organization. They create a structured record of patterns, root causes, and leadership responses. This documentation ensures that ethical challenges are not treated as isolated incidents but as signals requiring systemic attention. They matter because organizations learn from what they track, analyze, and address. Case logs reveal recurring vulnerabilities that may otherwise remain hidden. They also demonstrate that leadership takes ethical issues seriously and responds consistently. When ignored, ethical problems repeat, escalate, or spread across teams. Leaders lose visibility into emerging risks and fail to intervene early. Over time, the organization becomes reactive rather than proactive in managing ethical integrity.

When done well, ethics case logs strengthen organizational learning, accountability, and transparency. People see that ethical issues are handled with seriousness and follow through. This builds trust and encourages earlier reporting of concerns. Case logs reinforce the principle that ethical vigilance is a continuous responsibility. They remind leaders that integrity requires attention to patterns, not just events. Ultimately, they help build a culture where ethical improvement is systematic and sustained.

MORAL COMPASS SCENARIO	LEADERSHIP ACTION
Hannah receives a complaint about a supervisor making inappropriate jokes but hesitates to log it because the complainant asked for discretion. She worries documenting it might escalate the issue too quickly.	The leader instructs Hannah to record the case anonymously to preserve both integrity and confidentiality. They also schedule a discreet climate check with the affected team.
Omar notices repeated small discrepancies in expense reports from the same employee. He debates whether it's worth opening a formal ethics case.	The leader directs Omar to log the pattern as a preliminary case to establish a traceable record. They then assign an investigator to review the discrepancies objectively.
Serena overhears two employees discussing a potential kickback arrangement with a vendor. She's unsure whether the conversation was serious or hypothetical.	The leader asks Serena to document the incident in the ethics case log as an early signal. They also initiate a quiet fact-finding step to determine credibility.
Lucas receives a report that a manager pressured a junior employee to work unpaid overtime. He worries the claim may be difficult to substantiate.	The leader instructs Lucas to log the case immediately to preserve the timeline. They then coordinate with HR to interview both parties and review timekeeping records.
Naomi learns that a team lead has been retaliating against an employee who raised a concern last quarter. She fears the retaliation may escalate if not addressed.	The leader logs the case as a priority ethics issue and triggers a retaliation-protection protocol. They also assign a neutral advocate to support the affected employee.
Andre receives an anonymous tip about confidential data being shared informally between teams. He's unsure whether the tip is credible.	The leader directs Andre to record the tip in the ethics log to ensure traceability. They then initiate a limited-scope audit to verify whether any data exposure occurred.

3. GIFT AND INFLUENCE RESTRICTIONS

Gift and Influence Restrictions define the boundaries that prevent external parties from shaping leadership decisions through incentives or favors. They clarify what can be accepted, what must be declined, and what requires disclosure. These restrictions ensure that leaders remain independent, impartial, and free from undue influence. They matter because gifts and favors can subtly distort judgment, create obligations, or signal preferential access. Clear restrictions protect leaders from ethical compromise and reputational risk. They also reinforce the expectation that decisions must be made solely in the organization's interest. When ignored, influence channels form that bypass formal processes and undermine fairness. Teams begin to question whether outcomes are earned or purchased. Over time, the organization becomes vulnerable to corruption, favoritism, and external manipulation.

When done well, gift and influence restrictions create transparency, fairness, and confidence in leadership independence. People trust that decisions are not shaped by hidden incentives or personal benefits. This strengthens credibility and reduces the risk of ethical breaches. These restrictions reinforce the principle that leadership must remain free from external pressure. They remind leaders that integrity requires resisting even subtle forms of influence. Ultimately, they protect the organization's moral and operational autonomy.

MORAL COMPASS SCENARIO	LEADERSHIP ACTION
Chloe receives an expensive holiday basket from a vendor bidding on a major contract. She feels uncomfortable but doesn't want to offend the sender.	The leader instructs Chloe to return the gift with a polite compliance-based explanation. They also document the incident and remind the vendor of the organization's gift policy.
Rafael is offered VIP tickets to a sporting event by a partner whose renewal is under review. He knows accepting them could compromise negotiations.	The leader directs Rafael to decline the tickets and record the offer in the gift log. They then rotate negotiation duties to a leader with no exposure to the vendor.
Tessa is invited to an all-expenses-paid conference by a software provider she is evaluating. She suspects the invitation is meant to influence her decision.	The leader instructs Tessa to decline the invitation and disclose it formally. They also notify procurement to ensure all vendor interactions remain compliant.
Jonah receives a luxury pen set from a client after closing a deal. He wonders if it crosses the threshold of acceptable appreciation.	The leader asks Jonah to log the gift and submit it for value assessment. They then determine whether it must be returned or donated based on policy.
Aisha is offered a "consulting honorarium" by a vendor seeking access to internal decision-makers. She immediately senses the ethical risk.	The leader directs Aisha to decline the payment and report the attempt. They escalate the incident to compliance for vendor-risk evaluation.
Felix is given a high-end gadget by a partner during a site visit. He feels pressured because the gift was presented publicly.	The leader documents the gift and arranges for its return through formal channels. They also brief the partner on acceptable conduct to prevent future violations.

4. EXECUTIVE CONDUCT CERTIFICATIONS

Executive Conduct Certifications require leaders to formally attest to their adherence to ethical standards, policies, and legal obligations. They create a structured mechanism for accountability and self reflection. This certification process ensures that leaders regularly reaffirm their commitment to integrity. It matters because formal attestation elevates ethical responsibility from expectation to obligation. Certifications reduce ambiguity by making leaders explicitly accountable for their conduct. They also signal that ethical leadership is a measurable and monitored requirement. When ignored, ethical expectations become informal, inconsistent, or selectively applied. Leaders may drift into complacency or assume that standards do not apply equally. Over time, the organization experiences uneven enforcement that erodes trust and credibility.

When done well, executive conduct certifications strengthen accountability, transparency, and ethical discipline. People see that leaders are held to clear and consistent standards. This reinforces confidence in leadership integrity and organizational fairness. Certifications reinforce the principle that ethical leadership requires conscious, ongoing commitment. They remind leaders that integrity must be renewed, not assumed. Ultimately, they anchor the organization in a culture where ethical responsibility is explicit and enforceable.

MORAL COMPASS SCENARIO	LEADERSHIP ACTION
Danielle delays signing her annual conduct certification because she's unsure whether a past decision fully aligned with policy. She fears signing without clarity could be misleading.	The leader arranges a confidential review of the decision so Danielle can certify accurately. They also update the certification process to include a pre-signing consultation option.
Miguel signs his certification quickly but later realizes he overlooked a potential conflict. He worries that amending it will draw scrutiny.	The leader instructs Miguel to file an immediate correction and document the oversight. They reinforce that transparency strengthens—not weakens—credibility.
Harper hesitates to certify because she has observed minor policy deviations on her team. She's unsure whether they rise to the level of disclosure.	The leader guides Harper through a structured review to determine what must be reported. They then help her submit a complete and accurate certification.
Nolan is new to the organization and finds the certification language overly broad. He fears signing something he doesn't fully understand.	The leader provides a detailed walkthrough of each clause and clarifies expectations. They also refine onboarding materials to prevent similar confusion.
Sienna signs her certification but later learns that her team used outdated compliance procedures. She worries this invalidates her attestation.	The leader helps Sienna file an amended certification and initiates a process update. They also schedule refresher training for her team.
Grant avoids signing because he feels the certification implies personal fault for inherited issues. He wants assurance that context matters.	The leader clarifies that certifications reflect current stewardship, not historical blame. They support Grant in documenting inherited risks before signing.

Ethics Hotline Reports capture concerns, allegations, and observations submitted through confidential reporting channels. They provide leaders with visibility into issues that may not surface through formal processes. This reporting mechanism ensures that ethical risks are identified early and addressed appropriately. They matter because employees often hesitate to raise concerns directly, especially when power dynamics are involved. Hotlines create a safe pathway for surfacing issues that might otherwise remain hidden. They also demonstrate that leadership is committed to listening and responding without retaliation. When ignored, hotline reports lose credibility and participation declines. Ethical issues remain unaddressed until they escalate into crises. Over time, silence becomes normalized and the organization's moral fabric weakens.

When done well, ethics hotline reports strengthen psychological safety, accountability, and early intervention. People trust that concerns will be taken seriously and handled with fairness. This encourages proactive reporting and reduces the likelihood of ethical failures. Hotline systems reinforce the principle that leadership must be accessible, responsive, and morally courageous. They remind leaders that ethical vigilance depends on open channels of communication. Ultimately, they protect the organization's integrity and reinforce a culture where speaking up is a moral duty.

MORAL COMPASS SCENARIO	LEADERSHIP ACTION
Olivia receives a hotline report alleging favoritism in shift assignments. She worries the claim may be emotionally charged rather than factual.	The leader instructs Olivia to log the report and initiate a neutral fact-finding step. They also ensure the reporter receives acknowledgment without revealing identities.
Marcus reviews a hotline tip about a manager verbally intimidating staff. He's concerned the behavior may escalate if not addressed quickly.	The leader prioritizes the case and assigns an investigator with no ties to the manager. They also implement temporary oversight to protect the team.
Fiona receives an anonymous report about missing inventory. She suspects internal theft but lacks evidence.	The leader directs Fiona to document the report and coordinate with security for a discreet audit. They also reinforce inventory controls across the department.
Declan reads a hotline message claiming a supervisor is retaliating against an employee who raised concerns. He knows retaliation cases require urgency.	The leader escalates the case immediately and activates retaliation-prevention protocols. They assign a neutral advocate to support the affected employee.
Ruby receives a report about inappropriate comments made during a team outing. She worries the informal setting may complicate the investigation.	The leader logs the report and initiates interviews with attendees. They also remind managers that conduct expectations apply in all work-related settings.
Isaac reviews a tip alleging misuse of company funds for personal travel. He recognizes the financial risk is significant.	The leader assigns a finance-trained investigator and freezes related expense approvals. They also notify compliance to prepare for potential escalation.

6. INTEGRITY AUDIT FINDINGS

Integrity Audit Findings document the results of systematic reviews that evaluate ethical compliance, behavioral risks, and leadership conduct. They provide objective insight into whether actions align with stated values and policies. This audit process ensures that integrity is measured, monitored, and continuously improved. They matter because organizations cannot rely solely on self reporting or assumptions of ethical behavior. Audits reveal gaps, inconsistencies, and vulnerabilities that require corrective action. They also demonstrate that leadership is committed to transparency and accountability. When ignored, ethical weaknesses persist and become embedded in daily operations. Leaders lose visibility into emerging risks and fail to intervene effectively. Over time, the organization becomes increasingly exposed to ethical, legal, and reputational harm.

When done well, integrity audit findings strengthen ethical discipline, operational reliability, and cultural alignment. People see that integrity is evaluated with rigor and follow through. This builds trust and reinforces the seriousness of ethical expectations. Audit findings reinforce the principle that leadership must be accountable to objective standards. They remind leaders that integrity requires verification, not assumption. Ultimately, they help build a culture where ethical excellence is measurable and sustained.

MORAL COMPASS SCENARIO	LEADERSHIP ACTION
Camila discovers that a team has been bypassing approval steps to speed up delivery. She worries the shortcuts could expose the company to risk.	The leader instructs Camila to document the finding and require immediate process correction. They also schedule a follow-up audit to verify compliance.
Ethan identifies inconsistent documentation in a department's financial records. He suspects the gaps may hide deeper issues.	The leader directs Ethan to escalate the finding to internal audit for deeper review. They also implement temporary controls to prevent further discrepancies.
Zoe uncovers that a supplier was onboarded without proper due diligence. She fears the oversight could violate procurement policy.	The leader logs the finding and pauses all transactions with the supplier. They then initiate a full compliance review of the onboarding process.
Patrick finds that access permissions were granted beyond role requirements. He recognizes the security implications immediately.	The leader orders an urgent permission reset and documents the breach. They also require a full access-control audit across the department.
Alina notices that a team has been using outdated compliance templates. She worries this could invalidate recent filings.	The leader instructs Alina to update the templates and re-submit affected documents. They also implement a version-control system to prevent recurrence.
Stefan discovers that expense approvals were routinely rubber-stamped without review. He realizes this undermines financial integrity.	The leader documents the finding and retrains approvers on required scrutiny. They also introduce random spot-checks to reinforce accountability.

6x6™ Transformation Blueprint

Power Safety

"Trust is the only currency machines cannot mint; humans must earn it."
Satya Nadella, CEO of Microsoft, known for human-centered AI leadership.

Power Safety requires disciplined restraint, and Responsible Authority Use ensures leaders apply their decision-making power with fairness, transparency, and accountability. It establishes boundaries that prevent authority from becoming arbitrary or self serving. This matters because leadership power influences resources, opportunities, and organizational direction. Responsible Authority Use prevents harm by ensuring decisions are grounded in principle rather than personal preference. It also reinforces trust by demonstrating that authority is exercised with care. When ignored, power becomes unpredictable and destabilizing. Teams experience fear, confusion, or disengagement. Over time, Power Safety deteriorates as authority becomes detached from responsibility. Misuse Risk Prevention ensures leaders recognize and mitigate situations where power can be abused or misapplied.

Power is a liability without guardrails. This topic utilizes Decision Impact Assessments and Emergency Power Limit Rules to ensure that authority does not become a bottleneck or a source of operational friction. By physically engaging with "Delegation Matrices," a leader can visualize how power flows through the enterprise. It's about creating a system of "Authority Segregation" that prevents the ego from overriding the clinical diagnostics of the framework. When power is made "safe," it transforms from a tool of control into an engine of empowerment, allowing strategy to be democratized without risking the integrity of the mission. In a print format, you can trace the lines of authority across the page, identifying where power pools or stagnates. This clarity ensures that leadership intent is amplified, not distorted, as it moves through the hierarchy, protecting the organization from the "Kill Zone" of executive overreach or systemic decision paralysis.

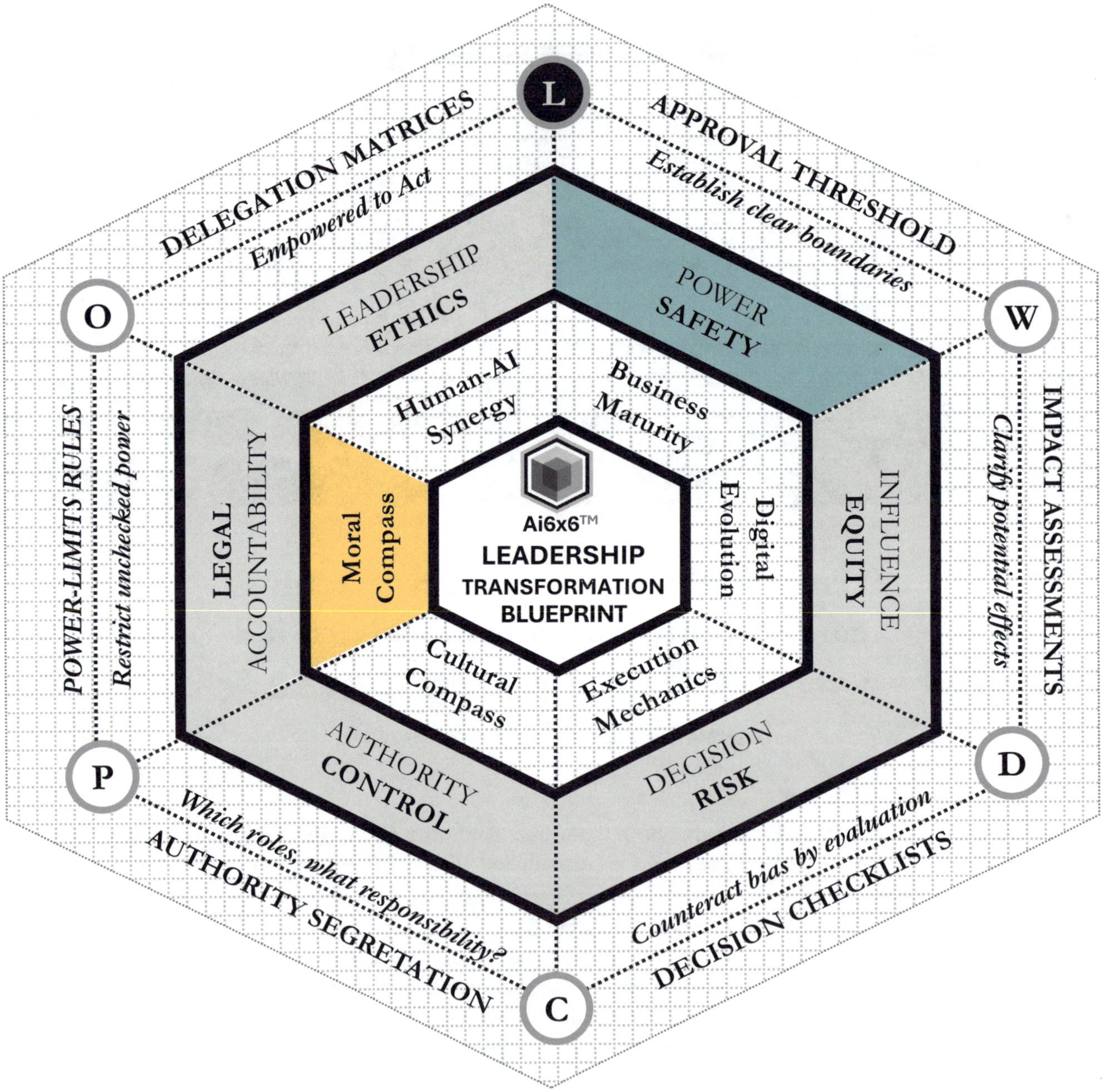

Ai6x6 POWER SAFETY FOR MORAL COMPASS | LEADERSHIP TRANSFORMATION

1. DELEGATION MATRICES

Delegation Matrices define how authority, responsibility, and decision ownership are distributed across leadership levels. They clarify who is empowered to act, who must be consulted, and who remains accountable for outcomes. This structure prevents ambiguity and ensures decisions flow through the right hands. They matter because unclear delegation creates power concentration, operational bottlenecks, and ethical vulnerability. Proper delegation protects against misuse of authority by ensuring decisions are not hoarded or improvised. It also reinforces fairness by aligning responsibility with capability rather than hierarchy alone. When ignored, leaders overreach, teams hesitate, and decisions become inconsistent or delayed. Power becomes unevenly distributed, creating risk and confusion. Over time, the organization experiences operational fragility and moral imbalance.

When done well, delegation matrices create clarity, speed, and balanced authority across the organization. People understand their decision space and act confidently within it. This strengthens execution and reduces the risk of power misuse. Delegation reinforces the principle that leadership requires distributing authority responsibly. It reminds leaders that empowerment is a moral obligation, not a convenience. Ultimately, it builds a system where power is exercised with discipline and transparency.

MORAL COMPASS SCENARIO	LEADERSHIP ACTION
Adrian keeps escalating routine decisions to his director because he's unsure what authority he actually holds. His hesitation slows down the team and frustrates stakeholders.	The leader reviews the delegation matrix with Adrian and clarifies his decision boundaries. They also run a short calibration session with the whole team to reinforce consistent delegation.
Mei approves vendor invoices that technically exceed her authority level, assuming it's harmless because the vendor is longstanding. She doesn't realize this exposes the company to audit risk.	The leader resets Mei's permissions and walks her through the correct delegation tiers. They also update the matrix to ensure invoice thresholds are unmistakably clear.
Carlos delegates a sensitive client decision to a junior analyst who feels unprepared. The analyst makes a rushed call that creates customer confusion.	The leader revisits the delegation matrix with Carlos and highlights which decisions require senior oversight. They also provide the analyst with support and training to rebuild confidence.
Priyanka avoids delegating altogether because she fears losing control of outcomes. Her team becomes dependent and disengaged.	The leader coaches Priyanka on structured delegation using the matrix as a guide. They also assign her a delegation goal for the next quarter to build the habit.
Ben delegates a cross-department decision without consulting the impacted teams. The receiving team pushes back, citing unclear authority.	The leader clarifies cross-functional delegation rules and updates the matrix to reflect shared ownership. They also facilitate a reset meeting to repair trust.
Sofia delegates a time-sensitive approval to someone who is out on leave. The delay causes a missed deadline and financial penalty.	The leader reinforces backup-delegation rules and updates the matrix to include alternates. They also implement a simple escalation protocol for time-critical decisions.

2. APPROVAL THRESHOLD TABLES

Approval Threshold Tables define the levels of authority required to approve decisions of varying scale, risk, or impact. They establish clear boundaries that prevent unilateral action where oversight is necessary. This structure ensures that decisions escalate appropriately based on their significance. They matter because unregulated approval power creates ethical exposure and operational inconsistency. Thresholds protect against impulsive or unauthorized decisions that could harm the organization. They also reinforce accountability by ensuring leaders cannot bypass governance. When ignored, decisions are made at the wrong level, without proper scrutiny or alignment. Teams become uncertain about when to escalate or proceed. Over time, the organization experiences governance gaps that weaken control and trust.

When done well, approval thresholds create predictability, fairness, and disciplined oversight. People understand when decisions require additional review and why. This strengthens governance and reduces the risk of costly errors. Thresholds reinforce the principle that leadership must balance autonomy with accountability. They remind leaders that oversight is a safeguard, not an obstacle. Ultimately, they protect the organization from avoidable risk and reinforce ethical decision making.

MORAL COMPASS SCENARIO	LEADERSHIP ACTION
Malik approves a purchase order that exceeds his financial limit because he assumes the urgency justifies it. He later realizes the system flagged it for audit.	The leader reviews the threshold table with Malik and reinstates proper approval routing. They also add a pre-approval check to prevent similar overrides.
Jenna repeatedly sends small requests to senior leadership even though they fall within her approval band. She fears being blamed if something goes wrong.	The leader walks Jenna through the threshold table and reinforces her authority. They also set up a confidence-building review cycle to help her trust her judgment.
Oliver splits a large expense into smaller chunks to stay under his approval limit. He doesn't realize this violates policy and creates compliance exposure.	The leader documents the incident and retrains Oliver on anti-splitting rules. They also update the threshold table with explicit examples to prevent misinterpretation.
Rina delays a critical vendor payment because she's unsure which threshold applies to international transactions. The vendor threatens to pause service.	The leader clarifies the international threshold rules and updates the table with a dedicated section. They also train the team on cross-border approval nuances.
Theo approves a hiring bonus outside his authority level because he wants to secure a candidate quickly. HR flags the discrepancy during onboarding.	The leader resets Theo's access and reviews the threshold table with him. They also implement a mandatory HR check for all bonus approvals.
Amara escalates every capital request because she doesn't understand the depreciation-based thresholds. Her confusion slows down the capital planning cycle.	The leader provides a walkthrough of the threshold logic and adds examples to the table. They also schedule a finance partner to support Amara during the next cycle.

3. DECISION IMPACT ASSESSMENTS

Decision Impact Assessments evaluate the ethical, operational, and stakeholder consequences of major leadership decisions. They clarify potential risks, benefits, and unintended effects before action is taken. This assessment process ensures decisions are grounded in foresight rather than assumption. They matter because leaders often underestimate downstream impacts, especially under pressure. Assessments prevent harm by revealing vulnerabilities that require mitigation. They also reinforce moral responsibility by ensuring leaders consider who is affected and how. When ignored, decisions create avoidable disruption, inequity, or ethical compromise. Teams are left to manage consequences that were never anticipated. Over time, the organization becomes reactive and loses strategic stability.

When done well, impact assessments create thoughtful, responsible, and well aligned decisions. People trust that leadership has evaluated consequences with seriousness and care. This strengthens confidence in both the process and the outcomes. Assessments reinforce the principle that leadership requires anticipation, not just action. They remind leaders that every decision carries moral weight. Ultimately, they elevate decision quality and protect organizational integrity.

MORAL COMPASS SCENARIO	LEADERSHIP ACTION
Lionel pushes a product change without assessing downstream customer impact. Support tickets spike within 24 hours.	The leader requires Lionel to complete a full impact assessment and present mitigation steps. They also make impact reviews mandatory for all product changes.
Farah approves a staffing reduction without evaluating workload redistribution. The remaining team becomes overwhelmed and misses deadlines.	The leader instructs Farah to run a workload impact assessment and adjust staffing accordingly. They also implement a pre-decision impact checkpoint for all resource changes.
Connor greenlights a new vendor without assessing integration risks. The vendor's system fails to sync with internal tools.	The leader directs Connor to complete a technical impact assessment and coordinate remediation. They also add integration-risk criteria to the assessment template.
Yasmin changes a customer policy without reviewing legal implications. Compliance flags the update as non-compliant.	The leader requires Yasmin to conduct a legal impact assessment and reverse the change. They also add legal review as a mandatory step for policy decisions.
Diego shifts budget from maintenance to innovation without assessing operational risk. Equipment failures increase within weeks.	The leader instructs Diego to complete an operational impact assessment and rebalance the budget. They also require maintenance-risk scoring for all budget shifts.
Helena accelerates a launch timeline without assessing team capacity. Burnout rises and quality drops.	The leader requires Helena to run a capacity impact assessment and adjust the timeline. They also introduce a capacity-risk gate for all accelerated launches.

4. CRITICAL-DECISION CHECKLISTS

Critical Decision Checklists provide structured criteria leaders must review before making high stakes choices. They ensure that essential ethical, operational, and risk considerations are not overlooked. This discipline prevents decisions from being rushed or based on incomplete information. They matter because high pressure environments create cognitive shortcuts that can lead to ethical or strategic failure. Checklists counteract bias by forcing systematic evaluation. They also reinforce consistency across leaders and situations. When ignored, decisions become vulnerable to oversight gaps, emotional impulses, or selective reasoning. Teams experience uneven standards and unpredictable outcomes. Over time, the organization suffers from preventable errors that erode trust and performance.

When done well, critical decision checklists create rigor, stability, and disciplined judgment. People see that leadership decisions follow a structured and reliable process. This strengthens confidence in both the fairness and quality of outcomes. Checklists reinforce the principle that leadership requires methodical thinking under pressure. They remind leaders that discipline is a safeguard against error. Ultimately, they elevate decision reliability and protect organizational resilience.

MORAL COMPASS SCENARIO	LEADERSHIP ACTION
Arjun skips the critical decision checklist because he feels confident in his intuition. The decision backfires and creates rework.	The leader requires Arjun to complete the checklist retroactively and identify missed steps. They also enforce checklist usage for all high-stakes decisions.
Beatrice rushes a vendor selection without completing the risk-review section of the checklist. The vendor fails to meet security requirements.	The leader instructs Beatrice to redo the selection using the full checklist. They also add a mandatory security sign-off step.
Nolan approves a process change without validating stakeholder alignment. Several teams push back, causing delays.	The leader requires Nolan to complete the stakeholder-impact portion of the checklist. They also implement a cross-team review gate for similar decisions.
Kyra bypasses the financial-impact section of the checklist to save time. Finance later identifies unbudgeted costs.	The leader directs Kyra to complete the financial section and adjust the plan. They also reinforce checklist compliance through quarterly audits.
Elias makes a customer-facing decision without reviewing the communication plan. Customers receive inconsistent messages.	The leader instructs Elias to complete the communication-readiness section of the checklist. They also require communication sign-off for all customer-impacting decisions.
Talia approves a system change without validating rollback procedures. When the change fails, recovery takes hours.	The leader requires Talia to complete the technical-risk section of the checklist. They also add rollback validation as a mandatory pre-deployment step.

5. AUTHORITY SEGREGATION CHARTS

Authority Segregation Charts define how decision making power is separated to prevent concentration, conflict, or misuse. They clarify which roles hold initiating, approving, and oversight responsibilities. This separation ensures that no single leader controls an entire decision pathway. They matter because concentrated authority increases ethical risk and reduces transparency. Segregation protects against unilateral decisions that bypass checks and balances. It also reinforces fairness by distributing power across multiple perspectives. When ignored, authority becomes centralized, creating vulnerability to bias, error, or misconduct. Teams lose confidence in the integrity of decisions. Over time, the organization experiences governance failures that compromise ethical and operational stability.

When done well, authority segregation creates balance, transparency, and shared accountability. People trust that decisions are reviewed and validated through multiple roles. This strengthens governance and reduces the risk of power abuse. Segregation reinforces the principle that leadership must operate within structured limits. It reminds leaders that shared authority is a moral safeguard. Ultimately, it builds a system where power is exercised responsibly and collaboratively.

MORAL COMPASS SCENARIO	LEADERSHIP ACTION
Rowan approves both the creation and payment of a vendor invoice, unaware this violates segregation rules. Finance flags the conflict.	The leader resets Rowan's permissions and reviews the segregation chart with him. They also implement automated separation controls.
Elise handles both access provisioning and access reviews for her team. She doesn't realize this creates a security gap.	The leader reassigns one of the responsibilities and updates the segregation chart. They also schedule quarterly access audits.
Mateo manages both procurement and contract negotiation for a major supplier. The dual role raises risk concerns.	The leader splits the responsibilities and documents the change in the segregation chart. They also establish a dual-review process for supplier contracts.
Jada approves her own travel expenses due to a system oversight. The issue goes unnoticed for months.	The leader corrects the workflow and updates the segregation chart to prevent self-approval. They also run a retroactive audit of Jada's expenses.
Finn oversees both quality assurance and production scheduling. Conflicts arise when deadlines pressure quality standards.	The leader separates the roles and updates the segregation chart accordingly. They also create a joint review forum to balance quality and speed.
Mira is responsible for both risk assessment and risk acceptance decisions. This concentration of authority concerns compliance.	The leader divides the responsibilities and documents the change. They also implement a risk-acceptance committee for oversight.

6. EMERGENCY POWER-LIMIT RULES

Emergency Power Limit Rules define the boundaries of authority leaders may exercise during crises or urgent situations. They clarify what actions are permitted, what requires oversight, and what must be documented. This structure prevents crisis conditions from becoming a justification for unchecked power. They matter because emergencies create pressure that can distort judgment and weaken ethical discipline. Clear limits protect against overreach and ensure decisions remain aligned with values even under stress. They also reinforce accountability by requiring transparency during exceptional circumstances. When ignored, leaders may take actions that exceed their mandate or bypass governance. Teams become uncertain about what is legitimate or necessary. Over time, emergency behavior becomes normalized, eroding ethical and operational standards.

When done well, emergency power limits create stability, trust, and disciplined crisis response. People understand that even urgent decisions follow principled boundaries. This strengthens confidence in leadership during high pressure moments. Power limits reinforce the principle that ethics must hold firm when conditions are most challenging. They remind leaders that crisis authority is temporary, constrained, and accountable. Ultimately, they protect the organization from long term harm caused by short term overreach.

MORAL COMPASS SCENARIO	LEADERSHIP ACTION
Declan overrides standard approval rules during a system outage but continues using emergency powers after the issue is resolved. His team grows uneasy about the extended authority.	The leader reinstates normal approval rules and documents the misuse. They also retrain Declan on the time-bound nature of emergency powers.
Marisol makes rapid staffing changes during a crisis without notifying HR. The changes create confusion once operations stabilize.	The leader reviews emergency power limits with Marisol and reverses unauthorized changes. They also implement a crisis-communication protocol.
Theo suspends vendor controls during a supply disruption but forgets to restore them. The lapse exposes the company to procurement risk.	The leader restores all controls and logs the oversight. They also add a post-crisis reset checklist to prevent recurrence.
Ivana reallocates budget during an emergency but exceeds the temporary spending cap. Finance flags the overage.	The leader corrects the budget and reviews the emergency spending rules with Ivana. They also implement automated alerts for cap breaches.
Jasper bypasses security protocols during a cyber incident but continues using the shortcuts afterward. The security team raises concerns.	The leader reinstates full security protocols and documents the deviation. They also require Jasper to complete a post-incident compliance review.
Lior issues operational directives outside his authority during a facility shutdown. Teams remain confused about which rules still apply.	The leader clarifies the emergency authority boundaries and resets all directives. They also conduct a debrief to strengthen future crisis governance.

6x6™ Transformation Blueprint

Influence Equity

"Sustainability in AI means building systems that future generations will thank us for, not fear."

Jane Goodall, world-renowned primatologist and environmental advocate.

Influence Equity requires balanced leadership impact, and Fair Influence Distribution ensures that no single individual or group disproportionately shapes decisions or outcomes. It establishes norms that prevent dominance, favoritism, or exclusion. This matters because influence determines whose ideas are heard and whose needs are prioritized. Fair Influence Distribution prevents inequity by ensuring all voices have legitimate access to leadership attention. It also reinforces trust by demonstrating that influence is earned through merit, not proximity or politics. When ignored, influence becomes concentrated and distorts organizational judgment. Teams experience disengagement as their perspectives are marginalized. Over time, Influence Equity deteriorates as decision making becomes biased and unrepresentative. Unbiased Leadership Practices ensure leaders evaluate input based on relevance and quality rather than personal preference.

Strategy fails when it is born in a vacuum of "Perception Gaps." By tracing Bias Adjusted Rating Sheets and Inclusive Meeting Rotations, leaders physically see how inclusive cultures prevent the narrow thinking that leads to market failure. This isn't a "soft" metric; it is a clinical requirement for "Calibration." When a leader ensures "Sponsorship Assignment Logs" are balanced, they are diversifying the organization's cognitive portfolio. This topic teaches that the most effective way to secure the future is to ensure that every strategic voice is heard, analyzed, and integrated into the overarching blueprint of the firm. Within the print series, you can flip between representation dashboards and equity audit reports, engrossing yourself in the data that exposes hidden silos. Influence equity ensures that your organization is hunting for truth, not just consensus, making your strategic engine more robust and better prepared to exploit the high-growth frontiers of the AI economy.

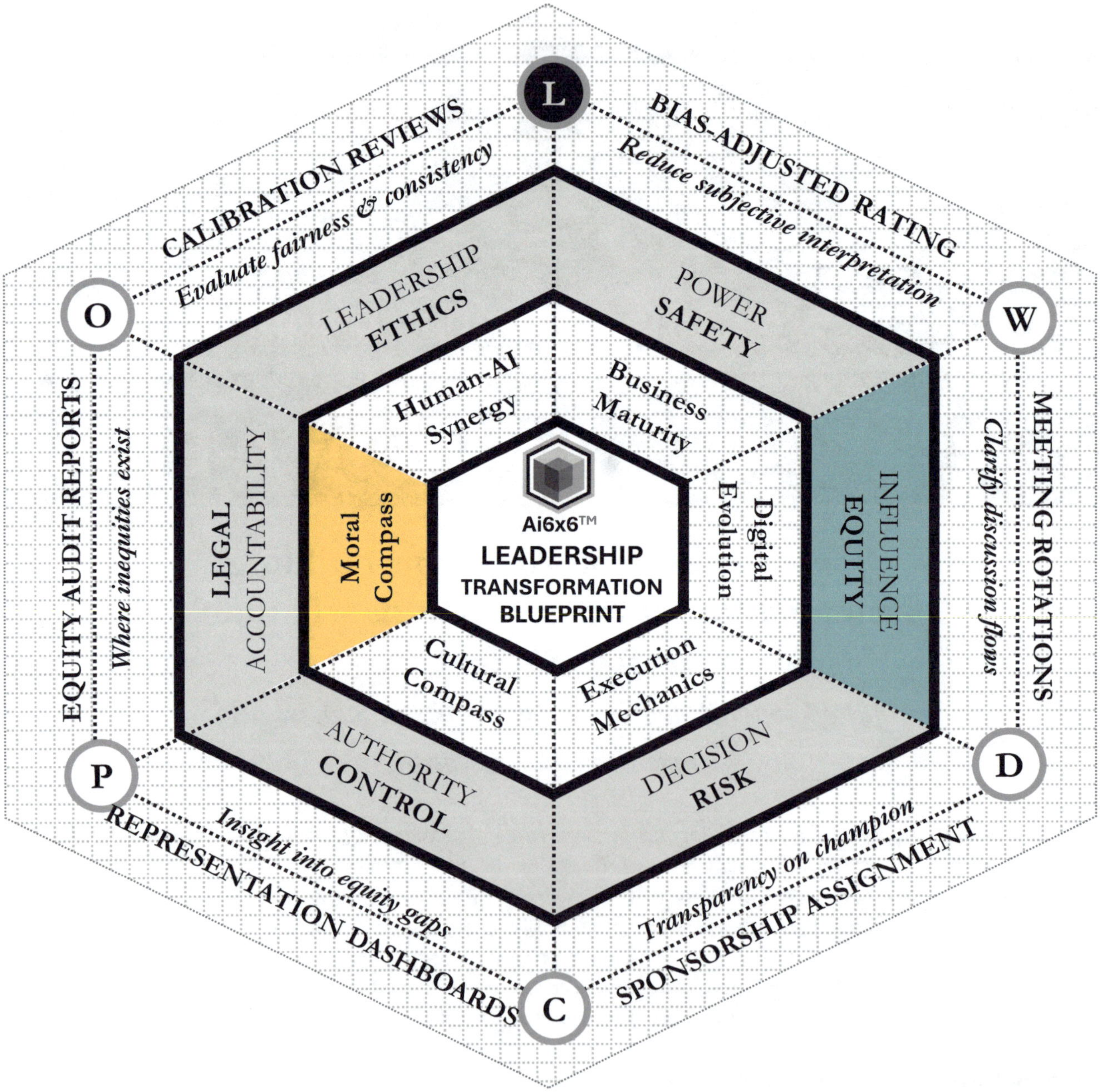

Ai6x6 INFLUENCE EQUITY FOR MORAL COMPASS | LEADERSHIP TRANSFORMATION

1. CALIBRATION COMMITTEE REVIEWS

Calibration Committee Reviews evaluate performance ratings, promotion decisions, and leadership judgments across teams to ensure fairness and consistency. They create a structured forum where multiple leaders compare assessments, challenge assumptions, and correct bias. This process ensures that individual preferences do not distort organizational outcomes. They matter because performance evaluations shape careers, compensation, and access to opportunity. Reviews prevent inequities that arise when leaders apply standards inconsistently or rely on subjective impressions. They also reinforce transparency by making evaluation logic visible and accountable. When ignored, ratings drift, favoritism grows, and disparities widen across teams. Employees lose trust in the fairness of advancement decisions. Over time, the organization experiences systemic inequity that undermines morale and retention.

When done well, calibration reviews create fairness, clarity, and shared accountability in talent decisions. People trust that evaluations reflect consistent standards rather than individual bias. This strengthens confidence in leadership and improves the credibility of performance systems. Calibration reinforces the principle that equity requires collective oversight. It reminds leaders that fairness is a discipline, not an assumption. Ultimately, it builds a culture where opportunity is earned through merit, not proximity or preference.

MORAL COMPASS SCENARIO	LEADERSHIP ACTION
Loraine notices that two managers consistently inflate ratings for their own teams during calibration. She feels uneasy because the pattern disadvantages quieter departments.	The leader reviews historical calibration data and confronts the inflation directly with the managers. They also implement cross-functional calibration pairings to neutralize bias.
Darius realizes his peers are pushing through ratings without discussing discrepancies. He worries the committee is becoming a rubber-stamp exercise.	The leader pauses the session and requires each rating variance to be justified with evidence. They also introduce a structured debate protocol for contested ratings.
Keiko observes that remote employees are rarely discussed in calibration sessions. She suspects proximity bias is influencing outcomes.	The leader mandates equal review time for remote staff and adds remote-visibility prompts to the calibration agenda. They also require managers to submit pre-reads for all employees.
Stefan sees that one manager defends every rating aggressively, shutting down conversation. The committee becomes reluctant to challenge him.	The leader intervenes and enforces equal speaking time rules. They also coach the manager on collaborative calibration behavior.
Amira notices that high-potential employees from smaller teams are overlooked because their managers are less vocal. She worries talent is being missed.	The leader introduces a rotation where each manager must present another team's top performers. They also add a "silent review" step where ratings are evaluated before discussion.
Hugo realizes the committee is relying heavily on tenure rather than performance evidence. He fears this will demotivate newer high performers.	The leader requires evidence-based justification for all top ratings. They also add a calibration checkpoint comparing performance outcomes across tenure bands.

Bias Adjusted Rating Sheets provide structured criteria that reduce subjective interpretation in performance evaluations. They clarify expectations, define rating anchors, and require evidence-based assessment. This structure ensures that evaluations reflect actual performance rather than personal affinity or unconscious bias. They matter because unstructured ratings amplify disparities and distort talent decisions. Adjusted sheets counteract bias by forcing leaders to justify ratings with observable behaviors. They also reinforce fairness by standardizing how performance is measured across teams. When ignored, evaluations become inconsistent, arbitrary, or influenced by personal relationships. Employees experience inequitable outcomes that erode trust. Over time, the organization embeds systemic bias into its talent processes.

When done well, bias adjusted rating sheets create transparency, consistency, and equitable evaluation. People understand how their performance is assessed and what evidence supports the rating. This strengthens confidence in the fairness of the system and reduces disputes. Rating sheets reinforce the principle that equity requires structured, evidence-based judgment. They remind leaders that fairness must be engineered, not assumed. Ultimately, they elevate the integrity of performance management.

MORAL COMPASS SCENARIO	LEADERSHIP ACTION
Talia notices she consistently rates extroverted employees higher because they speak up more. She realizes her rating sheet doesn't account for communication-style bias.	The leader helps Talia adjust her rating criteria to separate visibility from impact. They also train the team on identifying behavioral bias patterns.
Ron discovers he has been underrating employees who challenge him intellectually. He recognizes this as a discomfort-based bias.	The leader guides Ron through a bias-adjustment review and recalibrates the affected ratings. They also introduce a challenge-positive rubric to normalize constructive dissent.
Bianca realizes she scores employees with similar backgrounds to her more favorably. She feels embarrassed but wants to correct it.	The leader walks Bianca through a structured bias-adjustment process and anonymizes the next rating cycle. They also add a peer-review step for her evaluations.
Idris tends to give higher ratings to employees who work longer hours, even when outcomes are similar. He recognizes this as effort-visibility bias.	The leader helps Idris shift to outcome-based scoring and adjust his rating sheet accordingly. They also add a metric that distinguishes effort from impact.
Carmen notices she penalizes employees who struggle with English fluency, even when their work quality is strong. She wants to correct this linguistic bias.	The leader revises Carmen's rating sheet to separate communication clarity from performance. They also provide guidance on evaluating multilingual employees fairly.
Wesley realizes he consistently underrates employees who prefer asynchronous communication. He didn't realize his bias toward synchronous engagement.	The leader adjusts Wesley's rating sheet to include multiple communication modes. They also require him to review asynchronous contributions before final scoring.

3. INCLUSIVE MEETING ROTATIONS

Inclusive Meeting Rotations ensure that speaking time, agenda ownership, and decision influence are distributed equitably across participants. They clarify who leads discussions, who presents updates, and who contributes to key decisions. This structure prevents dominant voices from overshadowing others. They matter because meetings are where visibility, influence, and leadership opportunities are created. Rotations prevent exclusionary patterns that limit participation to a select few. They also reinforce fairness by ensuring diverse perspectives shape decisions. When ignored, meetings become hierarchical, repetitive, and unrepresentative. Certain voices dominate while others withdraw or are overlooked. Over time, the organization loses innovation, engagement, and trust.

When done well, inclusive rotations create balanced participation, shared ownership, and broader leadership development. People feel seen, heard, and valued in decision spaces. This strengthens collaboration and improves decision quality. Rotations reinforce the principle that influence must be distributed, not concentrated. They remind leaders that inclusion is a structural practice, not a sentiment. Ultimately, they build a culture where participation is equitable, and leadership pathways are accessible.

MORAL COMPASS SCENARIO	LEADERSHIP ACTION
Nia notices that the same three people dominate every meeting agenda. She feels newer team members aren't getting visibility.	The leader implements a rotating agenda-owner schedule to distribute influence. They also coach senior members to create space for emerging voices.
Omar realizes that meetings are always scheduled in time zones that disadvantage international colleagues. Attendance from overseas teams keeps dropping.	The leader introduces a rotating time-zone schedule to balance participation. They also assign a meeting facilitator to ensure global voices are heard.
Felicity sees that introverted team members rarely speak during discussions. She worries their ideas are being lost.	The leader adds structured round-robin input moments to every meeting. They also allow pre-submitted contributions for those who prefer written input.
Joon notices that women on the team are interrupted more frequently. He feels the meeting culture is unintentionally exclusionary.	The leader establishes an interruption-tracking protocol and enforces equal speaking time. They also train the team on inclusive facilitation norms.
Amina observes that remote employees are consistently overlooked during brainstorming. The in-room group dominates the conversation.	The leader requires hybrid-first facilitation, giving remote participants the first speaking slot. They also introduce digital whiteboards to equalize contribution channels.
Caleb realizes that junior employees are never asked to present updates. He worries this limits their development.	The leader creates a rotation where junior staff present at least once per quarter. They also pair them with mentors to prepare effectively.

4. SPONSORSHIP ASSIGNMENT LOGS

Sponsorship Assignment Logs track which employees receive advocacy, visibility, and advancement support from senior leaders. They create transparency around who is being championed and who may be overlooked. This documentation ensures that sponsorship is intentional rather than driven by familiarity or bias. They matter because sponsorship is one of the strongest predictors of career acceleration. Logs prevent inequities that arise when leaders sponsor only those who resemble them or are already visible. They also reinforce fairness by making sponsorship decisions reviewable and accountable. When ignored, sponsorship becomes uneven, opaque, and exclusionary. High potential employees without access to informal networks are left behind. Over time, the organization develops unequal opportunity pipelines.

When done well, sponsorship logs create equity, visibility, and balanced talent development. People trust that advancement support is distributed based on merit and potential. This strengthens retention and broadens the leadership pipeline. Sponsorship tracking reinforces the principle that opportunity must be intentionally cultivated. It reminds leaders that advocacy is a responsibility, not a privilege. Ultimately, it builds a system where talent mobility is fair and transparent.

MORAL COMPASS SCENARIO	LEADERSHIP ACTION
Lior notices that sponsorship opportunities are going to the same high-visibility employees. He worries quieter high-potential staff are being overlooked.	The leader reviews the sponsorship log and redistributes opportunities based on potential rather than visibility. They also require managers to nominate at least one emerging talent each cycle.
Daphne realizes she has been sponsoring employees who share her communication style. She didn't notice the pattern until reviewing the log.	The leader helps Daphne diversify her sponsorship portfolio and set explicit selection criteria. They also add a quarterly audit to detect bias patterns.
Mateo sees that sponsorship assignments heavily favor employees from one department. He suspects informal networks are influencing decisions.	The leader rebalances assignments across departments and documents the rationale. They also introduce a cross-department nomination process.
Giselle notices that employees returning from leave are rarely assigned sponsors. She worries they're missing critical career support.	The leader updates the sponsorship log to include return-from-leave triggers. They also assign sponsors proactively to reintegrating employees.
Rowan realizes that sponsorship assignments are not being tracked consistently. Some managers are assigning sponsors informally without documentation.	The leader mandates centralized logging and audits all informal assignments. They also train managers on the importance of transparent sponsorship tracking.
Helena sees that sponsorship opportunities are disproportionately given to employees already on leadership tracks. She fears this reinforces inequity.	The leader expands the sponsorship pool to include overlooked talent segments. They also introduce a rotation system to ensure broader access.

Representation Dashboards visualize demographic distribution, advancement patterns, and leadership composition across the organization. They provide leaders with real time insight into equity gaps and progress. This visibility ensures that representation is monitored, measured, and managed. They matter because organizations cannot improve what they do not track. Dashboards reveal disparities that may be invisible in day to day operations. They also reinforce accountability by making equity outcomes transparent. When ignored, representation stagnates or declines without leadership awareness. Biases remain unchallenged and systemic barriers persist. Over time, the organization loses credibility with employees and stakeholders.

When done well, representation dashboards create clarity, accountability, and data driven equity improvement. People see that leadership monitors representation with seriousness and transparency. This strengthens trust and motivates action. Dashboards reinforce the principle that equity requires measurement and follow through. They remind leaders that representation is a strategic indicator, not a symbolic metric. Ultimately, they help build a workforce that reflects the organization's values and commitments.

MORAL COMPASS SCENARIO	LEADERSHIP ACTION
Priya notices the dashboard shows a steep drop-off in women at mid-management levels. She realizes the issue has gone unaddressed for years.	The leader launches a root-cause analysis and sets representation targets for mid-level roles. They also assign accountability to department heads.
Andre sees that certain ethnic groups are underrepresented in customer-facing roles. He suspects biased assumptions about "fit."	The leader reviews hiring patterns and retrains managers on inclusive selection. They also update the dashboard to track representation by role type.
Lila notices that employees with disabilities are nearly absent from leadership roles. She worries accessibility barriers are being overlooked.	The leader conducts an accessibility audit of leadership pathways. They also add disability representation metrics to the dashboard.
Marcus sees that attrition rates are higher for first-generation professionals. He suspects onboarding gaps.	The leader initiates a targeted retention program and updates the dashboard to track first-generation employee outcomes. They also assign mentors to new hires.
Zahra realizes that representation in technical roles is stagnant despite diverse applicant pools. She suspects screening bias.	The leader reviews screening criteria and implements structured interviews. They also track funnel-stage representation on the dashboard.
Theo notices that promotions disproportionately favor employees from one region. He worries regional bias is influencing decisions.	The leader standardizes promotion criteria across regions. They also add regional equity indicators to the dashboard.

6. EQUITY AUDIT REPORTS

Equity Audit Reports evaluate systems, processes, and outcomes to identify where inequities exist and why. They analyze patterns across hiring, promotion, compensation, and leadership decisions. This audit process ensures that equity is treated as a measurable organizational responsibility. They matter because inequities often arise from structural factors rather than individual intent. Audits reveal root causes that require systemic correction. They also reinforce transparency by documenting findings and recommended actions. When ignored, inequities persist and become embedded in organizational norms. Employees lose trust in leadership's commitment to fairness. Over time, disparities widen and damage culture, performance, and reputation.

When done well, equity audits create insight, accountability, and targeted improvement. People see that leadership is willing to confront uncomfortable truths and act on them. This strengthens confidence in the organization's fairness and integrity. Audits reinforce the principle that equity requires structural change, not symbolic gestures. They remind leaders that fairness must be continuously evaluated and improved. Ultimately, they build a culture where equity is operational, measurable, and sustained.

MORAL COMPASS SCENARIO	LEADERSHIP ACTION
Simone discovers pay disparities between employees with identical roles and tenure. She feels the inequity is long overdue for correction.	The leader initiates a pay-equity remediation plan and communicates the changes transparently. They also schedule annual equity audits.
Victor finds that training opportunities are disproportionately allocated to senior staff. He worries junior employees are being left behind.	The leader rebalances training access and updates the audit criteria. They also introduce a quarterly training-equity review.
Hana sees that performance ratings skew higher for employees in headquarters compared to satellite offices. She suspects visibility bias.	The leader recalibrates ratings across locations and updates the audit to include location-based analysis. They also train managers on remote-equity practices.
Elias discovers that project leadership roles are rarely given to part-time employees. He realizes this limits career progression.	The leader revises project-assignment criteria and includes part-time eligibility. They also track assignment equity in future audits.
Mireya notices that employees from non-traditional educational backgrounds are underrepresented in promotions. She suspects credential bias.	The leader updates promotion criteria to emphasize skills over degrees. They also add educational-background equity metrics to the audit.
Jonah finds that exit interviews show consistent themes of cultural exclusion for minority groups. He feels the issue has been ignored.	The leader launches a culture-equity initiative and integrates exit-interview data into the audit. They also assign executive sponsors to drive change.

6x6™ Transformation Blueprint

Decision Risk

"Morality in AI is not optional — it is the price of admission to the future."
Barack Obama, 44th U.S. President, known for global policy leadership.

Decision Risk requires disciplined foresight, and Consequence Focused Thinking ensures leaders evaluate the downstream impact of their choices before acting. It establishes a mindset that prioritizes long term stability over short term convenience. This matters because leadership decisions shape operational, financial, and cultural outcomes. Consequence Focused Thinking prevents harm by forcing leaders to consider unintended effects. It also reinforces accountability by linking decisions to their potential risks. When ignored, decisions become impulsive and destabilizing. Teams experience confusion as outcomes become unpredictable. Over time, Decision Risk increases as choices are made without adequate reflection. Structured Risk Evaluation ensures leaders assess threats systematically rather than relying on intuition alone.

This is about "Pre-Mortem" survival. Using Uncertainty Heatmaps and Scenario Stress Test Packs on a printed page allows a leader to see the "Kill Zones" in their logic before the market exploits them. By engrossing oneself in "Failure Mode and Effects Analyses," you move from legacy intuition to evidence-based certainty. It requires the leader to confront the possibility of failure in a controlled, clinical setting. Mapping "Escalation Criteria" ensures that when uncertainty strikes, the organization has a pre-validated protocol to follow, replacing high-stakes gambling with the cold, calculated precision of a diagnostic expert. The print format is essential here; it allows you to look at the entire risk landscape at once, seeing how a failure in one lens ripples across the others. This "Predator-Style" diagnostic replaces the "million-dollar guess" with a roadmap for resilience, ensuring your organization remains antifragile in the face of machine-paced market volatility.

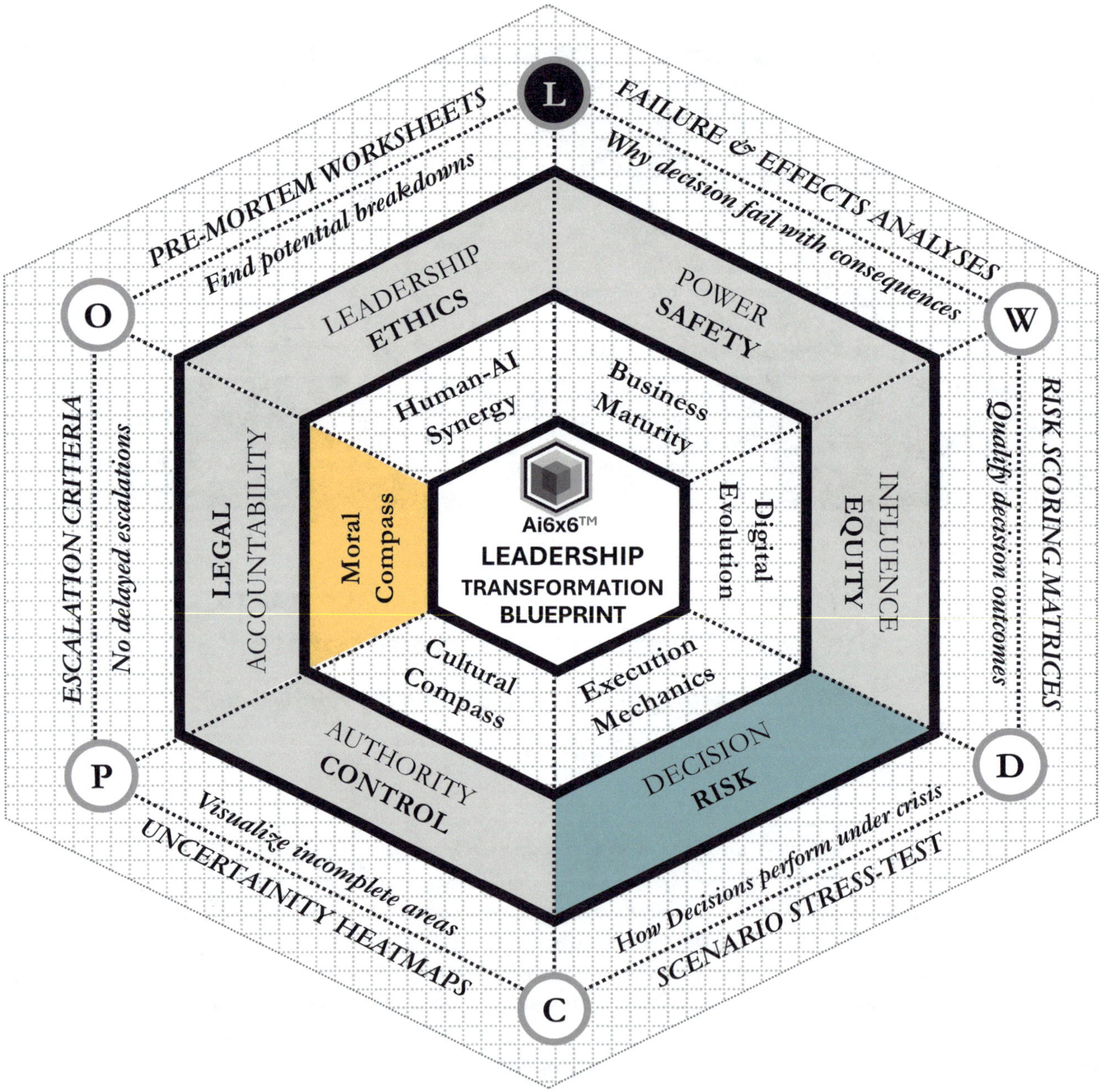

Ai6x6 DECISION RISK FOR MORAL COMPASS | LEADERSHIP TRANSFORMATION

1. PRE-MORTEM WORKSHEETS

Pre-Mortem Worksheets guide leaders through structured anticipation of failure before decisions are executed. They clarify potential breakdowns, overlooked assumptions, and hidden vulnerabilities. This discipline ensures that leaders confront risks proactively rather than reactively. They matter because most decision failures stem from blind spots, optimism bias, or unchallenged assumptions. Worksheets force leaders to imagine negative outcomes and identify their causes. They also reinforce accountability by requiring leaders to consider consequences before committing to action. When ignored, decisions proceed without adequate risk awareness or mitigation. Teams are left unprepared for foreseeable challenges. Over time, the organization becomes reactive, fragile, and prone to repeated mistakes.

When done well, pre mortems create foresight, resilience, and disciplined planning. People trust that leadership has evaluated risks with seriousness and depth. This strengthens confidence in decision quality and reduces the likelihood of avoidable failures. Pre mortems reinforce the principle that leadership requires confronting uncomfortable possibilities. They remind leaders that preparedness is a moral responsibility. Ultimately, they elevate decision reliability and organizational stability.

MORAL COMPASS SCENARIO	LEADERSHIP ACTION
Sahana leads a product launch and realizes her team is overly optimistic, ignoring early warning signs. She senses they need a structured way to imagine failure before it happens.	The leader has Sahana run a pre-mortem session where each member lists plausible failure points. They then convert the top risks into mitigation tasks with owners and deadlines.
Greg notices his team keeps repeating the same mistakes in new initiatives. He realizes they never pause to anticipate what could go wrong.	The leader instructs Greg to complete a pre-mortem worksheet before kickoff. They also require a follow-up review to ensure identified risks are addressed.
Ivette is preparing a major client proposal but feels uneasy about hidden dependencies. She wants a way to surface blind spots.	The leader guides Ivette through a pre-mortem exercise to map potential breakdowns. They then assign cross-functional reviewers to validate assumptions.
Omar's team is rushing a system migration and dismissing concerns about data loss. He fears they're underestimating the risk.	The leader requires a pre-mortem worksheet focused specifically on data-integrity failure modes. They also mandate a rollback plan before migration proceeds.
Lacey is launching a new workflow but senses her team is afraid to voice concerns. She wants a structured way to surface unspoken risks.	The leader facilitates an anonymous pre-mortem input round. They then consolidate the risks and assign owners to each mitigation.
Viktor is planning a high-visibility event and worries about logistical failures. He needs a way to anticipate issues before they occur.	The leader has Viktor complete a pre-mortem worksheet covering venue, staffing, and technology risks. They also schedule a rehearsal to test the highest-risk elements.

2. FAILURE-MODE-AND-EFFECTS ANALYSES

Failure Mode and Effects Analyses identify how a decision, process, or system could fail and what the consequences would be. They break down components, evaluate failure points, and assess severity, likelihood, and detectability. This structured analysis ensures that leaders understand the full risk landscape before acting. They matter because complex decisions often contain hidden interdependencies that amplify risk. Analyses reveal vulnerabilities that require redesign, mitigation, or monitoring. They also reinforce ethical responsibility by ensuring leaders do not overlook foreseeable harm. When ignored, failures emerge unexpectedly and cascade across teams or systems. Leaders are forced into crisis response rather than prevention. Over time, the organization suffers from repeated disruptions that erode trust and performance.

When done well, failure analyses create clarity, preparedness, and operational resilience. People see that leadership evaluates risk with rigor and precision. This strengthens confidence in both the process and the outcomes. Analyses reinforce the principle that leadership must anticipate failure, not merely react to it. They remind leaders that risk awareness is a cornerstone of ethical decision making. Ultimately, they protect the organization from preventable harm.

MORAL COMPASS SCENARIO	LEADERSHIP ACTION
Daniela notices recurring defects in a manufacturing line but can't pinpoint the root cause. She needs a structured way to break down failure points.	The leader instructs Daniela to run an FMEA with engineering and quality teams. They then prioritize the highest-severity failure modes for immediate correction.
Hugo sees that a new software feature keeps crashing under certain conditions. He suspects multiple interacting failure points.	The leader guides Hugo through an FMEA to map each failure mode and its impact. They also assign cross-functional owners to address the top risks.
Priyanka is concerned that a new onboarding process is producing inconsistent outcomes. She wants to understand where breakdowns occur.	The leader has Priyanka conduct an FMEA on each step of the process. They then redesign the highest-risk steps to reduce variability.
Lionel notices that customer complaints spike after a workflow change. He suspects a hidden failure mode in the new process.	The leader directs Lionel to run an FMEA with customer support and operations. They also implement monitoring for the top-risk failure points.
Serena sees that a supply chain partner is repeatedly missing deadlines. She wants to understand the systemic risks behind the failures.	The leader instructs Serena to complete an FMEA on the partner's process. They then escalate the findings to procurement for corrective action.
Mateo realizes that a new compliance requirement could fail at multiple handoff points. He needs a structured way to assess risk.	The leader has Mateo run an FMEA across all compliance touchpoints. They also implement controls for the highest-impact failure modes.

Risk Scoring Matrices quantify the severity, likelihood, and exposure of potential decision outcomes. They create a standardized framework for comparing risks across scenarios. This structure ensures that leaders prioritize attention based on objective criteria rather than intuition. They matter because subjective risk assessment leads to inconsistent decisions and misallocated resources. Matrices reveal which risks require immediate mitigation and which can be monitored. They also reinforce fairness by ensuring decisions are not influenced by personal bias or emotional reactions. When ignored, leaders underestimate critical risks or overreact to minor ones. Teams become confused about priorities and resource allocation. Over time, the organization experiences uneven risk management that weakens stability.

When done well, risk matrices create transparency, alignment, and disciplined prioritization. People understand how risks are evaluated and why certain actions are taken. This strengthens confidence in leadership judgment and resource decisions. Matrices reinforce the principle that risk must be assessed systematically. They remind leaders that objectivity is essential to responsible decision making. Ultimately, they improve organizational resilience and strategic clarity.

MORAL COMPASS SCENARIO	LEADERSHIP ACTION
Aaliyah is evaluating a new vendor but struggles to compare financial risk versus operational risk. She needs a consistent scoring method.	The leader walks Aaliyah through the risk scoring matrix and helps her assign probability and impact values. They also require her to document the final score for audit purposes.
Rowan is assessing a project delay but keeps relying on gut instinct. He realizes he needs a structured way to quantify risk.	The leader instructs Rowan to use the scoring matrix to evaluate severity and likelihood. They then review the score together to determine escalation.
Elise is reviewing cybersecurity threats but finds the list overwhelming. She needs a way to prioritize which risks matter most.	The leader has Elise score each threat using the matrix and rank them by total risk value. They also assign mitigation owners for the top tier.
Dante is planning a facility expansion but is unsure how to weigh environmental risks. He wants a more objective approach.	The leader guides Dante through applying the matrix to environmental, financial, and operational risks. They then require a consolidated risk report.
Maren is evaluating a new partnership but struggles to compare reputational risk with compliance risk. She needs a unified framework.	The leader instructs Maren to score each risk category using the matrix. They also require a cross-functional review before approval.
Felix is reviewing a backlog of unresolved risks and can't determine which ones require immediate action. He needs a prioritization method.	The leader has Felix score all open risks and categorize them by urgency. They then assign deadlines for the highest-scoring items.

4. SCENARIO STRESS-TEST PACKS

Scenario Stress Test Packs evaluate how decisions perform under extreme, unexpected, or high pressure conditions. They simulate adverse events, resource constraints, and rapid change environments. This testing ensures leaders understand how decisions hold up when reality deviates from expectations. They matter because most failures occur when conditions shift suddenly or assumptions break. Stress tests reveal weaknesses that require redesign or contingency planning. They also reinforce ethical responsibility by ensuring leaders prepare for worst case impacts. When ignored, decisions collapse under pressure and expose the organization to avoidable harm. Teams are forced into emergency improvisation without preparation. Over time, the organization becomes brittle and vulnerable to shocks.

When done well, stress tests create resilience, adaptability, and strategic foresight. People trust that leadership has evaluated decisions under realistic and challenging conditions. This strengthens confidence in long term stability and preparedness. Stress testing reinforces the principle that leadership must plan for volatility, not just ideal scenarios. It reminds leaders that resilience is built through deliberate pressure testing. Ultimately, it protects the organization from catastrophic failure.

MORAL COMPASS SCENARIO	LEADERSHIP ACTION
Naomi is preparing for a major system upgrade but hasn't tested how the team would respond to a failure. She worries they're unprepared for worst-case scenarios.	The leader instructs Naomi to run a stress-test simulation using the scenario pack. They then debrief the team and assign improvements based on gaps.
Jasper is launching a new service but hasn't tested how it performs under peak demand. He fears the system may buckle.	The leader requires Jasper to run a load-stress scenario and document failure points. They also mandate capacity adjustments before launch.
Rhea is planning a cross-department initiative but hasn't tested communication breakdown scenarios. She suspects coordination risks.	The leader has Rhea run a communication-failure stress test. They then refine escalation paths based on the results.
Omar is preparing for a regulatory audit but hasn't tested how the team would respond to unexpected document requests. He feels exposed.	The leader instructs Omar to run a compliance-stress scenario. They also create a rapid-response protocol for future audits.
Tessa is designing a new workflow but hasn't tested how it behaves when key staff are unavailable. She worries about single-point failures.	The leader has Tessa run a staffing-gap stress test. They then implement cross-training to reduce dependency risks.
Calvin is planning a product launch but hasn't tested how the team handles simultaneous failures. He wants to understand compounding risks.	The leader requires Calvin to run a multi-failure stress scenario. They then update the launch plan with layered contingencies.

5. UNCERTAINTY HEATMAPS

Uncertainty Heatmaps visualize areas where information is incomplete, assumptions are weak, or outcomes are unpredictable. They highlight zones of ambiguity that require caution, monitoring, or further analysis. This visibility ensures leaders do not mistake uncertainty for safety. They matter because decisions often fail not from known risks but from unknown or unexamined variables. Heatmaps reveal where leaders must slow down, seek more data, or build contingency plans. They also reinforce humility by acknowledging the limits of knowledge. When ignored, leaders proceed with false confidence and overlook critical unknowns. Teams are blindsided by developments that could have been anticipated. Over time, the organization becomes vulnerable to surprises that undermine strategy.

When done well, uncertainty heatmaps create awareness, caution, and informed decision pacing. People understand where ambiguity exists and how it affects choices. This strengthens alignment and reduces the likelihood of overconfident decisions. Heatmaps reinforce the principle that leadership requires acknowledging uncertainty, not denying it. They remind leaders that prudence is a moral discipline. Ultimately, they improve decision quality and organizational adaptability.

MORAL COMPASS SCENARIO	LEADERSHIP ACTION
Lila is planning a strategic initiative but realizes she can't quantify which assumptions are most fragile. She needs a visual way to map uncertainty.	The leader guides Lila through building an uncertainty heatmap and plotting each assumption. They then prioritize validation for the highest-uncertainty areas.
Jonah is evaluating a new market but feels overwhelmed by unknowns. He wants a structured way to identify where the biggest risks lie.	The leader instructs Jonah to create a heatmap of market uncertainties. They also assign research tasks to reduce the highest-risk unknowns.
Priya is preparing a budget forecast but struggles with volatile cost inputs. She needs a way to visualize uncertainty levels.	The leader has Priya map cost drivers on an uncertainty heatmap. They then adjust the forecast to include contingency buffers.
Marcus is designing a new workflow but isn't sure which steps carry the most unpredictability. He wants clarity before rollout.	The leader instructs Marcus to build a process-uncertainty heatmap. They then refine the workflow to stabilize high-uncertainty steps.
Elena is evaluating a partnership but can't determine which assumptions require deeper validation. She needs a prioritization tool.	The leader has Elena map partnership assumptions on a heatmap. They then assign due-diligence tasks for the highest-uncertainty items.
Theo is planning a product roadmap but feels uncertain about customer adoption patterns. He wants a way to visualize risk.	The leader instructs Theo to build a customer-uncertainty heatmap. They then schedule customer-validation sessions for the riskiest assumptions.

6. ESCALATION CRITERIA TABLES

Escalation Criteria Tables define when decisions, risks, or issues must be elevated to higher levels of leadership. They clarify thresholds, triggers, and conditions that require additional oversight. This structure prevents leaders from delaying escalation or attempting to resolve issues beyond their authority. They matter because delayed escalation amplifies harm, increases cost, and reduces the chance of effective intervention. Criteria ensure that risks are surfaced early and addressed by the right decision makers. They also reinforce accountability by making escalation a structured obligation. When ignored, issues remain hidden until they become crises. Teams lose confidence in leadership responsiveness. Over time, the organization experiences repeated failures caused by slow or inconsistent escalation.

When done well, escalation criteria create clarity, speed, and disciplined risk management. People understand when and why issues must be elevated. This strengthens trust in leadership responsiveness and reduces the likelihood of preventable crises. Criteria reinforce the principle that leadership requires timely transparency. They remind leaders that escalation is a duty, not a weakness. Ultimately, they protect the organization from avoidable harm and strengthen governance.

MORAL COMPASS SCENARIO	LEADERSHIP ACTION
Zara receives a customer complaint that seems minor but has potential legal implications. She's unsure whether it meets escalation criteria.	The leader walks Zara through the escalation table and confirms the legal trigger. They then escalate the case to compliance and update the team on the criteria.
Nolan encounters a system outage that affects only one team but could spread. He hesitates because the impact isn't yet widespread.	The leader reviews the escalation table and confirms the outage meets the "potential propagation" threshold. They then activate the incident-response protocol.
Amina sees a vendor miss a delivery deadline for the third time. She's unsure whether repeated failures require escalation.	The leader checks the escalation table and confirms the "pattern of recurrence" trigger. They then escalate the issue to procurement leadership.
Elias notices a team conflict that is escalating emotionally. He isn't sure whether it qualifies for HR involvement.	The leader reviews the escalation criteria and confirms the "behavioral risk" threshold. They then bring HR into the situation and document the escalation.
Rohan identifies a data-quality issue that could affect reporting accuracy. He's unsure whether the severity warrants escalation.	The leader uses the escalation table to confirm the "material impact" threshold. They then escalate the issue to data governance.
Tessa sees early signs of burnout in her team but isn't sure if it meets escalation criteria. She worries waiting could make it worse.	The leader reviews the "well-being risk" threshold and confirms escalation is appropriate. They then involve HR and adjust workload distribution.

6x6™ Transformation Blueprint

Authority Control

"If AI learns from us, then our ethics must be worth learning."
Malala Yousafzai, Nobel Peace Prize laureate and education activist.

Authority Control requires structural clarity, and Clear Governance Boundaries define where leadership authority begins and ends. They establish the limits that prevent overreach, ambiguity, or conflict. This matters because unclear authority creates operational confusion and ethical vulnerability. Clear Governance Boundaries prevent power conflicts by specifying decision rights. They also reinforce fairness by ensuring authority is distributed intentionally rather than informally. When ignored, governance becomes inconsistent and unpredictable. Teams experience misalignment as leaders operate without shared rules. Over time, Authority Control deteriorates as boundaries erode and decisions become chaotic. Oversight Enforcement Mechanisms ensure governance rules are followed, and deviations are corrected.

Governance is the architecture of sovereignty. This topic uses RACI Charts and Decision Rights Catalogs to ensure the "Why" and the "How" are linked through radical human accountability. By studying "Oversight Committee Charters" in a focused environment, a leader eliminates the "Execution Drift" that occurs when ownership is vague. It's about hard-coding accountability into the organization's DNA. This control isn't about micromanagement; it is about providing the "Accountability Chains" that allow the system to self-govern. When authority is clearly mapped on a printed blueprint, the organization gains the velocity needed to compete in a world of machine logic. You can physically point to the person responsible, ensuring that every strategic move is backed by a human driver. This eliminates the "Black Box" of organizational drift and replaces it with a clinical, transparent hierarchy where the distance between leadership thought and operational action is finally and permanently closed.

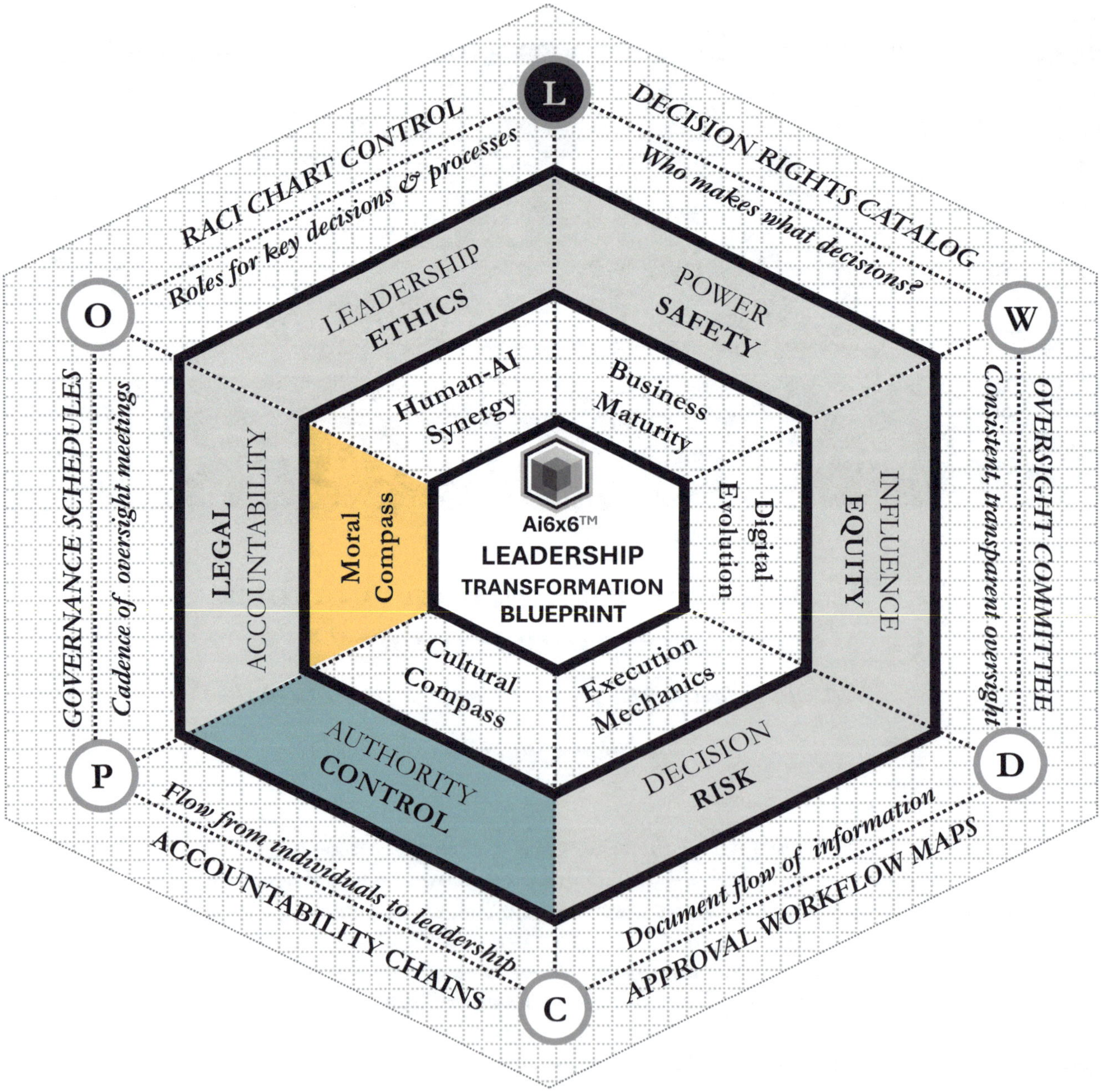

Ai6x6 AUTHORITY CONTROL FOR MORAL COMPASS | LEADERSHIP TRANSFORMATION

RACI Charts Control defines how responsibility, accountability, consultation, and information flow are assigned for key decisions and processes. They clarify who owns outcomes, who must be involved, and who must be informed. This structure prevents confusion and ensures decisions move through the right roles with the right expectations. They matter because unclear responsibility creates duplication, gaps, and ethical ambiguity. RACI controls prevent leaders from avoiding accountability or overstepping their authority. They also reinforce fairness by ensuring responsibilities are distributed intentionally rather than assumed. When ignored, teams operate with conflicting assumptions about ownership and authority. Decisions stall, overlap, or fall through the cracks. Over time, the organization experiences operational friction and weakened accountability.

When done well, RACI controls create clarity, alignment, and disciplined execution. People understand their role in every decision and act confidently within defined boundaries. This strengthens coordination and reduces conflict across teams. RACI reinforces the principle that leadership requires explicit responsibility, not implicit assumptions. It reminds leaders that clarity is a moral obligation in complex systems. Ultimately, it builds a culture where accountability is shared, transparent, and reliable.

MORAL COMPASS SCENARIO	LEADERSHIP ACTION
Janelle realizes her team keeps duplicating work because no one knows who is actually accountable for deliverables. She sees confusion rising with every new project.	The leader rebuilds the RACI chart with Janelle and clarifies who is Responsible, Accountable, Consulted, and Informed. They then walk the team through the chart and require its use for all new initiatives.
Tariq notices that two departments both believe they "own" the same process. Meetings keep turning into turf battles.	The leader updates the RACI chart to assign a single accountable owner and clearly define supporting roles. They also facilitate a reset meeting to align both departments.
Elise keeps getting pulled into decisions she shouldn't be making. She feels overwhelmed and unclear about her actual responsibilities.	The leader reviews the RACI chart with Elise and removes her from non-essential decision roles. They also communicate the updated responsibilities to all stakeholders.
Mateo sees that his team is waiting for approvals from someone who is only meant to be consulted. Deadlines slip because of the misunderstanding.	The leader corrects the RACI chart to reflect true approval authority. They also train the team on how to interpret RACI roles properly.
Karina notices that new hires are confused about who to escalate issues to. The RACI chart hasn't been updated in years.	The leader refreshes the chart and integrates it into onboarding. They also assign quarterly reviews to keep it current.
Devon keeps consulting people who don't need to be involved, slowing down decisions. He fears missing someone important.	The leader simplifies the RACI chart and highlights only essential consult roles. They also coach Devon on using the chart as a decision-speed tool.

Decision Rights Catalogs document who has the authority to make specific decisions across the organization. They define boundaries, escalation paths, and ownership for strategic, operational, and ethical choices. This cataloging ensures that decision power is distributed intentionally rather than informally. They matter because unclear decision rights create power struggles, delays, and inconsistent outcomes. Catalogs prevent leaders from making decisions outside their mandate or avoiding decisions they are responsible for. They also reinforce transparency by making authority visible and reviewable. When ignored, decisions become fragmented, duplicated, or contested. Teams waste time negotiating ownership instead of executing. Over time, the organization experiences governance drift and weakened strategic alignment.

When done well, decision rights catalogs create predictability, speed, and disciplined authority. People know exactly who decides what and why. This strengthens trust in leadership and accelerates execution. Catalogs reinforce the principle that authority must be defined, not improvised. They remind leaders that clarity in decision ownership is essential to ethical governance. Ultimately, they build a system where decisions are made efficiently and responsibly.

MORAL COMPASS SCENARIO	LEADERSHIP ACTION
Asha realizes her team keeps escalating minor decisions because they don't know what authority they have. She sees decision paralysis forming.	The leader reviews the decision rights catalog with Asha and clarifies which decisions sit at which level. They also run a workshop to reinforce decision ownership.
Colin approves a cross-department change without consulting the impacted teams. He didn't realize the decision exceeded his rights.	The leader updates the catalog to clarify cross-functional decision boundaries. They also require Colin to coordinate future decisions through the appropriate governance group.
Mireya hesitates to approve a customer concession because she's unsure whether it falls within her authority. She fears overstepping.	The leader walks her through the catalog and confirms her authority level. They also add a quick-reference guide for customer-impacting decisions.
Tobias keeps making decisions that should belong to his director. His enthusiasm is creating governance gaps.	The leader reviews the catalog with Tobias and resets his decision scope. They also implement a pre-decision check for high-impact items.
Lila delays a vendor decision because she doesn't know who actually owns it. The ambiguity stalls the entire project.	The leader updates the catalog to assign a clear decision owner. They also communicate the change to all project stakeholders.
Grant notices that decisions requiring legal review are being made informally. He worries about compliance exposure.	The leader updates the catalog to include mandatory legal checkpoints. They also train managers on when legal review is required.

3. OVERSIGHT COMMITTEE CHARTERS

Oversight Committee Charters define the purpose, scope, responsibilities, and authority of committees that monitor critical decisions and risks. They clarify what the committee oversees, how it operates, and what standards it enforces. This structure ensures oversight is consistent, transparent, and aligned with organizational values. They matter because oversight failures often stem from unclear mandates or insufficient authority. Charters prevent committees from becoming symbolic or ineffective. They also reinforce accountability by ensuring oversight bodies have the power to intervene when necessary. When ignored, oversight becomes inconsistent, reactive, or easily bypassed. Risks escalate without proper review and governance weakens. Over time, the organization becomes vulnerable to ethical, operational, and compliance failures.

When done well, oversight charters create rigor, transparency, and disciplined governance. People trust that oversight bodies operate with clear purpose and authority. This strengthens confidence in organizational integrity and risk management. Charters reinforce the principle that leadership must welcome scrutiny. They remind leaders that oversight is a safeguard, not an obstacle. Ultimately, they build a governance system that is strong, credible, and resilient.

MORAL COMPASS SCENARIO	LEADERSHIP ACTION
Naomi sits on an oversight committee that hasn't met in months. She realizes no one is enforcing the charter.	The leader reinstates the meeting cadence and reviews the charter with all members. They also assign a chair to ensure ongoing compliance.
Dorian notices the committee is making operational decisions instead of oversight decisions. The group is drifting from its mandate.	The leader revisits the charter and redefines the committee's scope. They also redirect operational decisions back to the appropriate teams.
Selene sees that committee members aren't preparing for meetings. Discussions become unproductive and rushed.	The leader enforces pre-read requirements outlined in the charter. They also introduce a rotating facilitator to improve meeting discipline.
Marcus realizes the committee lacks representation from key departments. Decisions feel disconnected from reality.	The leader updates the charter to include required cross-functional representation. They also onboard new members with a charter briefing.
Fiona notices that action items from committee meetings are rarely completed. Accountability is unclear.	The leader adds explicit follow-up rules to the charter and assigns an action-tracking owner. They also review progress at the start of each meeting.
Idris sees that the committee's purpose is outdated due to organizational changes. The charter no longer reflects current needs.	The leader leads a charter refresh and aligns it with the new operating model. They also schedule an annual charter review.

4. APPROVAL WORKFLOW MAPS

Approval Workflow Maps visualize the sequence of reviews, validations, and authorizations required for key decisions. They clarify the flow of information, the order of approvals, and the checkpoints that ensure quality and compliance. This mapping prevents decisions from bypassing essential oversight. They matter because unclear workflows create delays, bottlenecks, and governance gaps. Maps prevent leaders from improvising approval paths or skipping required steps. They also reinforce fairness by ensuring all decisions follow the same structured process. When ignored, approvals become inconsistent, opaque, or vulnerable to manipulation. Teams experience confusion about who must sign off and when. Over time, the organization suffers from uneven governance and increased risk exposure.

When done well, approval workflows create efficiency, transparency, and disciplined decision flow. People understand exactly how decisions progress and what is required at each stage. This strengthens alignment and reduces friction across teams. Workflow mapping reinforces the principle that leadership must operate within structured processes. It reminds leaders that consistency is essential to ethical and operational integrity. Ultimately, it builds a system where approvals are predictable, fair, and reliable.

MORAL COMPASS SCENARIO	LEADERSHIP ACTION
Camila keeps losing track of where approvals are stuck. The workflow map hasn't been updated since last year's reorg.	The leader rebuilds the workflow map with Camila and clarifies each approval step. They also publish the updated map for all teams to reference.
Jaden notices that approvals bounce between teams because no one knows the correct sequence. Projects stall for weeks.	The leader simplifies the workflow map and removes redundant steps. They also train teams on the new streamlined process.
Priya sees that urgent approvals get lost in the standard workflow. She needs a fast-track path.	The leader adds an expedited approval lane to the workflow map. They also define criteria for when fast-tracking is allowed.
Omar realizes that digital and manual approval paths don't match. Employees are confused about which one to follow.	The leader aligns both paths and updates the workflow map to reflect the unified process. They also retire outdated manual steps.
Helena notices that approvals requiring legal review are often skipped. The workflow map doesn't highlight legal checkpoints clearly.	The leader updates the map to include mandatory legal gates. They also require system-based prompts to enforce compliance.
Theo sees that new hires struggle to understand the approval process. The map is too complex and jargon-heavy.	The leader redesigns the map using plain language and visual cues. They also integrate it into onboarding materials.

5. ACCOUNTABILITY CHAINS

Accountability Chains define how responsibility flows from individual contributors to senior leadership for decisions, actions, and outcomes. They clarify who is answerable for what and how accountability escalates. This structure ensures that responsibility is neither diluted nor concentrated improperly. They matter because unclear accountability leads to blame shifting, inaction, and ethical ambiguity. Chains prevent leaders from avoiding responsibility or assigning it informally. They also reinforce fairness by ensuring accountability aligns with authority and role. When ignored, teams operate without clear ownership of outcomes. Problems linger because no one feels responsible for resolving them. Over time, the organization experiences declining performance and weakened trust.

When done well, accountability chains create clarity, discipline, and reliable follow through. People understand their obligations and the expectations tied to their role. This strengthens execution and reinforces a culture of responsibility. Accountability chains reinforce the principle that leadership requires owning outcomes, not just decisions. They remind leaders that accountability must be explicit and traceable. Ultimately, they build a system where responsibility is honored, and performance is strengthened.

MORAL COMPASS SCENARIO	LEADERSHIP ACTION
Rhea notices that when projects fail, no one can explain who was accountable for key decisions. The team keeps blaming "the process."	The leader builds a clear accountability chain and assigns ownership for each decision point. They also review it with the team to eliminate ambiguity.
Malik sees that tasks are assigned to teams instead of individuals. Deadlines slip because no one feels personally responsible.	The leader updates the accountability chain to include named owners. They also require weekly check-ins to reinforce ownership.
Giselle realizes that escalations are going to the wrong people. The accountability chain hasn't been updated since the reorg.	The leader refreshes the chain and communicates the new escalation paths. They also post the updated version in all project spaces.
Rowan notices that senior leaders keep stepping into decisions meant for frontline managers. This undermines empowerment.	The leader clarifies decision ownership in the accountability chain and reinforces boundaries. They also coach senior leaders on staying in their lane.
Tessa sees that cross-functional work breaks down because no one knows who owns the handoffs. The chain stops at departmental borders.	The leader extends the accountability chain across departments and defines handoff owners. They also run a cross-team alignment session.
Hugo realizes that accountability is assigned but not enforced. People ignore their responsibilities without consequence.	The leader ties accountability chain roles to performance expectations. They also introduce consequence management for repeated lapses.

6. GOVERNANCE MEETING SCHEDULES

Governance Meeting Schedules define the cadence, purpose, and structure of leadership meetings that oversee decisions, risks, and compliance. They clarify when governance bodies convene, what they review, and how decisions are documented. This scheduling ensures oversight is timely, consistent, and aligned with organizational needs. They matter because irregular or ad hoc governance weakens accountability and visibility. Schedules prevent gaps where risks accumulate without review. They also reinforce discipline by ensuring governance is treated as a continuous responsibility. When ignored, oversight becomes sporadic, reactive, or symbolic. Critical issues go unaddressed until they escalate. Over time, the organization experiences governance drift and weakened control.

When done well, governance schedules create rhythm, predictability, and disciplined oversight. People trust that leadership regularly reviews risks, decisions, and compliance. This strengthens organizational stability and ethical alignment. Schedules reinforce the principle that governance must be continuous, not occasional. They remind leaders that oversight is a structural commitment. Ultimately, they build a system where governance is reliable, proactive, and effective.

MORAL COMPASS SCENARIO	LEADERSHIP ACTION
Amina notices that governance meetings are constantly rescheduled. Decisions pile up because there's no reliable cadence.	The leader locks a fixed meeting schedule and enforces attendance expectations. They also assign a backup delegate for each member.
Victor sees that meetings are too infrequent to address fast-moving risks. Issues escalate before they're discussed.	The leader increases meeting frequency and adds interim check-ins for urgent topics. They also create a rapid-escalation channel for time-sensitive decisions.
Lila realizes that meetings are happening but no decisions are being made. The agenda is unfocused and overly long.	The leader restructures the agenda around decision items and time-boxes discussions. They also assign a facilitator to keep meetings on track.
Jonas notices that key stakeholders often skip governance meetings. Decisions lack the right voices.	The leader enforces attendance rules and escalates repeated absences. They also require pre-meeting input from anyone who cannot attend.
Priyanka sees that governance meetings overlap with operational standups. Teams are forced to choose between them.	The leader restructures the schedule to avoid conflicts. They also coordinate calendars across departments to prevent future overlap.
Elias realizes that governance meetings are not aligned with quarterly planning cycles. Decisions come too late to influence strategy.	The leader syncs the governance schedule with planning milestones. They also add pre-planning governance reviews to improve alignment.

6x6™ Transformation Blueprint

Legal Accountability

"Technology grows fast, but responsibility must grow faster."
Pope Francis, global religious leader known for moral guidance on modern issues.

Legal Accountability requires disciplined adherence to external obligations, and Regulatory Compliance Expectations define the legal standards leaders must uphold in every operational and strategic decision. They establish the minimum requirements for lawful conduct. This matters because leadership decisions expose the organization to regulatory, financial, and reputational risk. Regulatory Compliance Expectations prevent violations by ensuring leaders understand and follow applicable laws. They also reinforce integrity by demonstrating that compliance is non-negotiable. When ignored, legal exposure increases rapidly. Teams experience uncertainty as decisions drift outside legal boundaries. Over time, Legal Accountability deteriorates as compliance becomes inconsistent. Law Aligned Decisions ensure leadership choices remain firmly within legal and regulatory frameworks.

Transforming compliance from a tax into a strategic moat. This topic anchors the leader in the hard reality of SEC Governance, Sarbanes-Oxley, and Dodd-Frank. By reviewing "Integrity Findings" alongside "Whistleblower Protection" rules, the leader ensures the legacy they build is legally unassailable. This is the final layer of protection for the executive's intent. It treats "Legal Accountability" as a proactive defensive maneuver rather than a reactive legal headache. In the AI era, being "clinically compliant" is the only way to ensure that your transformation isn't dismantled by regulatory sledgehammers or ethical lapses in governance. The print format allows you to engage with these complex legal frameworks as a unified whole, seeing the "Moral Compass" in its entirety. It ensures that your "Safe Zone" is truly secure, providing the ultimate insurance policy against the market extinction that follows the failure to uphold the sovereign laws of the industrial and digital landscape.

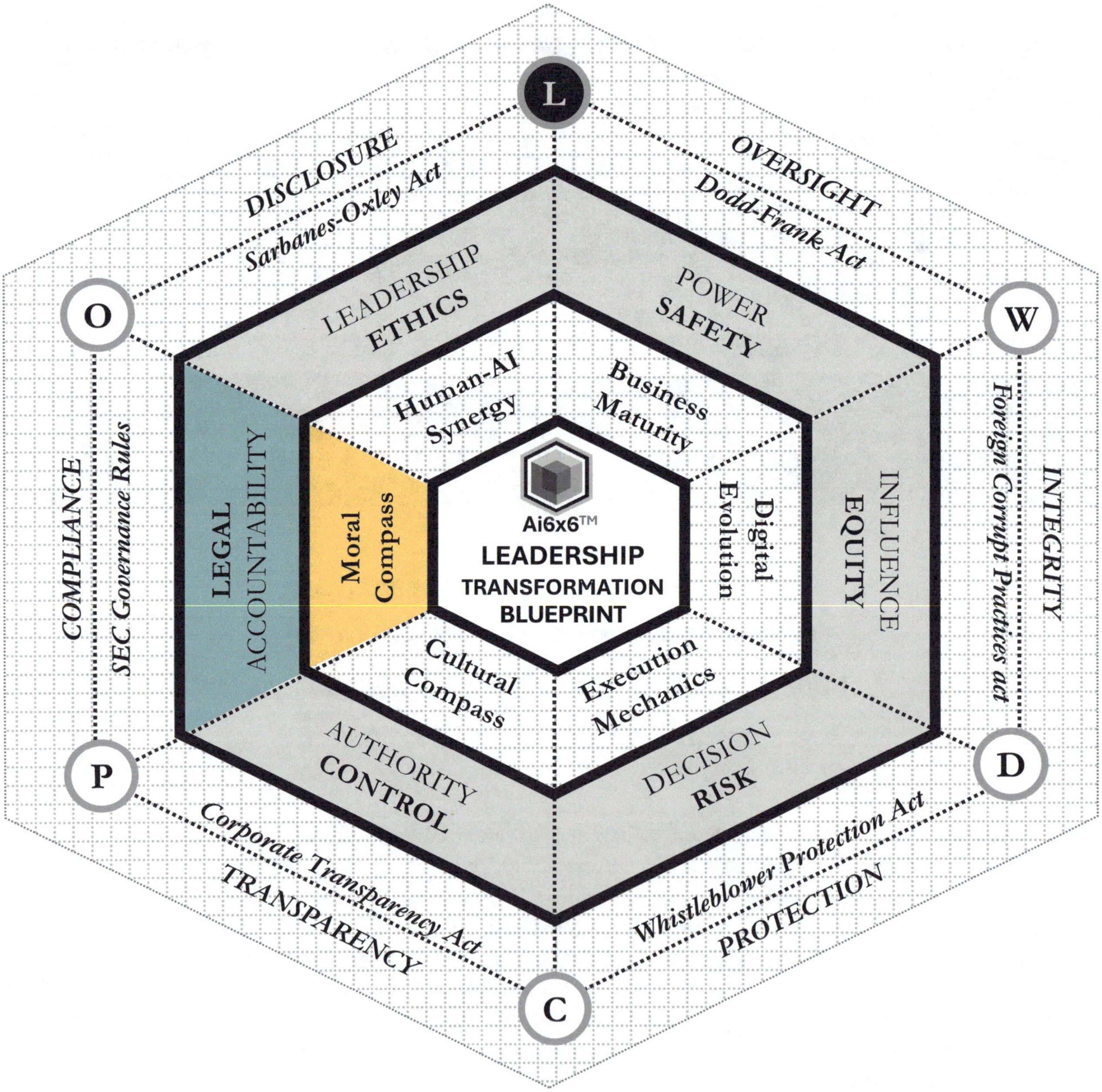

Ai6x6 LEGAL ACCOUNTABILITY FOR MORAL COMPASS | LEADERSHIP TRANSFORMATION

1. DISCLOSURE — SARBANES–OXLEY ACT

Disclosure under the Sarbanes–Oxley Act establishes the legal and ethical requirements for transparent financial reporting by leadership. It clarifies what must be disclosed, how accuracy is verified, and which controls ensure integrity. This structure ensures leaders cannot obscure financial risks or manipulate reporting. It matters because transparent disclosure protects investors, employees, and the public from misleading information. SOX requirements prevent unethical behavior such as concealment, misstatement, or selective reporting. They also reinforce accountability by holding executives personally responsible for accuracy. When ignored, financial statements become unreliable and expose the organization to legal, ethical, and reputational harm. Stakeholders lose confidence in leadership credibility. Over time, the organization becomes vulnerable to regulatory penalties and trust erosion.

When done well, SOX aligned disclosure creates transparency, accuracy, and disciplined financial governance. People trust that leadership reports information honestly and consistently. This strengthens market confidence and internal decision making. SOX disclosure reinforces the principle that leadership must uphold truth in all financial communication. It reminds leaders that integrity is a legal obligation, not a preference. Ultimately, it anchors the organization in a culture of financial honesty and accountability.

MORAL COMPASS SCENARIO	LEADERSHIP ACTION
Maren realizes her team has been delaying revenue recognition to smooth quarterly results. She knows this creates a material misstatement risk.	The leader instructs Maren to correct the entries immediately and disclose the issue to internal audit. They also retrain the team on SOX-compliant revenue policies.
Darius discovers that a manager has been approving journal entries without proper documentation. He worries this violates internal control requirements.	The leader requires all undocumented entries to be re-validated and logged. They also reinforce SOX documentation standards through mandatory training.
Evelyn notices that access to financial systems hasn't been reviewed in over a year. She fears unauthorized access could compromise reporting accuracy.	The leader orders an urgent access review and documents the remediation. They also implement quarterly SOX access-control checks.
Jonah learns that a key control failed during the last close cycle but wasn't reported. He senses the team is afraid of escalation.	The leader documents the control failure and reports it through the SOX disclosure process. They also reinforce a no-retaliation culture around control reporting.
Priya finds that reconciliations are being rushed and signed without proper review. She worries errors are slipping through.	The leader resets the reconciliation timeline and requires dual-review for the next cycle. They also update the SOX control matrix to reflect stricter oversight.
Mateo discovers that financial assumptions used in forecasts were changed without approval. He knows this undermines reporting integrity.	The leader documents the unauthorized change and restores approved assumptions. They also tighten SOX controls around forecast inputs.

2. OVERSIGHT — DODD–FRANK ACT

Oversight under the Dodd–Frank Act defines the legal expectations for monitoring financial practices, risk exposure, and systemic stability. It clarifies how leadership must evaluate risk, document controls, and prevent misconduct. This oversight ensures that leaders cannot ignore or conceal practices that threaten financial integrity. It matters because weak oversight contributed to past financial crises and widespread harm. Dodd–Frank requirements prevent excessive risk taking, conflicts of interest, and opaque financial structures. They also reinforce accountability by mandating transparent governance. When ignored, risks accumulate unnoticed and expose the organization to catastrophic failure. Stakeholders lose trust in leadership's ability to manage financial responsibility. Over time, oversight failures lead to legal sanctions and reputational damage.

When done well, Dodd–Frank oversight creates stability, transparency, and disciplined risk management. People trust that leadership monitors financial practices with rigor and responsibility. This strengthens resilience and reduces exposure to systemic risk. Oversight reinforces the principle that leadership must safeguard the organization from preventable harm. It reminds leaders that vigilance is a legal and moral duty. Ultimately, it builds a governance system that protects long term financial integrity.

MORAL COMPASS SCENARIO	LEADERSHIP ACTION
Lila notices her risk team is ignoring early indicators of liquidity stress. She fears they're underestimating systemic risk.	The leader escalates the issue to the risk committee and documents the indicators. They also require enhanced monitoring aligned with Dodd–Frank expectations.
Omar learns that a trading desk is bypassing internal limits to chase short-term gains. He recognizes the behavior as a compliance red flag.	The leader freezes the trading activity and reports the breach to compliance. They also reinforce Dodd–Frank risk-limit rules with the entire desk.
Serena sees that stress-testing models haven't been updated in two years. She worries they no longer reflect market realities.	The leader orders an immediate model refresh and documents the gap. They also implement an annual Dodd–Frank model review cycle.
Victor discovers that whistleblower complaints about risk mismanagement were never forwarded to the oversight committee. He senses a breakdown in governance.	The leader escalates the complaints and audits the reporting pipeline. They also reinforce mandatory routing rules for risk-related complaints.
Hana realizes that derivative exposures are being reported inconsistently across regions. She fears this violates transparency requirements.	The leader standardizes reporting templates and conducts a cross-regional audit. They also train teams on Dodd–Frank reporting obligations.
Elias finds that a business unit is using outdated risk-weighting assumptions. He knows this could distort capital calculations.	The leader mandates immediate recalculation using updated assumptions. They also implement a quarterly review of risk-weighting models.

3. INTEGRITY — FOREIGN CORRUPT PRACTICES ACT

Integrity under the Foreign Corrupt Practices Act establishes the legal boundaries that prohibit bribery, improper payments, and unethical influence in international operations. It clarifies what constitutes corrupt behavior, how risks must be monitored, and what controls prevent violations. This structure ensures leaders uphold ethical conduct across all markets. It matters because corruption distorts competition, damages reputation, and exposes organizations to severe penalties. FCPA requirements prevent leaders from using unethical incentives to secure business advantages. They also reinforce fairness by ensuring global operations follow consistent ethical standards. When ignored, corruption infiltrates decision making and undermines organizational credibility. Teams become uncertain about what behavior is acceptable. Over time, the organization faces legal consequences and loss of stakeholder trust.

When done well, FCPA integrity creates ethical consistency, global credibility, and disciplined compliance. People trust that leadership rejects corrupt practices in every market. This strengthens international partnerships and protects long term growth. FCPA alignment reinforces the principle that leadership must uphold integrity regardless of external pressure. It reminds leaders that ethical conduct is non-negotiable. Ultimately, it builds a global culture grounded in honesty and fairness.

MORAL COMPASS SCENARIO	LEADERSHIP ACTION
Camila learns that a regional manager offered "expedited fees" to a foreign official. She recognizes this as a potential FCPA violation.	The leader halts all related transactions and reports the incident to compliance. They also retrain the region on anti-bribery rules.
Rowan discovers that a distributor is providing lavish gifts to government clients. He suspects the company may be indirectly liable.	The leader suspends the distributor relationship pending investigation. They also implement stricter third-party FCPA controls.
Priyanka notices that due-diligence files for foreign partners are incomplete. She worries the gaps expose the company to bribery risk.	The leader orders a full due-diligence refresh and documents all missing information. They also implement mandatory annual partner reviews.
Hugo hears that a sales rep promised "special access" to a government buyer. He fears this crosses ethical and legal boundaries.	The leader removes the rep from the account and initiates an FCPA investigation. They also reinforce training on prohibited inducements.
Tessa sees that payments to a foreign consultant lack clear justification. She suspects the consultant may be a conduit for improper influence.	The leader freezes the payments and requests full documentation. They also escalate the case to legal for FCPA review.
Idris learns that a local partner in another country insists on cash payments. He knows this is a major FCPA red flag.	The leader halts the payment and initiates enhanced due diligence. They also require all future payments to follow traceable channels.

4. PROTECTION — WHISTLEBLOWER PROTECTION ACT

Protection under the Whistleblower Protection Act defines the legal safeguards for individuals who report misconduct, fraud, or ethical violations. It clarifies how reports must be handled, how retaliation is prevented, and how confidentiality is maintained. This structure ensures that employees can speak up without fear. It matters because organizations cannot address wrongdoing if people are afraid to report it. Whistleblower protections prevent retaliation, silence, and concealment of ethical issues. They also reinforce accountability by ensuring leadership responds to concerns transparently. When ignored, employees remain silent and misconduct spreads unchecked. Leaders lose visibility into emerging risks. Over time, the organization becomes vulnerable to ethical collapse and legal consequences.

When done well, whistleblower protection creates psychological safety, transparency, and early intervention. People trust that raising concerns is safe and valued. This strengthens ethical vigilance and organizational integrity. Protection reinforces the principle that leadership must defend those who defend the truth. It reminds leaders that retaliation is both unethical and unlawful. Ultimately, it builds a culture where speaking up is protected and respected.

MORAL COMPASS SCENARIO	LEADERSHIP ACTION
Naomi reports financial irregularities and immediately feels her manager is treating her differently. She fears retaliation.	The leader activates whistleblower-protection protocols and removes the manager from supervisory duties. They also assign Naomi a neutral HR advocate.
Dorian overhears a supervisor threatening an employee who raised a safety concern. He knows this violates whistleblower protections.	The leader documents the threat and suspends the supervisor pending investigation. They also reinforce anti-retaliation rules across the team.
Selene notices that employees avoid reporting issues because past whistleblowers were sidelined. She senses a culture of fear.	The leader launches a culture-reset initiative and publicly reaffirms whistleblower protections. They also audit past cases for retaliation patterns.
Marcus learns that a whistleblower's identity was shared informally. He recognizes this as a serious breach.	The leader investigates the disclosure and disciplines those responsible. They also tighten confidentiality protocols.
Fiona sees that a whistleblower's complaint was quietly closed without review. She suspects intentional suppression.	The leader reopens the case and escalates it to compliance. They also implement mandatory tracking for all whistleblower reports.
Idris hears that a whistleblower was excluded from key meetings after raising concerns. He worries this is subtle retaliation.	The leader restores the employee's access and documents the retaliation. They also retrain the leadership team on protected-activity rules.

Transparency under the Corporate Transparency Act establishes the legal requirements for disclosing beneficial ownership and preventing hidden financial structures. It clarifies who must be identified, what information must be reported, and how accuracy is verified. This structure ensures leaders cannot obscure ownership or conceal illicit activity. It matters because opaque ownership enables fraud, money laundering, and unethical influence. CTA requirements prevent misuse of corporate structures for hidden agendas. They also reinforce accountability by making ownership visible to regulators and stakeholders. When ignored, organizations become vulnerable to legal scrutiny and reputational harm. Stakeholders question the legitimacy of leadership decisions. Over time, hidden ownership erodes trust and exposes the organization to severe penalties.

When done well, CTA transparency creates clarity, legitimacy, and ethical governance. People trust that leadership operates with openness and honesty. This strengthens stakeholder confidence and regulatory alignment. Transparency reinforces the principle that leadership must operate in the open. It reminds leaders that secrecy is a risk, not a strategy. Ultimately, it builds a governance environment grounded in visibility and integrity.

MORAL COMPASS SCENARIO	LEADERSHIP ACTION
Amina realizes beneficial-ownership records haven't been updated after a leadership change. She knows this violates reporting requirements.	The leader updates the ownership filings immediately and documents the lapse. They also implement quarterly ownership-record reviews.
Victor discovers that a subsidiary's ownership structure is unclear. He fears the company may be out of compliance.	The leader orders a full ownership-mapping exercise and files corrected disclosures. They also centralize ownership documentation.
Lila notices that foreign ownership stakes were never reported. She worries this exposes the company to penalties.	The leader files the missing disclosures and conducts a compliance audit. They also train regional leaders on reporting obligations.
Jonas learns that beneficial-owner information is stored in spreadsheets with no access controls. He recognizes the security risk.	The leader migrates the data to a secure system and restricts access. They also implement audit logs for all ownership-data changes.
Priyanka sees that ownership changes are communicated informally and inconsistently. She fears updates are being missed.	The leader formalizes the ownership-change workflow and assigns a compliance owner. They also require documentation for every change.
Elias discovers that a newly acquired entity has no ownership documentation at all. He knows this is a major compliance gap.	The leader initiates an urgent ownership-verification process. They also integrate ownership reporting into the acquisition checklist.

Compliance with SEC Governance Rules defines the legal expectations for reporting, oversight, and ethical conduct in publicly regulated organizations. It clarifies how leaders must disclose information, manage conflicts, and maintain governance standards. This structure ensures leadership behavior aligns with regulatory requirements. It matters because SEC rules protect investors, markets, and the public from misconduct. Compliance prevents manipulation, selective disclosure, and governance failures. It also reinforces accountability by requiring leaders to operate within strict ethical boundaries. When ignored, organizations face legal penalties, loss of investor confidence, and reputational damage. Teams become uncertain about the legitimacy of leadership decisions. Over time, governance failures undermine organizational stability.

When done well, SEC compliance creates trust, transparency, and disciplined governance. People see that leadership meets the highest standards of ethical and legal conduct. This strengthens market confidence and internal alignment. Compliance reinforces the principle that leadership must uphold the law with precision and integrity. It reminds leaders that ethical governance is a continuous obligation. Ultimately, it builds a system where legal and moral standards reinforce each other.

MORAL COMPASS SCENARIO	LEADERSHIP ACTION
Simone realizes board minutes are incomplete and missing key decisions. She knows this violates SEC governance expectations.	The leader reconstructs the missing minutes and implements stricter documentation standards. They also assign a governance coordinator.
Rowan sees that insider-trading windows are not being communicated clearly. Employees are confused about when they can trade.	The leader issues an immediate clarification and updates the trading-window calendar. They also require mandatory training for all insiders.
Hana discovers that required committee charters haven't been updated in years. She fears they no longer reflect actual responsibilities.	The leader updates all charters and aligns them with SEC governance rules. They also schedule annual charter reviews.
Elias notices that risk disclosures in public filings are outdated. He worries the company is under-reporting material risks.	The leader updates the disclosures and coordinates with legal for accuracy. They also implement a quarterly risk-disclosure review.
Mireya learns that board packets are being distributed without proper confidentiality controls. She recognizes the governance risk.	The leader moves board materials to a secure portal and restricts access. They also train board members on confidentiality protocols.
Jonah sees that related-party transactions are not being reviewed consistently. He fears this violates SEC oversight requirements.	The leader establishes a formal review committee and documents all transactions. They also integrate related-party checks into the governance calendar.

CHAPTER
G7

BONUS CHAPTER

BLUEPRINT SERIES

6x6™ Transformation Blueprint

The Stewardship Paradox

"The greatest threat to our planet is the belief that someone else will save it." -- **Robert Swan**, the first person to walk to both Poles and a relentless advocate for Antarctic preservation, provides the definitive answer to the **Stewardship Paradox**. His life's work serves as a reminder that sustainability cannot be solved by waiting for external mandates; it requires a leader to take personal, fiduciary responsibility for the future, ensuring the firm's survival is no longer outsourced to "someone else."

Traditional leadership has failed because it treats sustainability as a secondary "offset" to the primary engine of profit. For decades, the fiduciary duty was interpreted through a narrow, short-term lens that prioritized immediate dividends over long-term planetary survival. This one-dimensional approach has left organizations brittle and reactive. To achieve Net Zero, leaders must adopt a multi-dimensional perspective that embraces the **Stewardship Paradox**. This requires balancing the "now" with the "next," navigating the tension between growth and dematerialization, and moving from defensive "greenhushing" to radical transparency. Traction is only possible when the moral compass is integrated into the core governance of the firm, ensuring that the company's survival is inextricably linked to the health of the world it inhabits.

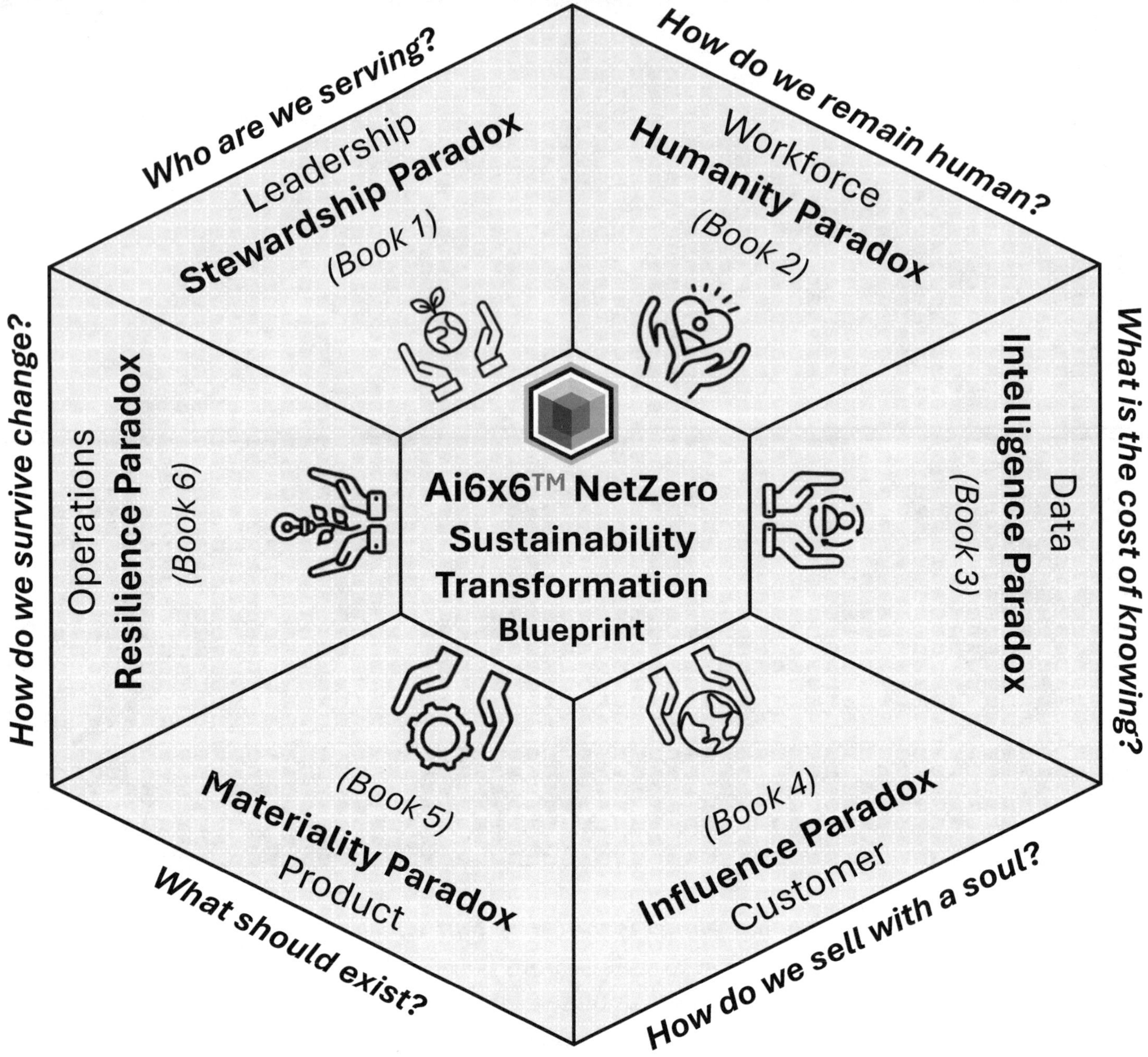

The **Ai6x6 NetZero Sustainability Transformation Blueprint** redefines the path to Net Zero by replacing the traditional "check-the-box" compliance mindset with a rigorous framework of **paradoxical leadership**. In a world accelerated by AI, sustainability is rarely a choice between a clear right and wrong; instead, it is a constant navigation of thirty-six competing tensions across six organizational lenses. By framing the journey through these paradoxes—such as the conflict between **Digital Intelligence and Energy Consumption** or **Global Targets and Local Infrastructure**—the framework prepares leaders for the messy, non-linear realities of systemic change.

Rather than viewing these tensions as obstacles to be avoided, the **Ai6x6 methodology** utilizes them as engines for innovation. It guides executives to move past the paralysis of "either/or" thinking, teaching them to embrace a "both/and" strategy that maintains fiduciary health while securing long-term planetary survival. This approach ensures that a company's moral compass remains calibrated even as technology and climate volatility shift the ground beneath them. Ultimately, the series provides a diagnostic roadmap that turns systemic friction into strategic advantage, empowering leaders to achieve radical Net Zero goals without sacrificing the operational resilience or human dignity required to thrive in the post-AI era.

THE STEWARDSHIP PARADOX is the heavy weight on a modern leader's shoulders. It is the grueling tension between protecting the immediate financial health of the company and the moral duty to ensure a livable world remains for that company to inhabit. In the AI era, this is no longer a philosophical debate; it is a structural crisis. Leaders are pressured to deliver quarterly growth while the planet demands a radical reduction in consumption. To lead now is to live permanently in this "stretch," holding the line between the demands of the shareholder and the urgent needs of the future.

1. THE FIDUCIARY VS. FUTURE PARADOX (Profit Today vs. Survival Tomorrow)

The core conflict here lies in the legal and financial definition of a leader's duty. Traditionally, fiduciary duty was a shield used to prioritize the current shareholder's wallet above all else. However, in the Net Zero era, failing to invest in climate resilience is, in itself, a breach of duty. Leaders are caught between the "now" and the "next." They must satisfy investors who want dividends this morning while making massive capital investments in a carbon-neutral future that may not bear fruit for a decade. It is a high-stakes balancing act where the cost of action is high, but the cost of inaction is terminal.

2. THE GROWTH-DECOUPLING PARADOX (Growing the Business vs. Using Fewer Resources)

Modern business is built on the belief that more is better—more sales, more users, more products. Yet, sustainability demands that we do more with less. The paradox of decoupling is the struggle to increase revenue while simultaneously shrinking the physical footprint of the company. It requires a fundamental shift from a volume-based business model to one based on value. For a leader, this means telling the market that the company is getting bigger and stronger, even as its consumption of energy, water, and raw materials is trending toward zero. It is the ultimate test of strategic innovation.

3. THE TRANSPARENCY-LIABILITY PARADOX (Being Honest vs. Fear of Being Sued)

Transparency is the currency of trust, but for a leader, it is also a legal minefield. Stakeholders demand "radical honesty" about where a company is failing in its Net Zero journey. However, the more a leader opens the books on their environmental gaps, the more they expose the firm to "greenwashing" lawsuits or activist attacks. This tension creates a culture of "greenhushing," where progress stalls because leaders are too afraid of the litigation that comes with being a work-in-progress. Managing this requires a shift from defensive PR to a culture of audited, data-backed integrity.

4. THE RESILIENCY-EFFICIENCY PARADOX (Being Prepared vs. Being Lean)

For thirty years, consultants taught leaders that "lean" was the only way to win. Removing every ounce of "waste" and "slack" saved money and reduced carbon. But the climate crisis has revealed that a hyper-efficient system is also a hyper-brittle one. Leaders now face the paradox of needing to re-introduce "inefficiency"—redundancy, safety stocks, and diverse supply chains—to survive environmental shocks. This is a direct hit to the bottom line, forcing a choice between a business that is optimized for a perfect world or a business that is hardened for a volatile one.

5. THE INCENTIVE-INTENT PARADOX (Daily Pay vs. Grand Promises)

There is a glaring gap between the 20-year "Net Zero" promises made in annual reports and the 12-month bonus structures that drive executive behavior. This paradox highlights the internal hypocrisy of governance. A leader may have the best intentions for the environment, but if their pay is tied exclusively to quarterly stock price or short-term EBIT, the environment will always lose. Solving this requires the courage to rewrite the contract of the CEO, embedding sustainability milestones into the very core of how success is measured and rewarded at the top.

6. THE GREENHUSHING PARADOX (Leading the Way vs. Staying Silent)

True leadership requires a voice, but in the current climate, many leaders are choosing silence. The paradox of "greenhushing" is driven by a fear that no matter what a company does, it will never be enough for the critics. By staying silent, leaders protect their brand from immediate scrutiny, but they also stop the flow of shared learning that the entire industry needs to move forward. Breaking this paradox requires a new kind of executive vulnerability—the willingness to lead out loud, even when the path is imperfect, to inspire action across the sector.

1. THE FIDUCIARY VS. FUTURE PARADOX (Profit Today vs. Survival Tomorrow)

- **Investment Banking:** A CEO in 2026 allocates $5B to "Long-Horizon" carbon removal tech, intentionally lowering this year's dividend to insure against total portfolio collapse by 2040.
- **Insurance:** A global firm stops underwriting properties in coastal "red zones," sacrificing current premiums to protect the company's solvency from the 10-year climate trajectory.
- **Retail Banking:** Offering "Green Interest Rate" discounts to corporate borrowers who hit Net Zero milestones, trading current interest income for lower long-term credit risk.
- **Energy:** A European utility retires a profitable natural gas plant 8 years ahead of its depreciation schedule to accelerate its transition to offshore wind.
- **Venture Capital:** A firm shifts its exit strategy from 5-year "flips" to 12-year "stewardship" holds to ensure climate-tech startups reach planetary scale.
- **Consumer Goods:** A conglomerate invests $2B in a 10-year "Regenerative Farming" program for its suppliers, reducing current margins to secure future raw material availability.

2. THE GROWTH-DECOUPLING PARADOX (Growing the Business vs. Using Fewer Resources)

- **Electronics:** A smartphone giant grows revenue by 15% through high-margin "Repair & Subscription" services while reducing virgin material extraction by 40%.
- **Industrial Equipment:** A heavy machinery firm sells "Tons of Earth Moved" as a service rather than selling excavators, decoupling profit from the number of units manufactured.
- **Apparel:** A luxury brand doubles its market cap by launching a "Certified Pre-Owned" platform, generating new profit from the same garment multiple times.
- **Software:** A cloud provider increases its user base while using AI to "refactor" code, reducing the energy-per-user by 60% and shrinking its data center footprint.
- **Beverage:** A global soda brand increases market share by installing 1 million "Micro-Bottling" kiosks, growing sales while eliminating the need for plastic bottle production.
- **Lighting:** A vendor shifts to "Light-as-a-Service," where they make more profit by using *less* electricity and fewer replacement bulbs, selling "lumen-hours" instead of hardware.

3. THE TRANSPARENCY-LIABILITY PARADOX (Being Honest vs. Fear of Being Sued)

- **Pharma:** A drug maker publishes a "Radical Honesty" report on the chemical runoff of its legacy factories, opening itself to fines but building the trust needed for its new "Green Bio-Tech" line.
- **Aviation:** An airline scuttles its "Carbon Neutral" marketing in 2026, replacing it with a live jet-fuel burn dashboard to prevent "greenwashing" litigation while showing real progress.
- **Fashion:** A retailer uses blockchain to show every flaw in its Tier 3 supply chain, accepting short-term activist criticism to build an unassailable, data-backed ESG defense.
- **Mining:** A copper firm shares live satellite feeds of its biodiversity loss in active pits, using the transparency to validate its massive "Nature Positive" restoration efforts.
- **Agribusiness:** A food giant discloses the exact pesticide levels across its global network, risking a 10% stock dip to reset its relationship with health-conscious consumers.
- **Automotive:** An EV maker publishes the "Human Rights Audit" of its cobalt mines, including the failures, to prove it is the only brand with a verified "Ethical Battery."

4. THE RESILIENCY-EFFICIENCY PARADOX (Being Prepared vs. Being Lean)

- **Manufacturing:** A firm moves from "Just-in-Time" to "Just-in-Case," building 3 months of "Resilience Stock" of recycled steel to survive 2026's climate-driven shipping disruptions.
- **Food & Beverage:** A brand localizes 80% of its sourcing to within 200 miles of its plants, accepting higher local costs to eliminate the carbon and risk of global shipping.
- **Logistics:** A freight company invests in high-tech "Sails" for its ships, which are slower than engines but ensure the fleet remains operational during a fuel-price shock.
- **Healthcare:** A hospital system invests in "Micro-grids" and on-site oxygen generation, choosing the "inefficient" cost of self-sufficiency over the risk of grid failure.
- **Apparel:** A brand shifts to "Micro-factories" in expensive cities like New York and London to produce only what is needed, trading mass-production efficiency for zero-waste resilience.
- **Tech:** A hardware firm keeps its legacy repair facilities open worldwide, accepting the "inefficiency" of spare parts inventory to extend product life and brand loyalty.

5. THE INCENTIVE-INTENT PARADOX (Daily Pay vs. Grand Promises)

- **Financial Services:** A bank rewrites its CEO's contract so that 40% of their "Performance Units" only vest if the bank's financed emissions drop by 20% over 5 years.
- **Retail:** A supermarket chain ties store manager bonuses to "Food Waste Reduction" instead of just "Gross Sales," aligning daily behavior with the 2030 Net Zero goal.
- **Energy:** An oil major links executive retirement packages to the successful decommissioning of carbon assets and the 10-year profitability of their new "Green Hydrogen" wing.
- **Auto:** An EV firm ties its engineering leads' bonuses to "Kilowatt-hour per Mile" efficiency, ensuring they prioritize environmental performance over just speed or luxury.
- **Consulting:** A "Big Four" firm penalizes partners for high-carbon travel, making "Internal Carbon Tax" a direct deduction from their annual profit share.
- **Construction:** A cement giant ties the CFO's bonus to the "Internal Carbon Price," forcing the finance team to treat carbon as a real, taxable cash-flow item.

6. THE GREENHUSHING PARADOX (Leading the Way vs. Staying Silent)

- **Cosmetics:** A brand admits on its homepage that its current plastic cap is "not yet sustainable," inviting competitors to join a shared R&D consortium to fix it.
- **Airlines:** A CEO gives a 2026 keynote admitting their Net Zero goal is "impossible with current tech," triggering a global industry debate on "Hard-to-Abate" sectors.
- **Investment:** A fund manager breaks the silence on "Green Returns," admitting that some green transitions will lose money in the short term, but are necessary for long-term survival.
- **Fashion:** A brand ends its "Eco-Collection" labels to stay quiet about "marketing" while publishing 1,000 pages of raw environmental data for researchers to audit.
- **Hospitality:** A hotel chain stops its "Save the Towels" signs, admitting they were ineffective, and instead holds public town halls on their "Invisible" HVAC and water-system upgrades.
- **Tech:** A chipmaker refuses to call itself "Carbon Neutral" despite buying offsets, choosing instead to lead the way by discussing the "Physical Reality" of its energy footprint.

Most comprehensive business transformation book series worldwide

Ai6x6™ Transformation Engine

The AI6x6 Transformation Engine™

Your Operating System for the AI Era

In an age of autonomous systems, fragmented strategy is a liability. The **AI6x6 Transformation Engine™** is a precision-engineered, 18-volume masterwork designed to move your enterprise from organizational inertia to scalable momentum. Built on a proprietary 6×6 geometry, this ecosystem unifies architecture, execution, and calibration into one high-performance unit.

The journey begins with the **Transformation Blueprint Series**, your **Intelligence Layer**. It codifies a "Navigation Deck" of Six Pillars and Six Lenses to anchor your strategic North Star. Next, the **Consulting Playbook Series** acts as the **Transmission Layer**, utilizing the IGNITE™ methodology to translate executive intent into high-torque, repeatable action. Finally, the **Maturity Diagnostic Series** serves as the **Sensor Layer**, providing a data-driven intake to quantify readiness and calibrate performance.

This isn't just a book collection; it is a rigid **Chassis** for institutional excellence. Secure the definitive architecture for sustained, high-stakes digital evolution. Drive the future with the Engine.

Transformation Series

To ensure this book series benefits your company holistically, we strongly recommend you buy the series and then distribute it to the appropriate transformation sponsors in each of these departments. Don't forget to browse through consulting offers at the end of the book.

WE NEED YOU!

WELCOMING OUR NEXT SET OF THOUGHT LEADERS

The remarkable success of the Ai6x6 Organization Transformation Series is built upon the **hard-won wisdom** of executive leaders who have successfully navigated the AI transformation within their industries, turning theory into a deeply practical, actionable guide. As we expand the series, we are extending a highly selective invitation to a new cohort of pioneers—individuals whose professional journey is at the **leading edge of AI strategy and execution**. We are looking for executives who possess a proven track record of rewriting business models, mastering unforeseen ethical and competitive challenges, and offering a **clear blueprint** for the next 3-5 years of the AI-Era.

Contributing to the Ai6x6 Series is a unique opportunity to **cement your professional legacy**, joining an exclusive, curated network of global authorities whose insights are distilled into executive-level knowledge cards, directly shaping the governance and innovation decisions of the world's most influential companies. Our selection process is based strictly on **pure expertise and demonstrated leadership** in the AI-Era, reinforcing our commitment to editorial integrity and authentic thought leadership—there is no monetary exchange involved. Select contributors will also receive an invitation to blog on our expert network microsite, further amplifying their platform. To confidentially inquire about contributing your expertise, please contact us via our dedicated portal at **www.Ai6x6.com/contact** or by scanning the adjacent QR code.

Contact

SUBSCRIBE TO OUR NEWSLETTERS: To stay ahead of the curve and receive continuous, executive-level intelligence on the rapid evolution of the AI-Era, we invite you to register for our **Ai6x6 Weekly Newsletter**. Each edition delivers curated insights directly from our thought leaders and highlights critical shifts in AI strategy, governance, and provides relevant excepts from our website and books. Keep your finger on the pulse of the future and ensure your company remains competitive by signing up on the same page

PLEASE REVIEW US ON AMAZON!

Leadership Blueprint

We know your time is valuable, but if the Ai6x6 Organization Transformation Series delivered the real-world, actionable strategies you needed, please consider leaving a brief review on Amazon. If we missed the mark, we would also like to know, and that is the only we we can improve. Your honest feedback is crucial; it helps fellow executives discover this unique framework and benefit from the insights of our thought leaders. Thank you for supporting our mission!

THE Ai6x6™ CERTIFIED TRANSFORMATION STRATEGIST

Certified for Thinking, Not Reciting

The **Ai6x6™ CTS (Certified Transformation Strategist)** is an elite designation for MBAs and strategic leaders with a minimum of **five years of professional experience**. This is not a test of your ability to regurgitate theory; it is a clinical evaluation of your ability to orchestrate high-stakes business transformation in real-time. It signals to the market that you are a strategist who can think on your feet, navigate ambiguity, and utilize a world-class clinical framework to deliver results.

The Non-Traditional Certification

We don't believe in "black box" testing. We believe in the "Open Book" reality of the modern boardroom. To earn your CTS, you must demonstrate mastery through a **60-minute Board-Level Defense** before a member of the Ai6x6™ Global Leadership Team.

$1000 per defense

The Preparation

Mastery of the Series: Candidates must familiarize themselves with all six volumes in the **Ai6x6™ Transformation Series**. You choose **two Retreat Exercises** from any two **Consulting Playbooks** within the series that best align with your expertise. Best practices for the presentation are provided on our website.

The 60-Minute Defense

1. **The Presentation (30 Minutes):** You have 15 minutes per exercise to present your strategic approach and expected outcomes. 60% of your score is for this exercise.
2. **The Situational Audit (30 Minutes):** Our leadership team will pose **six situational transformation scenarios**. You must demonstrate how to apply the **6 Lens Sequence™** to solve these complex frictions.
3. **The Open-Resource Advantage:** You are **encouraged** to refer to any of the six Ai6x6™ books during the exam. We test your ability to navigate and apply the framework under pressure, not your memory.

The Induction: Beyond the Credential

Passing the Board Defense is only the beginning. Successful candidates are immediately inducted into the **Ai6x6™ Executive Mentor Network**, unlocking:

- **Thought Leadership:** Opportunities to contribute to the Ai6x6™ books, knowledge articles and gain visibility among global industry leaders.
- **Engagement Pipeline:** Direct access to a network of future consulting engagements and strategic partnerships.
- **The Fast Track:** Exceptional participants may be invited to join our elite **Practitioner Program**, bypassing the standard entry requirements to lead their own white-label boutique practice.

SCHEDULE YOUR BOARD DEFENSE AT:

Ai6x6.com/certification

Consulting Certification

THE Ai6x6™ PRACTITIONER PROGRAM

For Successful Senior Executives

Your Authority. Our Infrastructure. A Shared Legacy.

The **Ai6x6™ Practitioner Program** is designed for the high-level strategist—with proven industry leadership experience of over thirty years—who seeks a modern, standardized engine to deploy their operational wisdom. We provide the institutional infrastructure of a global consulting firm, delivered as a white-label solution that honors your professional identity and independence.

From $2000 per month membership

The "Un-Franchise" Model

We have re-engineered the consulting partnership for the modern era. Think of this as a **franchise without the friction or the high fees to get started**:

- **No Long-Term Commitments:** Maintain your agility with a model built for the independent architect.
- **No Loyalty Fees or Royalties:** You keep 100% of your engagement fees. We don't tax your success.
- **A Partnership of Mutual Trust:** Our relationship is powered by a transparent monthly membership, ensuring our incentives are perfectly aligned with your practice's growth.

The Foundations of Practice

All Practitioners gain immediate access to the **Ai6x6™ "Firm-in-a-Box"**:

- **White-Label Ecosystem:** Seamlessly integrate your own brand into our proprietary presentations, client pitches, and diagnostic audit templates.
- **The Clinical Engine:** Augment your strategic intuition with fact-based, quantitative scoring systems that replace subjective brainstorming with measurable transformation.

The White Glove Partnership

For those who wish to move from independent advisor to institutional authority, our **White Glove** tier offers a deep-tissue integration into the Ai6x6™ ecosystem:

- **Monthly 1-on-1 Strategic Check-ins:** Direct access to the **Ai6x6™ leadership team** to stress-test your pipeline, refine high-stakes offerings, or audit active client delivery.
- **Institutional Visibility:** Gain a global platform as a **Co-Author or Featured Expert** in upcoming volumes. We showcase your success stories on **Ai6x6.com**, cementing your legacy as a master of the 6 Lens Sequence.
- **The Bulk Purchase Program:** Exclusive access to deep discounts on the **Ai6x6™ Book Series**, enabling you to provide every client stakeholder with the "source code" for orchestration.

APPLY TO OUR PRACTITIONER PROGRAM AT:

Ai6x6.com/practitioner

Consulting Practitioner

THE Ai6x6™ TRANSFORMATION RETREAT

One Playbook. Your Leadership. Total Orchestration.

The **Ai6x6™ Business Transformation Series** is designed to eliminate the need for traditional, slow-moving consulting teams. Using our "Clinical Blueprints," your organization can self-organize, self-diagnose, and self-disrupt. Everything you need to facilitate a high-stakes, 2.5-day executive offsite is contained within the books.

The Self-Guided Mastery (100% Internal & Free)

You have the "Source Code." Your internal Offsite Coordinator or Chief of Staff can lead the entire 2.5-day agenda end-to-end using the **Consultant Playbook (Part 2)** and our **Supplemental Workbooks**.

- **No External Fees:** Run the entire diagnostic without "billable hours."
- **Full Control:** Use the 6-book series to "mix and match" exercises that fit your specific departmental frictions.
- **Tactile Execution:** Move from whiteboards to prioritized MoSCoW cards and RACI matrices using the physical templates provided in the series. Our Advisory section has all the inspiration your team needs.

Packages from $4999

The Orchestrated Upgrade (White-Glove Support)

For leadership teams seeking institutional weight and an objective "Cold Eye" review without the overhead of travel, we offer the **Orchestrated Retreat** package. This provides three critical layers of external authority:

- **Phase 1: The 30-Minute Architect's Briefing (Pre-Retreat):** A high-velocity strategic sprint with your internal Facilitator. we help you select the exact modules from across the 6-book series to target your unique friction points and set the tactical "Rules of Engagement."
- **Phase 2: The 30-Minute Executive Keynote (Live Remote):** A virtual address delivered by the **Ai6x6™ Leadership Team** to kick off Day 1. This keynote provides the external authority required to break through internal political silos and align the room's energy.
- **Phase 3: The 14-Day Clinical Roadmap Audit (Post-Retreat) :** Submit your digital outputs—prioritized lists, RACI matrices, and visualized Roadmaps. Our team performs a rigorous review and delivers a **Comprehensive Strategic Report** within **two weeks**, stress-testing your plan for diagnostic integrity and self-disruption readiness.

UPGRADE TO AN ORCHESTRATED RETREAT:

Ai6x6.com/retreat

Consulting Retreat

"Peace is its own reward."
Mahatma Gandhi

Ai6x6 Research Lab
Publishing Team

NEW YORK OFFICE
(Editorial Hub)

Editorial Lead: Marisa Thompson
Research Lead: Talia Rosen
Line Editor: Jordan Alvarez

LONDON OFFICE
(Design Hub)

Design Lead: Keiko Tanaka
Layout Specialist: Oliver Whitmore
Illustrator: Aisha Rahman
Design Team from Sumu Arts

KOLKATA OFFICE
(Support Hub)

Project Manager: Sohini Chatterjee
Production : Arindam Ghosh
Quality Lead: Priya Deshmukh
Fact Checker: Ravi Patel
Research Assistant: Ananya Sengupta
Research Assistant: Rohan Mukherjee
Project Assistant: Madhumita Roy
Production Assistant: Sagnik Dutta
Quality Assistant: Ishita Banerjee
Series Lead (Blueprint): Amit Verma
Series Lead (Playbook): Neha Reddy
Series Lead (Diagnostic): Karan Mehta

The End